THE NATURE AND BEHAVIOR OF GRAIN BOUNDARIES

The Metallurgical Society of AIME Proceedings published by Plenum Press

---

1968–Refractory Metal Alloys: Metallurgy and Technology
Edited by I. Machlin, R. T. Begley, and E. D. Weisert

1969–Research in Dental and Medical Materials
Edited by Edward Korostoff

1969–Developments in the Structural Chemistry of Alloy Phases
Edited by B. C. Giessen

1970–Corrosion by Liquid Metals
Edited by J. E. Draley

1971–Metal Forming: Interrelation Between Theory and Practice
Edited by A. L. Hoffmanner

1972–The Nature and Behavior of Grain Boundaries
Edited by Hsun Hu

A Publication of the Metallurgical Society of AIME

THE NATURE AND BEHAVIOR OF GRAIN BOUNDARIES

A Symposium held at the TMS-AIME Fall Meeting in Detroit, Michigan, October 18-19, 1971

Edited by
Hsun Hu
Research Laboratory
United States Steel Corporation
Monroeville, Pennsylvania

PLENUM PRESS • NEW YORK-LONDON • 1972

Softcover reprint of the hardcover 1st edition 1972

Library of Congress Catalog Card Number 72-81907

ISBN 978-1-4757-0183-8 ISBN 978-1-4757-0181-4 (eBook)
DOI 10.1007/978-1-4757-0181-4

Plenum Press is a division of Plenum Publishing Corporation
227 West 17th Street, New York, N.Y. 10011

United Kingdom edition published by Plenum Press, London
A division of Plenum Publishing Company, Ltd.
Davis House (4th Floor), 8 Scrubs Lane, Harlesden, London,
NW10 6SE, England

PREFACE

In view of the dramatically increased interest in the study of grain boundaries during the past few years, the Physical Metallurgy Committee of The Institute of Metals Division of The Metallurgical Society, AIME, sponsored a four-session symposium on the NATURE AND BEHAVIOR OF GRAIN BOUNDARIES, at the TMS-AIME Fall Meeting in Detroit, Michigan, October 18-19, 1971. The main objectives of this symposium were to examine the more recent developments, theoretical and experimental, in our understanding of grain boundaries, and to stimulate further studies in these and related areas. This volume contains most of the papers presented at the Symposium. It is regrettable that space limitations allow the inclusion of only four of the unsolicited papers, in addition to thirteen invited papers. The papers are grouped into three sections according to their major content: STRUCTURE OF GRAIN BOUNDARIES, ENERGETICS OF GRAIN BOUNDARIES, and GRAIN BOUNDARY MOTION AND RELATED PHENOMENA.

Grain boundaries, or crystal interfaces, have been of both academic and practical interest for many years. An early seminar on "Metal Interfaces" was documented in 1952 by ASM. The Fourth Metallurgical Colloquium held in France, 1960, had a broad coverage on "Properties of Grain Boundaries". More recently the Australian Institute of Metals sponsored a conference on interfaces, with the proceedings being published by Butterworths in 1969. The most recent symposium was the "International Conference on the Structure and Properties of Grain Boundaries and Interfaces" held at the IBM Watson Research Center in August 1971. In contrast to these previous conferences, the present Symposium was devoted solely to the nature and behavior of grain boundaries or interfaces between crystals of the same phase. In consideration of the high degree of complexity of grain boundaries and of the rapid expansion of research findings in this area, the specific scope of the Symposium was considered appropriate and timely.

Much of the recent progress in the knowledge of grain boundaries has arisen from successful applications of new techniques.

These include electron transmission microscopy, computer calculation or simulation and, to a still limited extent, field-ion microscopy. These have greatly facilitated the direct examination or determination of the structure of grain boundaries. A great deal of information on the nature of grain boundaries has also been deduced from investigations of grain boundary properties (such as energy, mobility, diffusion, anisotropy, etc.). The introduction of disclination models to grain boundaries of metals, and the concept of phase transformations of grain boundaries may stimulate future studies.

Despite these recent developments, our current knowledge of grain boundaries is inadequate. As is well known, grain boundaries are vacancy sinks, and are highly susceptible to impurity segregation which have profound effects on the properties of boundaries. Additional studies of the interaction of these foreign species with grain boundaries, and their influence on the properties or behavior of the boundaries would be of great interest. Most investigations on the structure of grain boundaries have heretofore dealt principally with equilibrium or minimum energy configurations, but their observations may not represent the characteristics of a boundary in motion. The solution of these and other problems must await further developments.

Several members of the Committee, in particular, N.S. Stoloff, J.F. Breedis and M.F. Ashby, assisted in the planning of the Symposium. J.W. Cahn made helpful suggestions regarding the program. The four sessions of the Symposium were capably managed by the Session Chairmen: K.T. Aust, M.F. Ashby, G.F. Bolling and P.G. Shewmon. To all these members sincere appreciation is hereby extended. The cooperation of the authors and the enthusiasm of the Plenum Press, which made this publication possible, are sincerely acknowledged.

Hsun Hu
Program Chairman and Editor

Research Laboratory
U.S. Steel Corporation
Monroeville, Pennsylvania 15146

March 1, 1972

LIST OF CONTRIBUTORS

T. R. Anthony, Research and Development Center, General Electric Company, Schenectady, New York

R. W. Balluffi, Department of Materials Science and Engineering, Cornell University, Ithaca, New York

J. R. Beeler, Jr., North Carolina State University, Raleigh, North Carolina

M. Biscondi, Ecole Nationale Superieure des Mines de Saint-Etienne, Saint-Etienne, France

G. H. Bishop, Army Materials and Mechanics Research Center, Watertown, Massachusetts

R. D. Bourquin, Battelle-Northwest, Richland, Washington

G. A. Bruggeman, Army Materials and Mechanics Research Center, Watertown, Massachusetts

H. E. Cline, Research and Development Center, General Electric Company, Schenectady, New York

R. E. Dahl Jr., WADCO Corporation, Richland, Washington

F. M. d'Heurle, IBM Thomas J. Watson Research Center, Yorktown Heights, New York

A. Gangulee, IBM Thomas J. Watson Research Center, Yorktown Heights, New York

M. E. Glicksman, Naval Research Laboratory, Washington, D. C.

C. Goux, Ecole Nationale Superieure des Mines de Saint-Etienne, Saint-Etienne, France

E. A. Grey, Department of Metallurgical Engineering, Illinois Institute of Technology, Chicago, Illinois

Edward W. Hart, Research and Development Center, General Electric Company, Schenectady, New York

W. H. Hartt, Florida Atlantic University, Boca Raton, Florida

G. Hasson, Ecole Nationale Superieure des Mines de Saint-Etienne, Saint-Etienne, France

G. T. Higgins, Department of Metallurgical Engineering, Illinois Institute of Technology, Chicago, Illinois

R. J. Horylev Department of Materials Science, University of Southern California, Los Angeles, California

Hsun Hu, Research Laboratory, United States Steel Corporation, Monroeville, Pennsylvania

Y. Komem, Technion-Israel Institute of Technology, Haifa, Israel

P. Lagarde, Ecole Nationale Superieure des Mines de Saint-Etienne, Saint-Etienne, France

J. Levy, Ecole Nationale Superieure des Mines de Saint-Etienne, Saint-Etienne, France

J. C. M. Li, Department of Mechanical and Aerospace Sciences, University of Rochester, Rochester, New York

Y. C. Liu, Scientific Research Staff, Ford Motor Company, Dearborn, Michigan

K. Lücke, Institut für Allgemeine Metallkunde und Metallphysik, Technische Hochschule, Aachen, Germany

L. E. Murr, Department of Materials Science, University of Southern California, Los Angeles, California

J. P. Nielsen, New York University, New York

B. B. Rath, McDonnell-Douglas Research Laboratory, St. Louis, Missouri

R. Rixen, Institut für Allgemeine Metallkunde und Metallphysik, Technische Hochschule, Aachen, Germany

F. W. Rosenbaum, Institut für Allgemeine Metallkunde und Metallphysik, Technische Hochschule, Aachen, Germany

R. Rosenberg, IBM Watson Research Center, Yorktown Heights, New York

L. P. Stone, Anaconda American Brass, Waterbury, Connecticut

R. A. Vandermeer, Metals and Ceramics Division, Oak Ridge National Laboratory, Oak Ridge, Tennessee

C. L. Vold, Naval Research Laboratory, Washington, D. C.

G. R. Woolhouse, University of Cambridge, Cambridge, England

CONTENTS

THE NATURE AND BEHAVIOR OF GRAIN BOUNDARIES

STRUCTURE OF GRAIN BOUNDARIES

STRUCTURE OF GRAIN BOUNDARIES. THEORETICAL DETERMINATION AND EXPERIMENTAL OBSERVATIONS

G. Hasson, M. Biscondi, P. Lagarde, J. Levy and C. Goux

Ecole Nationale Superieure des Mines de Saint-Etienne

Saint-Etienne, France

ABSTRACT

In order to determine the structure and the energy of grain bounderies in pure metals a relaxation method based upon a minimum free energy criterion has been devised. The principle of the method is very simple; however, its correct application requires various precautions.

This method has been employed to determine the structure and the energy of a number of grain boundaries in aluminium where the boundary planes are either symmetrical or non-symmetrical with respect to the two lattices.

"Special" boundaries having been defined, an investigation has been conducted on boundaries having orientations close to those of the "Special boundaries". The presence of discontinuity lines in these boundaries, which are sometimes visible in the electron microscope, has been discussed.

The concept of grain boundary dislocations has been refined. In particular, a distinction has been made between "intersection dislocations" and "structural dislocations". An extension of the Burgers vector and Burgers circuit concepts has been proposed.

INTRODUCTION

In order to interpret the properties of grain boundaries the importance of knowing the structure of these boundaries is obvious. Unfortunately even now this structure can be determined only by very indirect methods. Electron microscopic or field ion microscopic observations often yield results which lack precision, and even in the most favourable cases interpretation is uncertain. The structures can be calculated on the basis of theoretical considerations. It is then necessary to make a number of assumptions which aim principally at simplifying and increasing the feasibility of the calculation. In this respect the advent of the computer has brought about considerable progress. The possibility of carrying out complex calculations allows one to limit the assumptions to a few, and their physical justification may be improved.

The present method has been applied to planar grain boundaries in fcc metals; however, it can be extended quite easily to more general cases.

When two crystals are separated by a boundary so that their lattices assume a twin position of low twin index, i.e. when the number of coincident lattice points is high, the calculation is relatively fast. It is noteworthy that one can include in this category the majority of boundaries whose characteristics appear to be of great interest.

A difficulty arises if the lattices are nearly, but not exactly, twin-related in orientation. In these cases, it would be very difficult to carry out the calculations because of the great number of atoms involved in a relatively wide area of the boundary plane, although such calculations can, in principle, still be performed. Boundaries separating such lattices commonly exhibit characteristic lines upon electron microscopic examinations. Hence these boundaries may be of considerable importance in structural studies.

In the present paper we intend to discuss these problems in more detail. We shall first summarize our method of calculations which will be illustrated by a few characteristic examples. We shall then examine the structural features of the boundaries and their interpretations. Finally, we shall try to analyze the nature of the grain boundary discontinuities, and discuss these defects as to what extent such defects may be related to classical dislocations. These studies will all be pertained to the case of fcc pure metals.

THEORETICAL CALCULATION OF THE STRUCTURE AND OF THE ENERGY OF GRAIN BOUNDARIES

General Principle

Let us consider in Fig. 1 two crystals I and II separated by

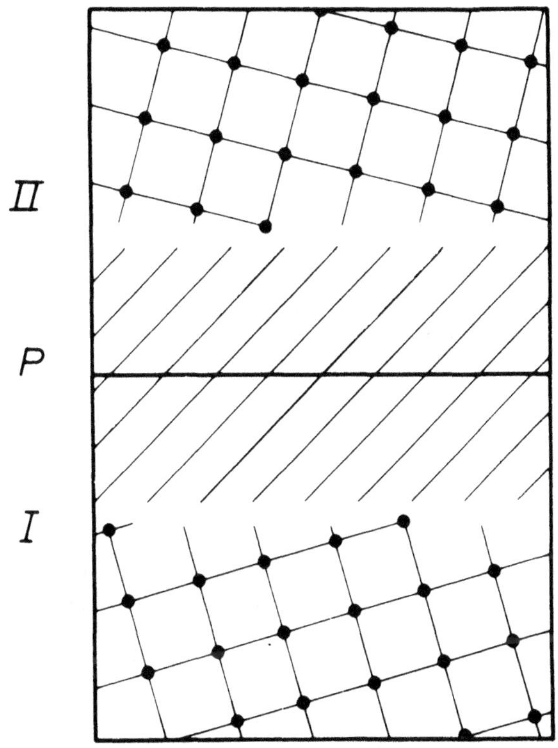

Fig. 1 - Schematic illustration of the existence of undistorted zones of crystals I and II near the grain boundary.

a plane P. One can assume that the structure of each crystal far enough from P is not influenced by the presence of the other so that the positions of the atoms far from P are exactly known. In Fig. 1 the positions are indicated by points representing the centers of the atoms.

In order to determine the structure in the intergranular region, we shall seek that atomic configuration which corresponds to the lowest possible free energy of the entire system.

In order to do this, let us place a certain number of atoms in arbitrary positions between crystals I and II (Fig. 2). E_0 is the free energy corresponding to the whole configuration of all the atoms. Let us now consider atom A in its initial position A_0. Moving this atom in the vicinity of A_0, may yield two possibilities:

- the energy of the system may be a minimum, expressed as E_0, when A is in its initial position A_0. In this case A will remain at A_0.
- there may be no energy minimum for A_0. In this case atom A will be shifted to a new position A_1 corresponding to an energy E_1, lower than E_0.

It is worth noting that this process does not necessarily mean that the new position A_1 corresponds to an energy minimum. Further, it would be very useful if one could determine A_1 in such a way that the convergence of the calculation be the fastest one. Unfortunately there seems to be no general way to achieve this.

After considering atom A, the same relaxation process is applied successively to all the atoms which have been placed in the intergranular region. All the atoms are again subjected to the same process in a second run starting with A and so on.

The successive values of the energy E_0, E_1 ... form a monotonically decreasing sequence. As this sequence is evidently bounded, it has a limit E which is a minimum value of the energy and corresponds to an equilibrium structure of the whole group of atoms.

Practical Conditions of the Calculation

The foregoing method is very simple in principle. However a number of precautions are necessary for its correct application. These precautions have been summarily discussed elsewhere(1) so we need not systematically reexamine them here. We shall restrict ourselves to the questions which arise in the following paragraphs.

Definition of the "calculation zone". The volume of the intergranular region in which the "mobile" atoms, subjected to the relaxation process, are contained is defined by a certain area of the plane P (Fig. 2), and the distance between the practically undistorted zones of crystals I and II. This volume we shall call the "calculation zone".

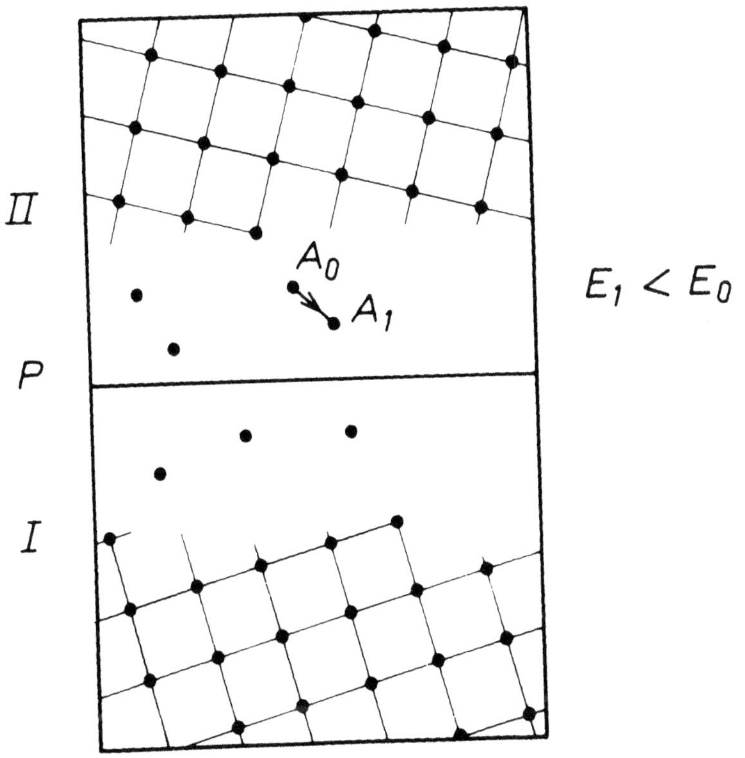

Fig. 2. Principle of the relaxation calculation.

The distance between the crystals may be chosen without serious difficulty. The greater it is, the more precise and the longer will be the calculation.

The determination of the region to be considered in plane P is much more difficult. First of all, it is in practice necessary to take into account only those cases where the direction of plane P corresponds to a plane cell which is common to the two crystallographic lattices I and II. However this condition does not introduce any severe limitation because it is satisfied in those cases which are evidently of the greatest interest. In fact, any relative crystal orientation may be chosen close to an orientation relationship of the crystals which meet this requirement.

In the plane P the calculation zone will be defined by the cell which is common to both crystals. The volume of the calculation zone will be a minimum if one chooses an elementary cell. However it may well be that a multiple cell leads to a lower value of the energy of the system. In such a case it is, in principle, this cell which should be considered.

Number N of atoms. If N is the number of mobile atoms situated in the calculation zone, the parameters to be considered are the 3 N coordinates of these atoms. However, parameter N itself is not initially known. It is necessary to determine the value of N which leads to the lowest possible value of E.

One might think that by changing the relative positions of the crystals I and II, the volume in which the N atoms are enclosed (i.e. the shape of the calculation zone) would be modified. This would yield the best conditions independently of the initial value of N. However, it is generally impossible to do this.

Let us consider, for example, the particular case of a grain boundary of zero misorientation (Fig. 3). In this case the calculation must lead to the equilibrium structure of a single crystal. However, as can be visualized clearly in Fig. 3, if 7 mobile atoms instead of 8 were placed in the calculation zone between I and II it would be impossible to obtain the correct structure no matter what relative positions might be chosen for I and II.

Translation vector T. Five parameters are required to define macroscopically the orientation of a grain boundary. To define the same boundary on an atomic scale requires three more parameters.

In fact, Fig. 4 shows that with reference to the position of O_{II} in the lattice of crystal I, the relative position of both crystals is different. The three parameters can be chosen as the coordinates of a vector such as:

$$T = O_I \, O_{II}$$

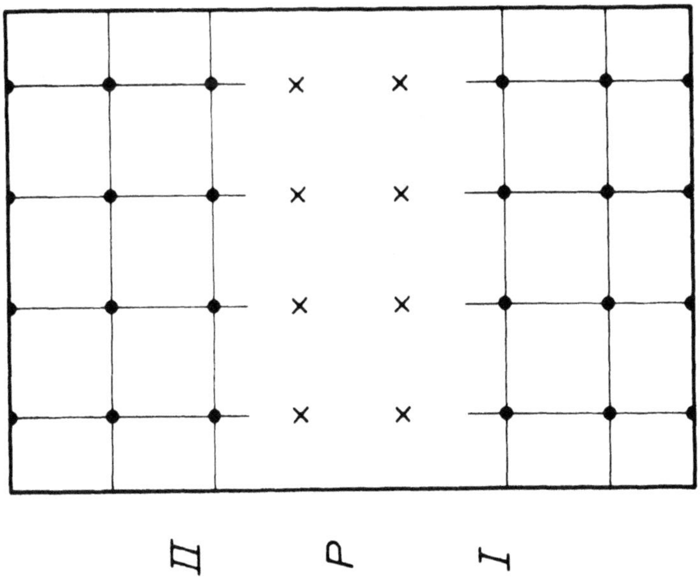

Fig. 3. Example showing the need for correctly choosing the value of N.

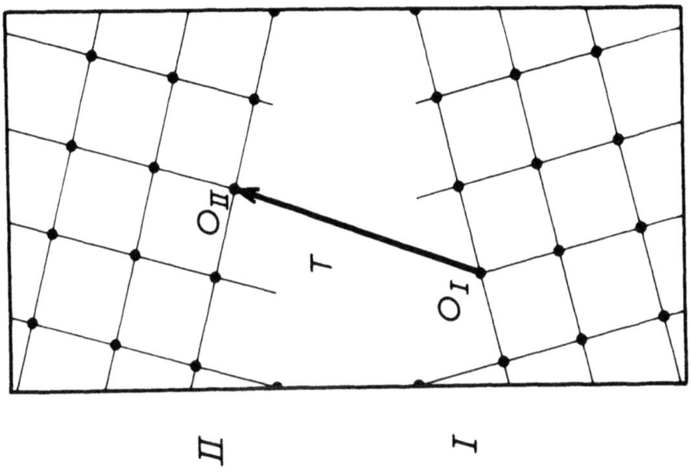

Fig. 4. Definition of vector T.

It is clear that all the possible relative positions of the crystals will be obtained if the end point O_{II}, of vector T is allowed to move inside one elementary cell of a fixed crystal assuming an arbitrary position of crystall II or, equivalently, if O_I moves inside an elementary cell of crystal I.

The coordinates of vector T are thus three parameters which have to be determined in order to arrive at the least possible value of E.

Results Concerning Various Kinds of Boundaries

The foregoing method has been applied to many grain boundaries and in particular to those in pure aluminium. In order to evaluate the energy E, a Morse potential, chosen for the case of aluminium, has been applied using the constants proposed by Cotterill and Doyama[2].

Examples of symmetrical tilt boundaries. Fig. 5 shows an example of a symmetrical [100] tilt boundary. This boundary is a (013) twin boundary, the misorientation being 36°52'. The points and the crosses represent the projections of the centers of atoms on two successive (100) planes. Apart from very small displacements of the atoms in the [100] direction, the structure is entirely determined by these projections.

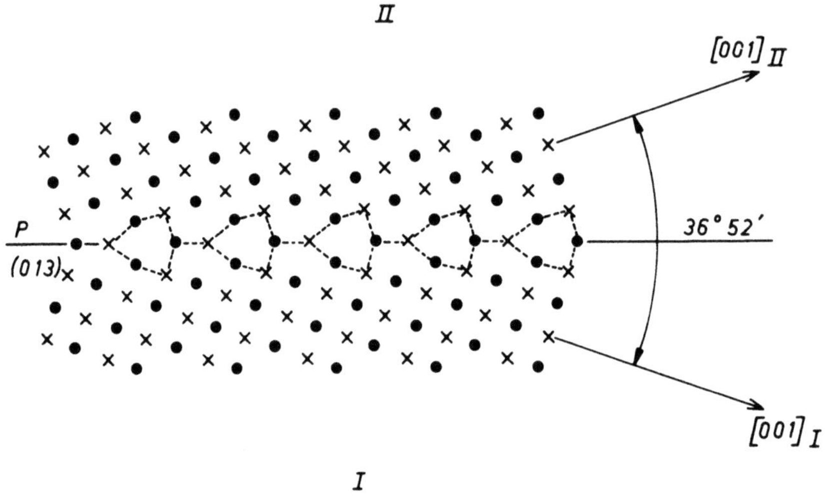

Fig. 5. Structure of a (013) twin boundary.

The curve in Fig. 6 shows the energy of the symmetrical [100] tilt boundaries. The (013), (012) and (023) twin boundaries are characterized by cusps in the curve but their energy values do not significantly differ from those of the other high angle boundaries.

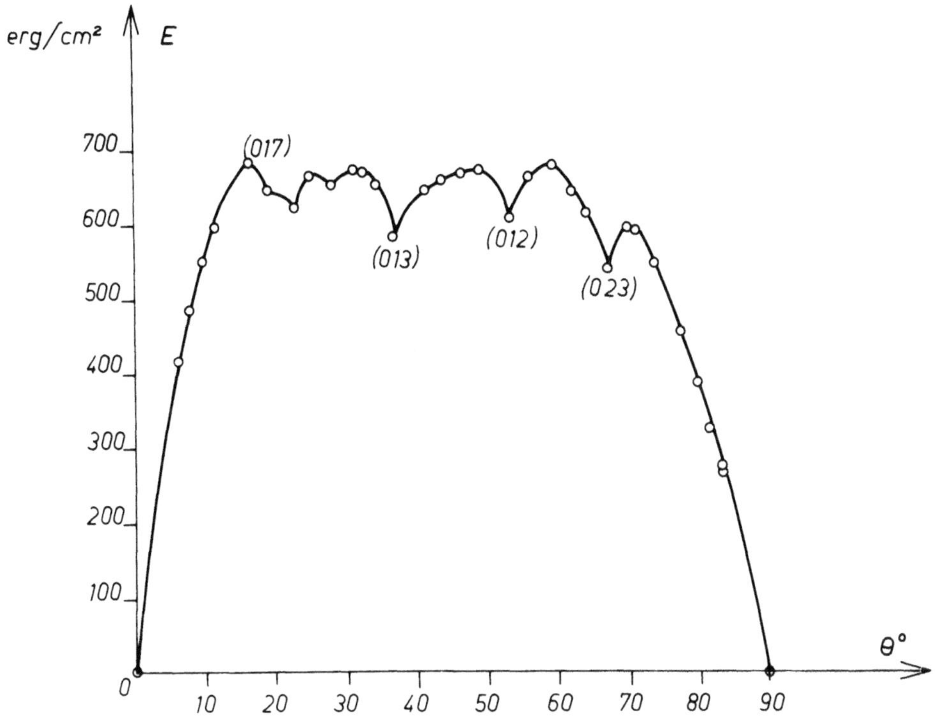

Fig. 6. Energy of symmetrical tilt boundaries about [100] in aluminium.

Figures 7 and 8 are equivalent to Figs. 5 and 6 but are for [110] tilt boundaries. In Fig. 7 the points and the crosses represent the projections of atom centers on two successive [110] planes. Figure 8 shows, as expected, that the (111) twin boundary has a very low energy. Obviously the calculated value is inferior to the real one; and the origin of this discrepancy relates to the kind of potential used. The (113) twin boundary also has a very low energy. In spite of its low twin index the (112) twin boundary does not have a low energy.

Figures 9 and 10 relate to [111] tilt boundaries. In order to define the boundary structure it is necessary to indicate the projections of atom centers on three successive (111) planes. The points, the crosses, and the triangles correspond respectively to

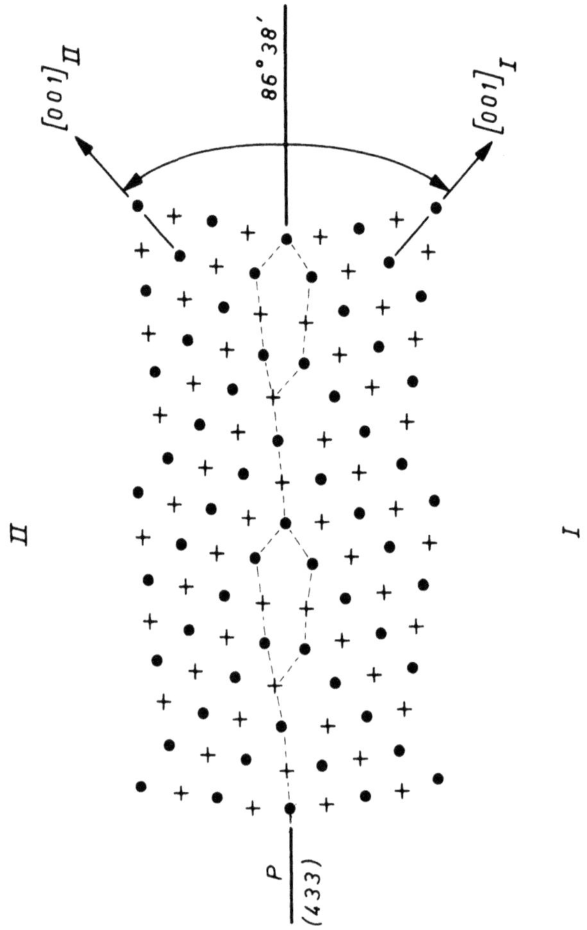

Fig. 7. Structure of a (334) twin boundary.

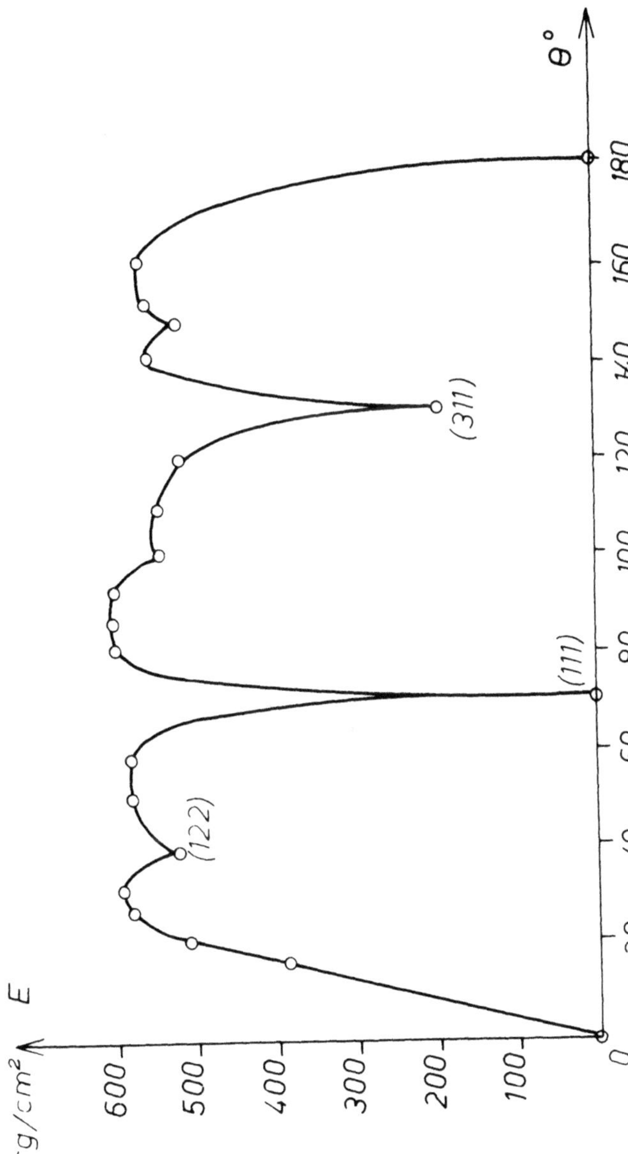

Fig. 8. Energy of symmetrical tilt boundaries about [110] in aluminium.

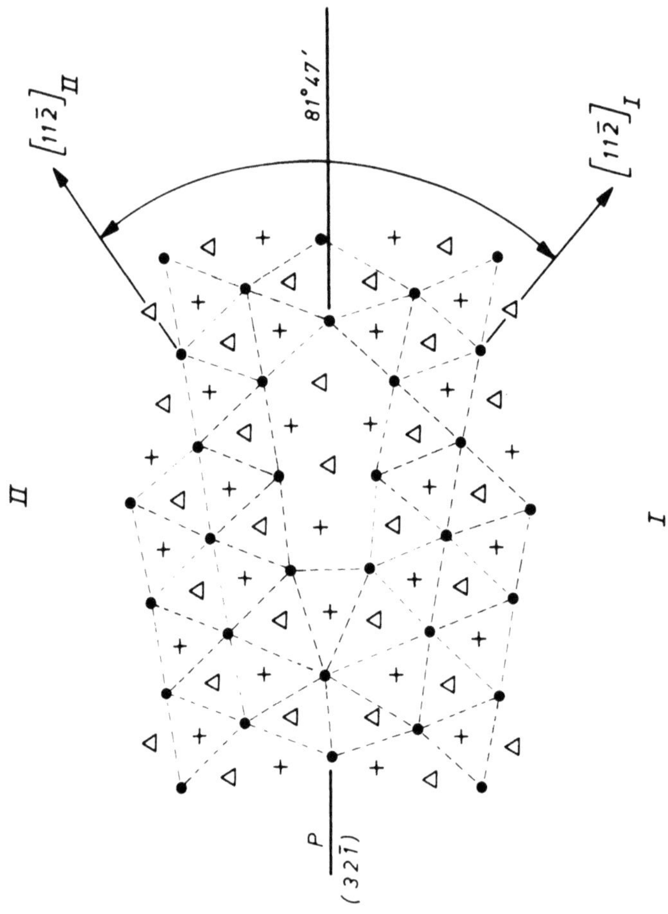

Fig. 9. Structure of a (123) twin boundary.

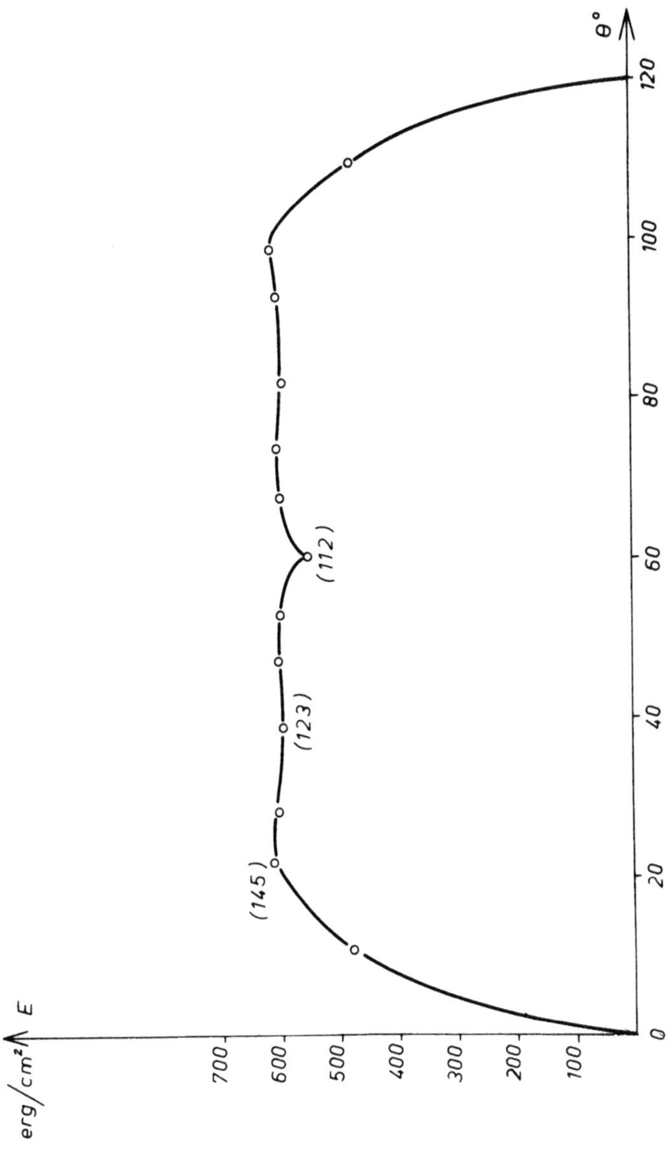

Fig. 10. Energy of symmetrical tilt boundaries about [111] in aluminium.

these three categories of atoms. In the curve of Fig. 10, the (112) twin boundary stands out because of a small decrease in its energy.

Non-symmetrical boundaries. Actually the structures of the boundaries shown in the above examples could well have been determined relatively accurately without using the present method which necessitates a computer(3)(4). The usefulness of the method resides in cases where the structures of the boundaries cannot be determined by relying on simple geometrical reasoning. This is generally the case when the boundary plane is not placed in a symmetrical position with respect to both lattices. Figure 11 represents such a case, and this can be treated without difficulty using the present method.

For lattice orientations such as shown in Fig. 11, the influence of the orientation of the boundary plane upon its energy may be calculated(1).

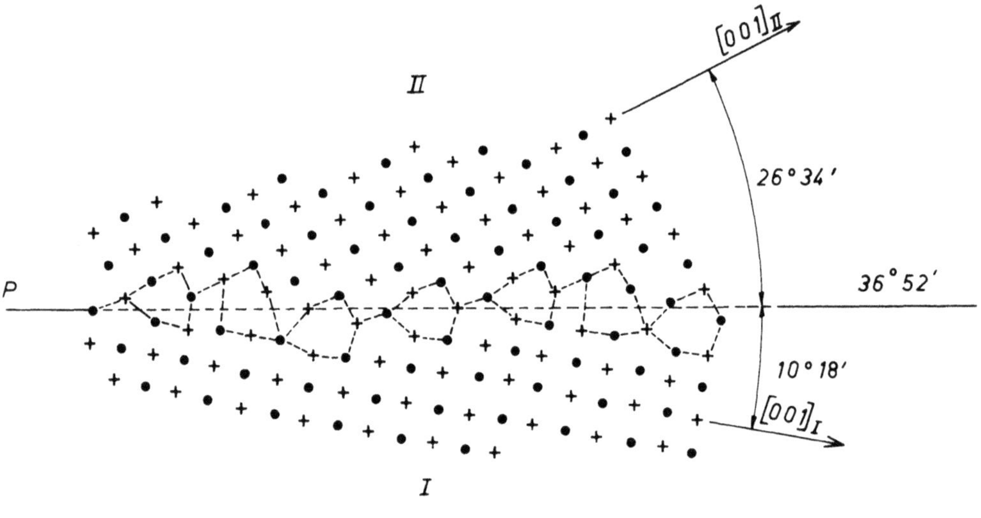

Fig. 11. Structure of an asymmetrical tilt boundary about [100].

A note concerning coincidence grain boundaries. Grain boundaries in which the boundary plane is a lattice plane common to both crystals are usually referred to as "coincidence grain boundaries" (of course, one assumes that the lattice points are placed in equivalent positions in both crystals). It has often been assumed that the corresponding disposition of the crystals has a minimum energy. Actually the value of the T vector which leads to a minimum in the energy does not generally correspond to this coincidence, even though the real structure is not greatly different from the coincidence structure.

A note concerning the structure of symmetrical boundaries. The structure of a boundary placed symmetrically between two crystals does not have to be symmetrical. Mathematically, the existence of an asymmetrical structure only implies the existence of an equivalent structure placed symmetrically with respect to the boundary plane P. This condition is evidently fulfilled if the crystals outside the boundary zone are in symmetrical positions with respect to P.

THE PROBLEM OF BOUNDARIES "CLOSE TO A SPECIAL BOUNDARY"

Definition of the Boundaries under Consideration

Let us consider the case of a (013) twin boundary in Fig. 5. Such a boundary can be referred to as being special in the sense that

- both crystals have in common a particular axis, in this case the [100]

- the misorientation about the [100], which is $36^o52'$, is the particular value for which the two crystals are in a twin position.

A "special" boundary will be characterized either by the first of these properties or by both.

A boundary designated as "close" to a special boundary will exhibit the following characteristics:

- either the misorientation, θ, about the common [100] axis will be close to $36^o52'$; the angular difference ($\theta - 36^o52'$) shall be called ε.

- or the two equivalent axes $[100]_I$ and $[100]_{II}$ will not exactly coincide, whereas the misorientation may or may not equal to the special value; the angle between $[100]_I$ and $[100]_{II}$ shall be called δ.

It is clear that the (013) twin boundary just presented is merely an example, and that the idea may be extended to other special boundaries without difficulty. In general, the double periodicity in the boundary plane may be defined by two translation vectors, V_1 and V_2, chosen being the shortest possible ones. A special boundary will then be characterized by the fact that at least one of these vectors will be very short. In other words the plane P will contain at least one row of highly densely packed atoms common to both lattices.

Importance of Considering Boundaries "Close to Special Boundaries"

Simple geometrical considerations have led one to expect that discontinuity lines comparable with dislocations might exist in such boundaries(3). Investigations with the electron microscope effectively revealed such discontinuities(5-7). However, their characteristics do not correspond to the first expectations. In order to truly understand intergranular structures, it is therefore important that such structures should be interpreted correctly.

Misorientation Close to a Special Value

Description of the case under investigation. Figure 12a schematically represents a single crystal in which plane P can be considered as the plane of a grain boundary of zero misorientation about an axis perpendicular to the plane of the figure (Fig. 12b). If the misorientation increases from zero to a small value ε, it is well known that the corresponding boundary is composed of a family of parallel edge dislocations (Fig. 12c).

The (012) twin boundary in Fig. 13a may be compared with the boundary of zero misorientation, since in the plane P, the plane cell common to both lattices is very small in both cases, thus the density of coincident atom sites is very high. If the misorientation, θ, were to be increased by a small amount ε, one might wonder whether the structure would be modified by the introduction of linear defects comparable to the dislocations which might eventually be visible in the electron microscope. Actually, this last assumption has been contradicted clearly by experimental observations, since no discontinuities of this kind have been revealed(5,6).

Grain boundaries of small misorientations. In order to solve the preceding question it is pertinent to make remarks concerning grain boundaries of small misorientations.

Let us assume, for example, that $\theta = 1^\circ$ (or, more accurately, $1^\circ 0' 18''$).

As shown in Fig. 14, the spacing, D, between the edge dislocations agrees with the formula of Read and Shockley:

$$D = 57\ b$$

Now, if θ is allowed to assume a value which is exactly double that of the first one, the same formula gives:

$$D = 28.5\ b$$

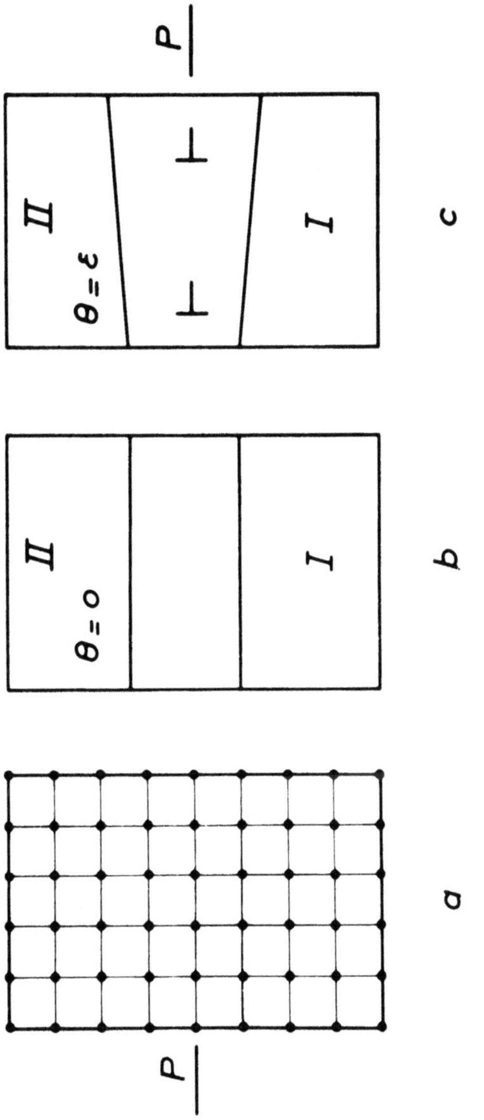

Fig. 12. Effect of introducing a small misorientation between two parts of a single crystal.

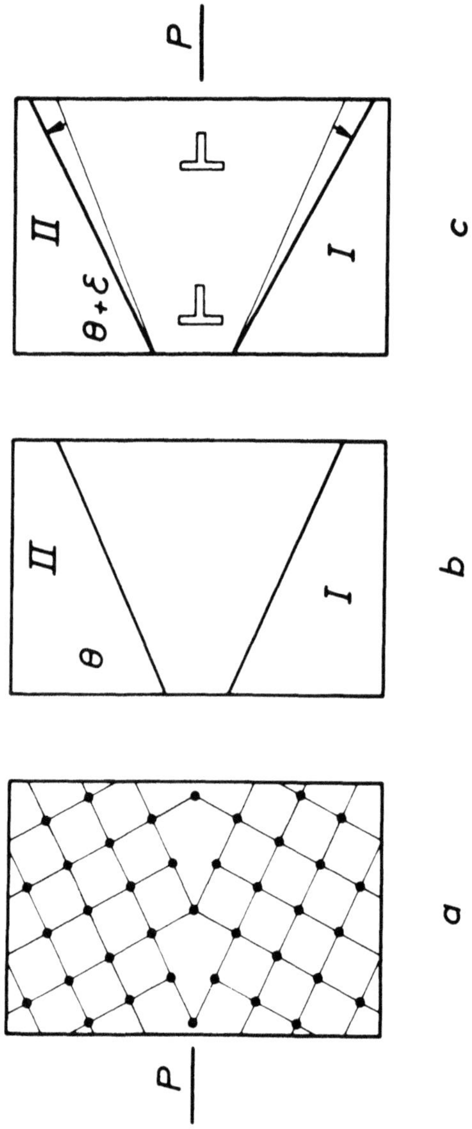

Fig. 13. Possible effect of slightly changing the misorientation of a twin boundary.

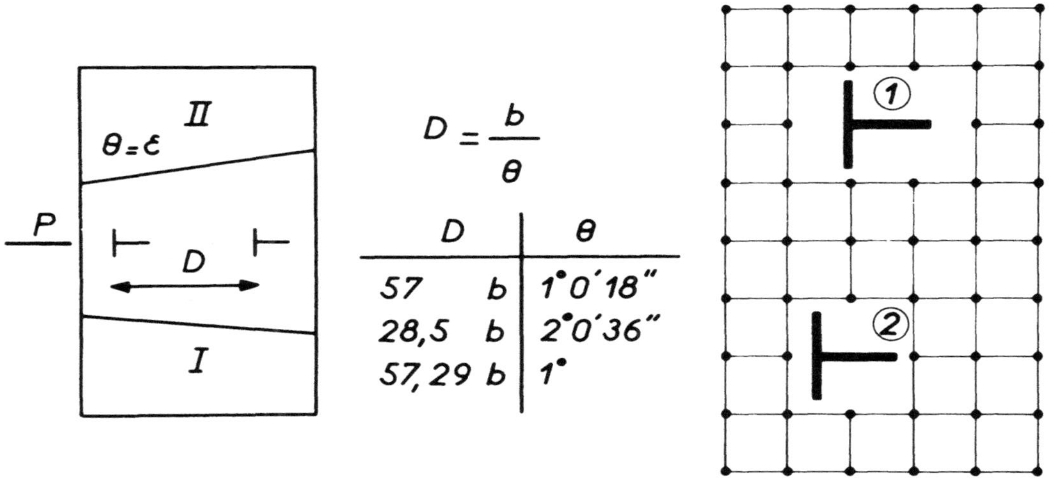

a b

Fig. 14. Figure showing the difficulty resultig from a strict application of the Read and Shockley formula.

If this formula were strictly followed, it would imply that the arrangements of successive dislocations were quite different, as shown in Fig. 14b. This is certainly not the case. In fact, there is little doubt that the boundary is composed of dislocations, alternately 28 b and 29 b apart. However:

- 28 b corresponds to a misorientation of $2^{\circ}3'$ ($2^{\circ}2'47''$)
- 29 b corresponds to a misorientation of $1^{\circ}59'$ ($1^{\circ}58'34''$)

It follows that for each value of misorientation between $1^{\circ}59'$ and $2^{\circ}3'$ the boundary is composed of dislocations 28 b or 29 b apart. If θ is close to $1^{\circ}59'$ the distance between two successive dislocations will be 29 b in general and in a few cases 28 b; the proportion of the latter will increase with θ. The number and distribution of the 28 b and 29 b spacings along the grain boundary could easily be determined.

High angle boundaries. The rule which has just been established for small angle boundaries can be generalized without difficulty. Let us consider for example a boundary with a common [100] axis, the misorientation varying between:

- $\theta_1 = 36^{\circ}52'$ corresponding to the (013) twin boundary in Fig.5, and
- $\theta_2 = 53^{\circ}08'$ corresponding to the (012) twin boundary in Fig. 15.

The same "pattern" appears in both boundaries. Yet two "patterns" are separated by a narrow strip of (001) plane if $\theta = \theta_1$ whereas they are joined together if $\theta = \theta_2$.

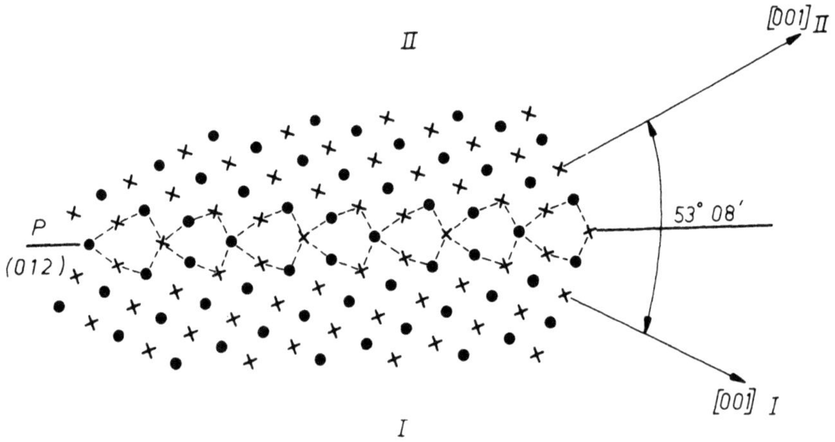

Fig. 15. Structure of a (012) twin boundary.

The structure of any boundary for which $\theta_1 < \theta < \theta_2$ is obvious. The same "patterns" are present; the number of "patterns" which are joined together or which are separated by a strip of (001) plane depends upon the value of θ with respect to θ_1 and θ_2.

This rule has been established through a direct determination of the structure of the (0,5,11) twin boundary which is represented in Fig. 16; in this case $\theta = 48^o54'$. There is only one strip of (001) plane for two patterns. This boundary has been chosen as an example because it does not require excessively long calculations; the extension to other boundaries is easy. However, in general, it is necessary to know the values θ_1 and θ_2 which limit θ. Without entering into the details of this question we will only mention the results concerning symmetrical [100] tilt boundaries. For a given θ, the values of θ_1 and θ_2 obey the following rules:

- if $\theta < 53^o8'$

  (θ_1 corresponds to a (0,1,n) twin boundary

  (θ_2 corresponds to a (0,1,n+1) twin boundary

- if $\theta > 53^o8'$

  (θ_1 corresponds to a (0,n,n+1) twin boundary

  (θ_2 corresponds to a (0,n+1,n+2) twin boundary

These rules allow one to immediately determine θ_1 and θ_2. Moreover, as in the case of the small angle boundaries, it is possible to specify the number and the distribution of the "patterns" corresponding to θ_1 and θ_2 respectively.

It is noted that if $\theta > 53^o8'$ the "patterns" of the structure are separated by strips of (011) planes as can be seen in Fig. 17 which shows a (034) twin boundary.

Appearance of the boundaries under consideration in the electron microscope. It has been noted previously that the discontinuity lines appearing in a boundary which is "close to a special boundary" do not generally arise from the misorientations being not exactly equal to the θ value for the special boundary. Actually discontinuity lines exist in the proposed model but these lines merely correspond to a transition between analogous structures. As for the "patterns" themselves they can be revealed in the electron microscope only if their mutual spacing is sufficiently large; that is to say, only for low values of θ.

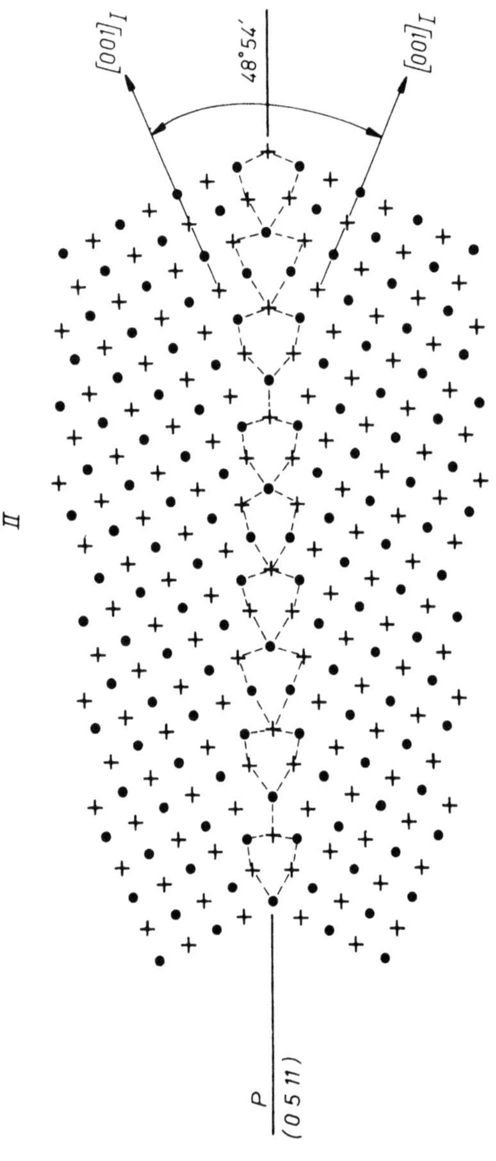

Fig. 16. Structure of a boundary intermediate between the (013) and (012) twin boundaries.

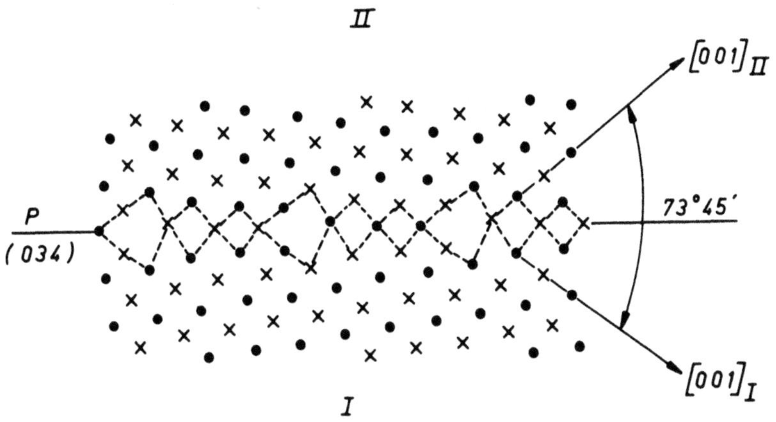

Fig. 17. Structure of a (034) twin boundary: presence in the boundary of zones of fit composed of (011) strips.

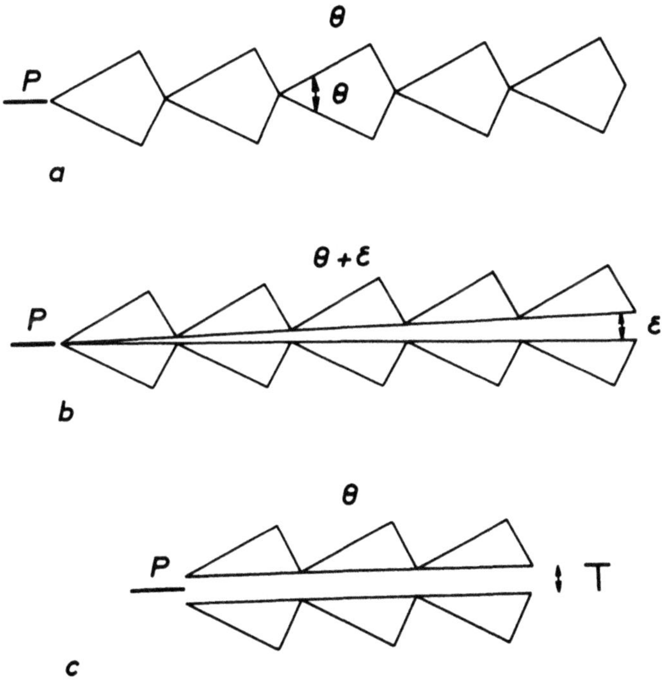

Fig. 18. Proposed method for studying boundaries "close to special boundaries".

General Method

The method as just described was for tilt boundaries having a common [100] axis. It can be extended to other types of boundaries. It is also possible to derive a more general method with which boundaries having two equivalent crystallographic axes of high density and being misoriented by a small angle δ can be studied.

Let us consider Fig. 18a which represents schematically a (012) twin boundary. If the misorientation is increased by a small angle ε, the new position of the crystals corresponds to that shown in Fig. 18b which suggests the following hypothesis.

With ε being small, one can consider that a small region of the boundary is obtained by separating the "patterns" in the structure by an amount T. (Fig. 18c). These patterns fit exactly when $\varepsilon = 0$.

There is a double interest in this hypothesis. First, it avoids the calculation of a structure having a long period; in fact the structure shown in Fig. 18c corresponds to that for the special boundary of angle θ, and the length of the period in the plane of the figure is independent of T; then a computer of limited capacity can be used. Of course, in order to know all the structure along the grain boundary it is necessary to determine the structures corresponding to all possible values of T.

A second advantage is that this hypothesis can easily be utilized to fit all the cases of boundaries "close to special boundaries", no matter what the modification of their orientation may be, in particular if two equivalent axes such as $[100]_I$ and $[100]_{II}$ make a small angle δ.

The method which emanates from this hypothesis has been used for the case already discussed of a boundary of misorientation $\theta + \varepsilon$, where θ corresponds to a special misorientation. The reliability of this method has been demonstrated by these results.

The Case when $[100]_I$ and $[100]_{II}$ make a Small Angle δ

Appearance of the corresponding boundaries in the electron microscope. Figure 19 shows the appearance of an oriented grain boundary of aluminium when examined in the electron microscope. It is a boundary close to a (013) twin boundary; the observation was made under bright field conditions. The fringes, which are approximately parallel and regularly spaced, are elements of the structure. They are parallel to the general [100] direction common to both lattices. The appearance and the spacing of these fringes do not depend upon the value of the misorientation.

Actually the $[100]_I$ and $[100]_{II}$ axes do not exactly coincide: it has been shown that these lines result from lack of coincidence and that their spacing, d, is inversely proportional to the angle δ' of their projections on the (013) plane. It has been found that $d = \frac{a}{2\delta'}$, where a is the lattice parameter of aluminium and δ' is the misorientation angle in radians.

Figure 20 is the electron micrograph using dark field conditions. One can easily recognize the family of parallel lines which appeared when using bright field illumination; however the photograph reveals a second family of lines which run perpendicularly to the former ones. Figure 21 schematically reproduces the general aspect of the photograph: the lines of the second system do not cross those of the first one and assume a zig-zag arrangement.

It is not surprising that the two methods are not equally efficient in revealing the structural elements. However, the difference is particularly pronounced in this case.

Method of investigation of the lines appearing in the electron microscope. Following the preceding remarks, no lines should be visible in the perfect (013) twin boundary, which is represented in Fig. 22a. Some lines will appear if crystal II is rotated with respect to crystal I by a small angle δ' about the normal to the (013) plane, as shown in Fig. 22b.

Figure 23a represents the (013) plane of crystal I with its periodic lattice of rectangular cells. The equivalent lattice in crystal II is obtained by rotating the former lattice by an angle δ' about O. If one considers the $(abcd)_I$ cell, the equivalent $(abcd)_{II}$ cell is obtained by the same rotation. As δ' is a small angle, one can assume that $(abcd)_I$ and $(abcd)_{II}$ are made equivalent by a translation of vector T (Fig. 23b). Then the structure can be calculated using the general method described in an earlier section.

We are thus induced to determine the structures resulting from a translation of vector T, imposed upon crystal II with respect to crystal I. For each value of T we obtain a structure which represents the true structure in a small region of the boundary plane.

One must now point out an important feature. With increasing values of T, the distances between atoms, which were originally close neighbours, will increase. It is to be expected that the bonds between such atoms will break at a certain value of T, so that the structure will exhibit a discontinuity for this value of T. However even if simple geometrical reasonings make it possible

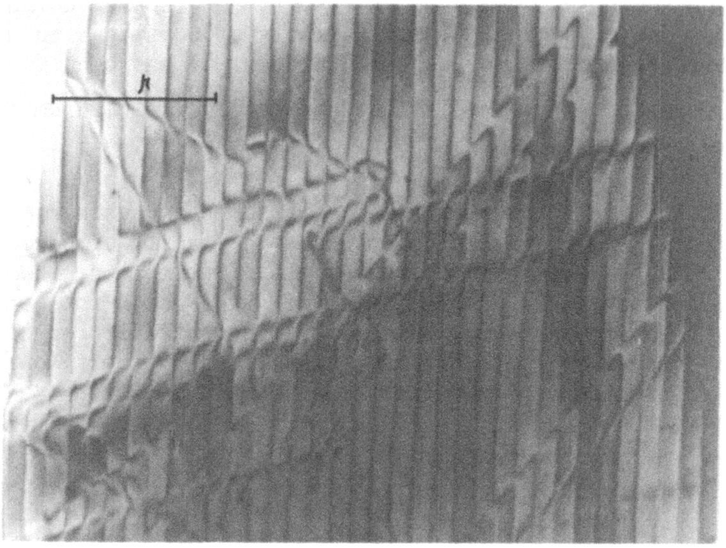

Fig. 19. Structure of a boundary "close to a (013) twin boundary"as examined under bright field illumination.

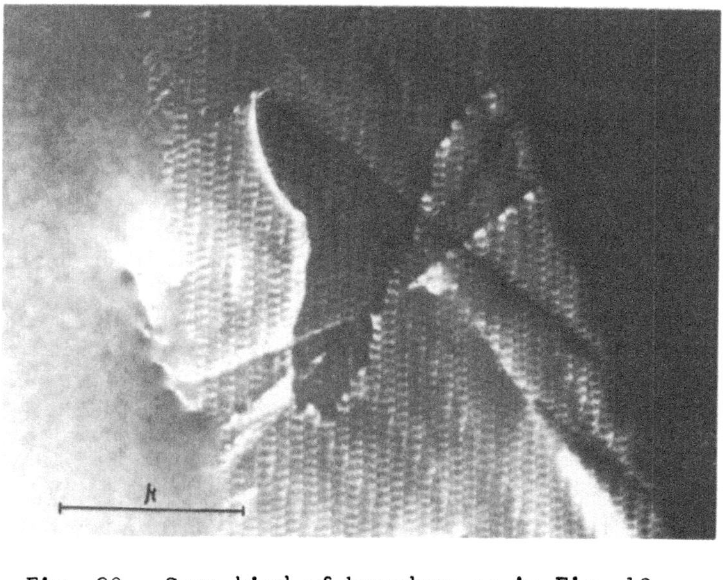

Fig. 20. Same kind of boundary as in Fig. 19 under dark field illumination.

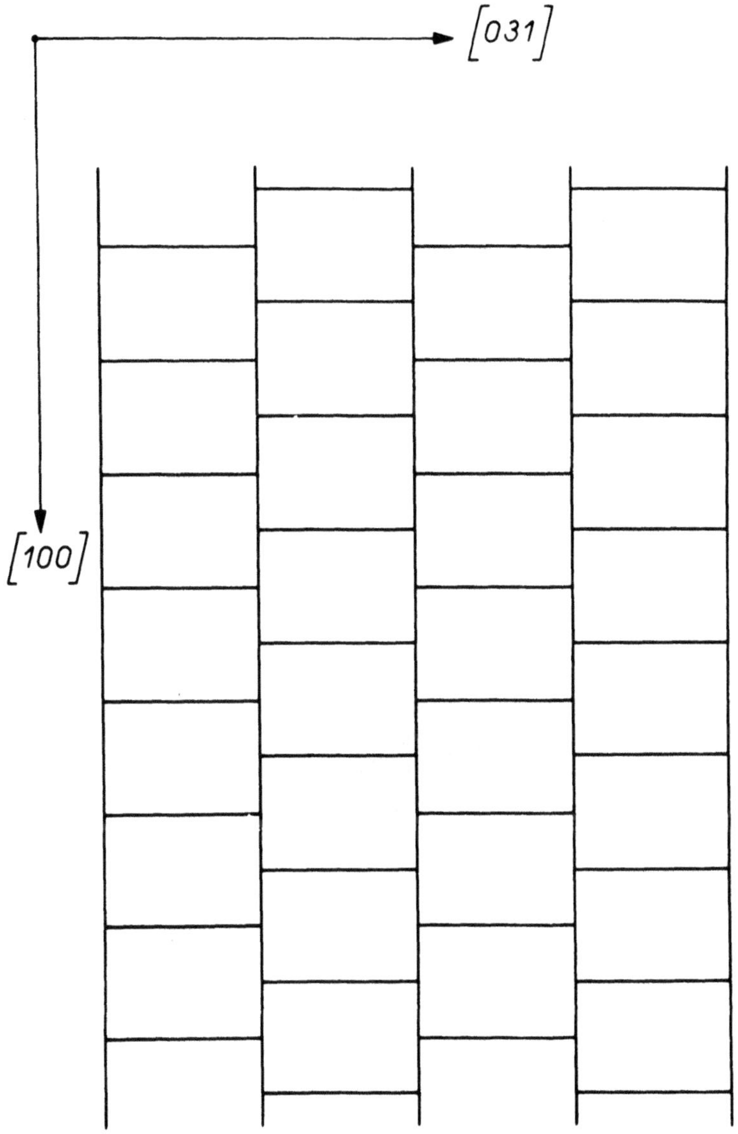

Fig. 21. Schematic representation of Fig. 20.

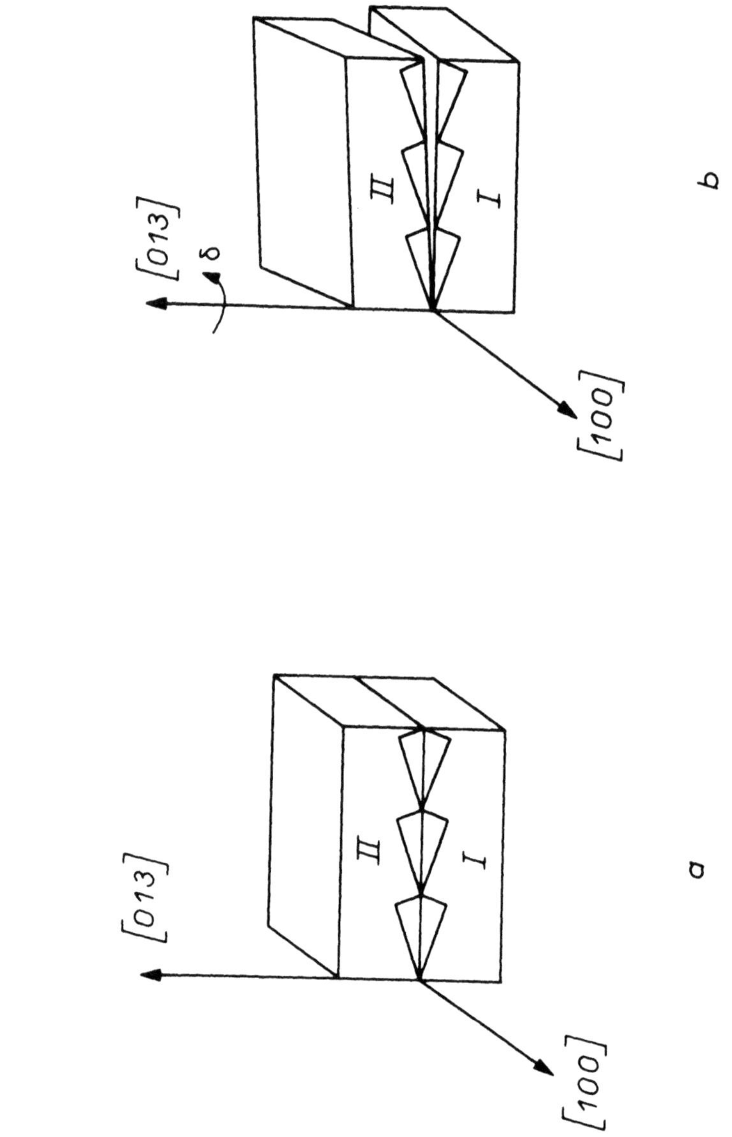

Fig. 22. Effect of a slight rotation about a twin axis.

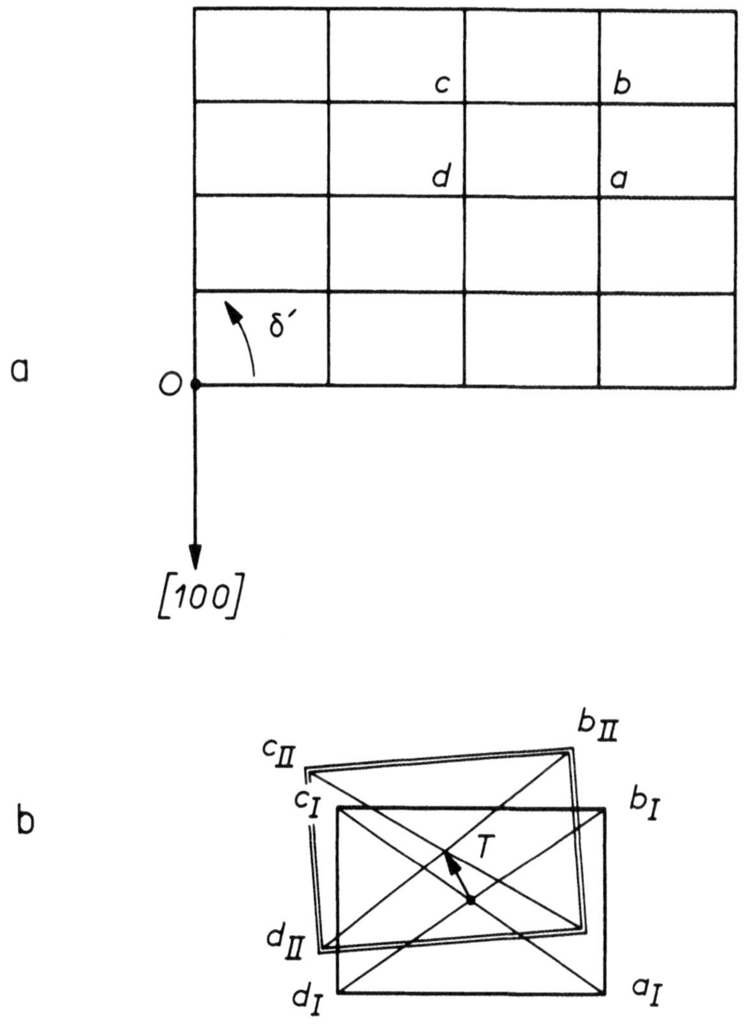

Fig. 23. Application of the method of Fig. 18 to the case of Fig. 22.

to account for these discontinuities, their exact nature can only be determined by the calculation of the structures for values of T near its critical value. Moreover, it is difficult to predict whether these discontinuities will show sufficient contrast in the electron microscope to make them visible.

Interpretation of the lines visible in the electron microscope. The calculation described in the preceding paragraphs has been performed for a boundary close to a (013) twin boundary. Despite the fact that this calculation has not yet been completed, it has already given some useful information.

Figure 24 again reproduces the shape of the lines which have been described previously. In the center, O, of a cell like abcd, one can assume that T = 0 so that the structure is the structure of

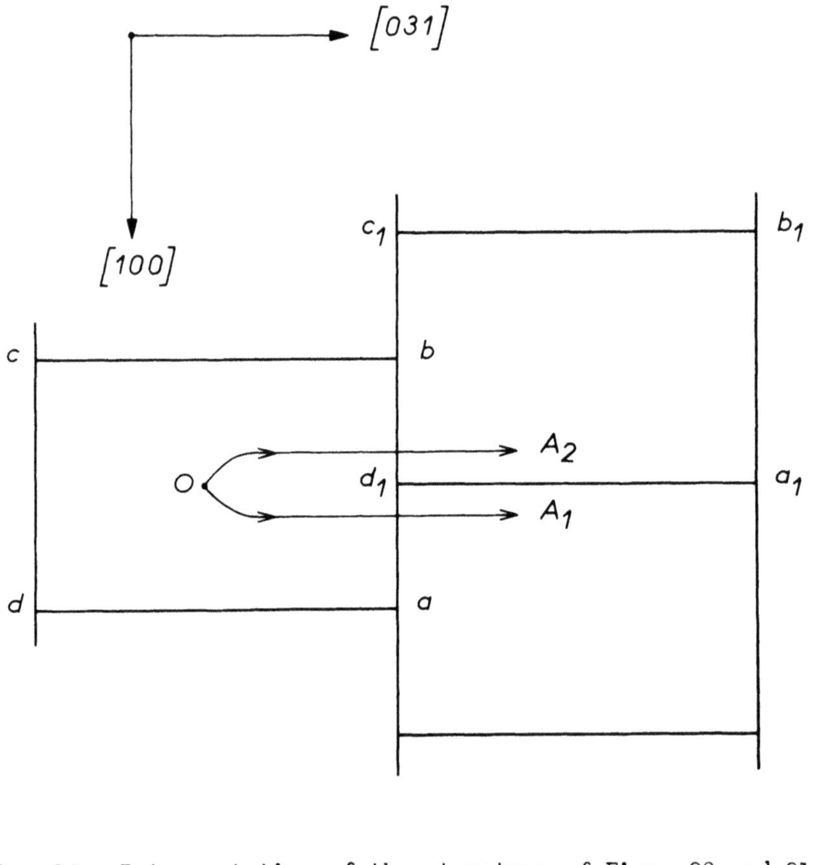

Fig. 24. Interpretation of the structure of Figs. 20 and 21.

the (013) twin boundary. At another point O_1, vector T is obviously:

- perpendicular to OO_1
- proportional to the length of OO_1 so that:

$$T = \delta' \times OO_1$$

Now if O_1 moves from O along a straight line parallel to (031), the vector T remains parallel to [100]. The calculation shows that there is a discontinuity of the structure at $T = \frac{a}{2}$, a being the lattice parameter.

This discontinuity corresponds to the uninterrupted fringes in the photographs of Fig. 19 and 20. Since $T = \delta' \times OO_1$, it follows that the spacing, d_1, of the lines is $d_1 = \frac{0.5\,a}{\delta'}$. This value agrees with electron microscopical observations.

If we now examine the structures which are encountered from O along two very close paths A_1 and A_2, it happens that the discontinuity across ab yields two different structures which exhibit a discontinuity across a_1d_1. This explains the occurrence of the second system of lines in Fig. 20.

Similarly it is possible to examine the structures from O along a line parallel to [100]. The calculation shows that a discontinuity occurs when $T \simeq \frac{a}{3}$. From this, it follows that the line spacing is $d_2 \simeq \frac{a}{3\delta'}$. The shape of the cells is determined by the ratio $\frac{d_1}{d_2} \simeq 1.5$. This value agrees well with the ratio measured from Fig. 20.

Thus it is clear that calculation of the structures gives a simple explanation of the observations obtained with the electron microscope. It is noteworthy that the discontinuity lines which appear under these conditions, and which are justified by the above reasoning, cannot be considered unquestionably as being dislocations because the nature of the discontinuities in the boundary and in the bulk crystal is different. This question will now be discussed in some detail.

REMARKS CONCERNING THE "GRAIN BOUNDARY DISLOCATIONS"

When observing grain boundaries in the electron microscope various kinds of lines or fringes may appear. Some of them correspond to elements of the structure which are effectively present in the boundary, whereas others like thickness fringes result only from purely optical phenomena. The structural elements which yield visible lines appear to be similar to dislocations in the bulk crystal so that they are often called "grain boundary dislocations".

Classification

The actual state of our knowledge concerning these grain boundary dislocations allows one to distinguish between several such types.

"Intersection" dislocations. We call "intersection dislocations" those discontinuities which form in the surface of a grain boundary when lattice dislocations of one crystal moves into the boundary. Such dislocations are visible in Fig. 19 where they disturb the arrangement of the parallel fringes causing the formation of jogs.

Clearly these dislocations are merely accidents in the grain boundary surface and are independent of the equilibrium structure of the boundary. One can assume that a suitable heat treatment, able to restore the true equilibrium structure of the grain boundary would make them disappear.

Structural dislocations. Among the dislocations which are real components of the boundary structure two groups can be distinguished depending upon whether the boundary under consideration is a low angle or a high angle boundary.

(1) Dislocations in low angle boundaries. A grain boundary placed between two crystals which are relatively slightly misoriented is a "sub-boundary". It is well known that, in the general case, a sub-boundary consists of a wall of dislocations which do not differ from the dislocations which can exist in either crystal.

It is worth noting however, that the dislocations which are present in a sub-boundary prepared artificially, that is by imposing upon the crystals a deternined orientation relationship, would not be stable if they were not associated in a wall[5,6]. This is the case for the boundary in Fig. 25 which is composed of a wall of edge dislocations whose Burgers vectors are parallel to a

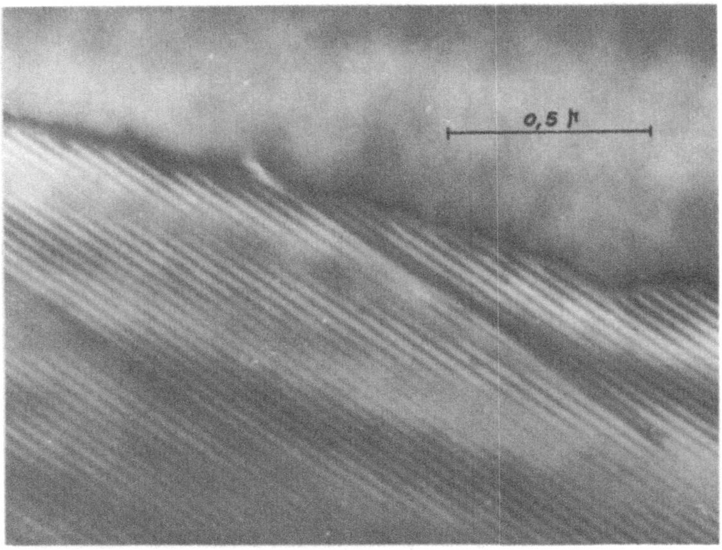

Fig. 25. Symmetrical tilt boundary of 1° misorientation about [100] in aluminium.

[001] direction. Such dislocations have never been seen in the interior of a crystal. The same applies to the boundary in Fig. 26 which consists of two groups of screw-dislocations perpendicular to each other. The Burgers vectors of these two groups of dislocations are perpendicular to two [011] type directions, respectively.

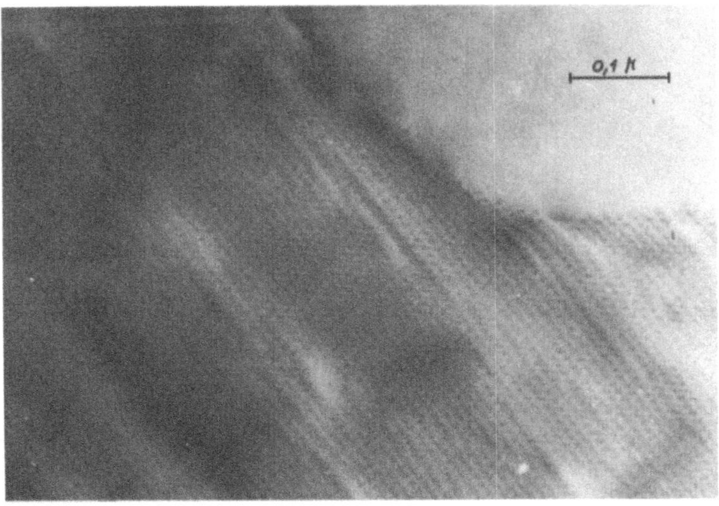

Fig. 26. Twist boundary of 1° misorientation about [100] in aluminium.

(2) Dislocations in high angle boundaries. In the general case, a high angle boundary has a chaotic structure. No two atoms in the boundary regions, for instance two neighbouring atoms, ever have the same environment; and it can be said that in a certain way the whole structure consists of very closely-spaced discontinuities. It is not possible to say what type of dislocation might be found in such a structure. In fact it is only in those boundaries which we define as being "close to special boundaries" that it is possible to define such dislocations.

In order to discuss this point more thoroughly the general method described earlier will be used again. The basic boundary is a "special boundary" whose periodicity in plane P is defined by two vectors V_1 and V_2, one of these being very short. In a boundary close to this special boundary the position of the crystals at any point is characterized by a vector T, comparable with the vector T of Fig. 18. Now, there is also in this boundary a double periodicity defined by two vectors V_1' and V_2', one or both of these vectors being long. One has:

$$V_1' = k_{1,1} V_1 + k_{1,2} V_2$$

$$V_2' = k_{2,1} V_1 + k_{2,2} V_2$$

where the coefficients, $k_{i,j}$, are whole numbers. Figure 27 represents the OACB cell related to the vectors V_1' and V_2'.

For each lattice point of the planar lattice defined by the OACB cell, the structure is the same as at O and

$$T = k_1 T_1 + k_2 T_2$$

k_1 and k_2 being whole numbers, T_1 and T_2 being the values of T in A and B respectively.

Let us now consider the structure of OA at a point P. If P moves from O, the length of vector T increases. Obviously the modification of the structure cannot be a continuous function of T; for a certain threshold value of T_1, T_1', at a point P_1, the structure will abruptly change. Beyond P_1 it may well happen that the structure changes continuously up to point A. In this case, the discontinuity in P_1 can be characterized simply by T_1. A discontinuity line L_1 must go through P_1; L_1 can be considered as equivalent to a dislocation line of Burgers vector T_1.

However, nothing prevents several discontinuities from taking place along OA at P_1, P_2, etc.. In this case lines such as L_1, L_2, etc., can always be considered as dislocation lines, yet a Burgers vector can be defined only for the whole group of lines crossing the segment OA.

Let us now consider the path $Q_1 Q_2$ in Fig. 27 in which Q_1 and Q_2 can move to O and A, respectively, without crossing a discontinuity line. This path can be characterized by the discontinuity T_1. Such a path can be thought of as being equivalent to a Burgers circuit.

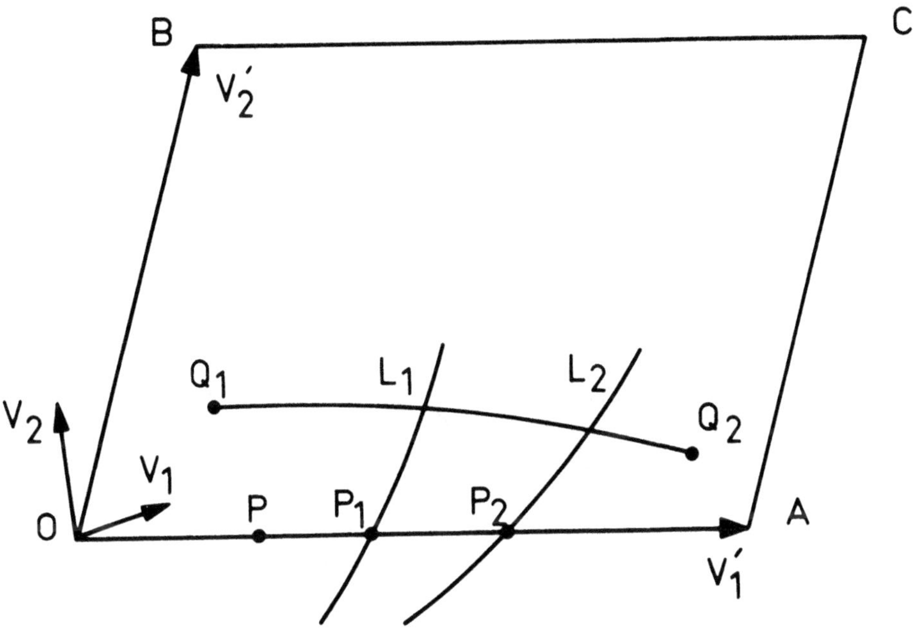

Fig. 27. Grain boundary discontinuity lines.

Let us now consider a closed circuit composed of segments such as $Q_1 Q_2$, it can be characterized by a sequence of discontinuities T_i, T_j, etc., which are analogous to T_1. Clearly,

$$\Sigma T_i = 0$$

This relationship being satisfied, the arrangement of the dislocations can assume various patterns. Figure 28 shows such an arrangement, although this is not meant to correspond to any real structure.

Of course, it will be difficult to predict if any particular discontinuity line will be visible during experimental observation with, for instance, an electron microscope.

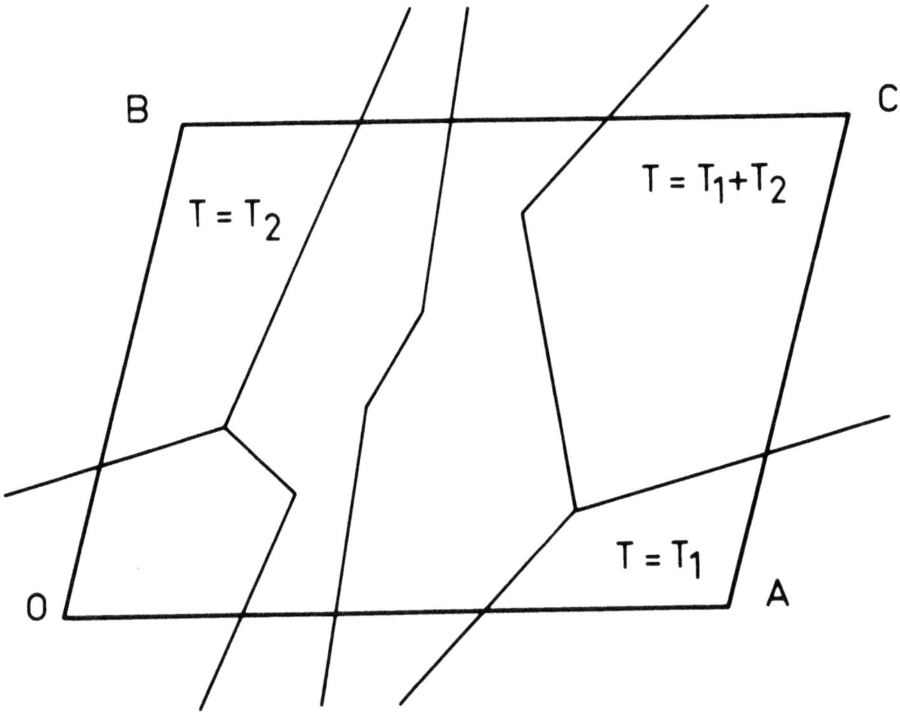

Fig. 28. T vectors in a boundary "close to a special boundary".

Note concerning the general boundary. The foregoing considerations have made it possible to define, in those boundaries which we call "close to special boundaries", dislocation lines which will in fact be visible under certain observation conditions. A general boundary can always be obtained by modification of a special boundary. It will differ from these boundaries, which we have been considering, on the following points:

- the dislocation density is high.
- each dislocation corresponds to a small discontinuity.

These features show clearly why nothing particular is visible in a general boundary when it is observed in the electron microscope.

CONCLUSION

A relaxation calculation based on a minimum free energy criterion has been devised in order to determine the structures and the energies of grain boundaries in pure metals. The correct use of the method requires that all the parameters, which are necessary in order to define the structure, are allowed to vary. This is particularly the case for the number N of atoms and the vector T.

This method has been applied particularly to aluminium. The structures and the energies of symmetrical [001], [011] and [111] tilt boundaries have been determined. The influence on these of a non-symmetrical orientation of the boundary plane has been calculated.

An investigation of boundaries defined as "close to special boundaries" has enabled one to obtain a deeper understanding of certain discontinuities in the structure of grain boundaries. It has been shown why some of these discontinuities are visible in the electron microscope whereas others are not.

The concept of grain boundary dislocations has been examined and a distinction has been drawn between "intersection dislocations" and "structural dislocations". The former occur by the intersection of a glide surface of one crystal with the boundary. They are not true components of the boundary. The latter, (structural dislocations) appear clearly only in boundaries "close to special boundaries". Under certain conditions one is allowed to define a discontinuity vector equivalent to a Burgers vector; and a concept analogous to a Burgers circuit has been proposed.

ACKNOWLEDGMENT

It is a pleasure for us to thank Doctor I. G. DAVIES for his invaluable help in preparing the translation in English of the French text.

REFERENCES

1. International Conference on Grain boundaries and interfaces. I.B.M. - August 23-25, 1971 (to be published).

2. R. M. J. Cotterill and M. Doyama, Phys. Rev., 137, 3A, 994, (1965).

3. C. Goux, Mem. Sci. Rev. Met., 58, 661 and 769, (1961).

4. G. H. Bishop and G. A. Bruggeman, (Paper of the "Army Materials and Mechanics Research Center"- Watertown, Massachusetts).

5. J. Levy and C. Goux, Mem. Sci. Rev. Met., 64, 663, (1967).

6. J. Levy, Thesis, Paris, 1968.

7. B. Michaut, Thesis, Grenoble, 1971.

ON GRAIN BOUNDARY DISLOCATION CONTRAST IN THE ELECTRON MICROSCOPE†

R. W. Balluffi*, G. R. Woolhouse**, and Y. Komem***

* Department of Materials Science & Engineering, Cornell University, Ithaca, New York 14850.
** Now at the University of Cambridge, Cambridge, England.
*** Now at the Technion-Israel Institute of Technology, Haifa, Israel.

ABSTRACT

Grain boundary dislocations (GBD's) may be studied conveniently in the electron microscope using specimens which are prepared by welding together two single crystal films having predetermined orientations to produce a thin-film bicrystal slab containing any type of grain boundary. The boundary is then examined at normal incidence. Problems associated with the imaging of GBD's in both low and high angle boundaries of this type are considered. The following main topics are discussed from both the theoretical and experimental standpoints: (1) the strain field diffraction contrast produced by isolated screw and edge GBD's under relatively simple conditions where no more than two strong beams exist in the upper crystal, and no more than four strong beams exist in the lower crystal; (2) the effects of superimposed Moiré patterns which are formed when more than one beam enters the objective aperture; (3) the diffraction contrast produced by regular arrays of closely spaced GBD's and the problem of resolving individual GBD's in such arrays; and (4) evidence that regular arrays of GBD's act as two-dimensional diffraction gratings which produce extra diffraction spots which may be of use in the direct imaging of finely spaced arrays.

† This work was supported by the U. S. Atomic Energy Commission under Contract AT(30-1)3504. Additional support was received from the Advanced Research Projects Agency through the technical facilities of the Materials Science Center at Cornell University.

INTRODUCTION

Electron microscope studies of grain boundary dislocations (GBD's) have been carried out in our laboratory during the last few years using thin-film bicrystal gold specimens possessing the geometry shown in Fig. 1. The bicrystal slabs have been prepared by welding evaporated single crystal films (Crystal 1 and Crystal 2) together face-to-face under controlled conditions according to a technique which has been described elsewhere[1-8]. By varying the orientations of both Crystals 1 and 2, a variety of low and high angle grain boundaries of twist, tilt and mixed type has been obtained. These boundaries have been examined in the electron microscope at approximately normal incidence (electron beam approximately parallel to z).

Two general types of GBD's have been observed in such boundaries with this technique:

1) Intrinsic GBD's. These dislocations are present in regular planar networks and are part of the equilibrium structure of many low and also high angle boundaries. The regular arrangement in the networks guarantees that the long range strain fields vanish. So far, such networks have been observed when the crystal

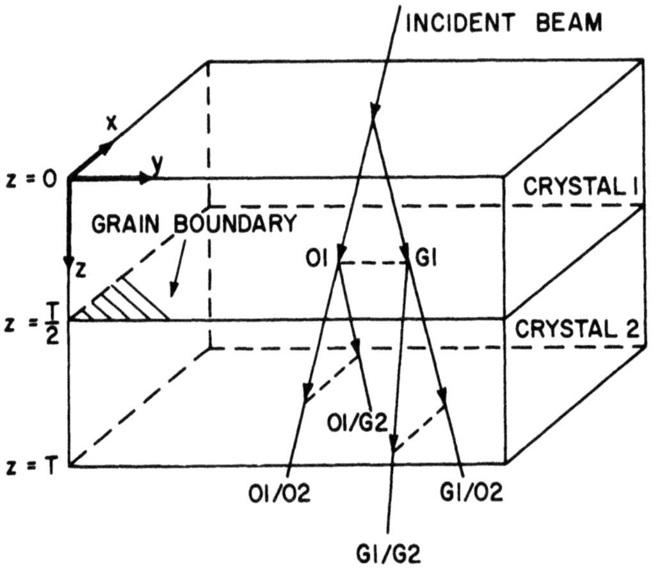

Fig. 1. Geometry of diffraction in bicrystal specimen.

misorientation has been close to a singular misorientation corresponding to a high density of coincidence sites in the unrelaxed boundary.

2) Extrinsic GBD's. These dislocations are extra GBD's which happen to be present and which are not part of the equilibrium structure of the boundary. These dislocations are generally distributed in a rather disorganized fashion, and their strain fields do not cancel at large distances. Such dislocations could be formed, for example, by plastic deformation.

A review of the relevant experimental results has been given recently[8]. The purpose of the present paper is to carry out a general analysis and review of the contrast effects which are expected to result from GBD's in bicrystal specimens of the present type.

DIFFRACTION CONTRAST DUE TO GBD's AND CRYSTAL MISORIENTATION IN BICRYSTAL SPECIMENS

General Problem

We begin with a discussion of the GBD diffraction contrast which may be expected in the general case in the present specimens* Such contrast may be quite complicated due to the possibility that either, or both, of the misoriented crystals may have several Bragg reflections excited. Proper account must then be taken of all the electron beams which are present due to diffraction and double diffraction, etc., in the two crystals. In addition, Moiré patterns may be produced, if it turns out that two or more beams exit from the lower surface of the bicrystal in nearly the same direction and are allowed to enter the objective aperture of the microscope. As is well known[9], interference between such beams generally produces a regular fringe (Moiré) pattern which under certain conditions may be rather similar to an array of intrinsic GBD's[10]. In the completely general case, therefore, the total diffraction contrast observed may arise from a complex superposition of dislocation strain field scattering effects in the various beams which are excited together with possible Moiré interference effects between the various beams.

---

* We point out in the final section of the present paper that some of the GBD contrast obtained in the case where the GBD's are present in the form of a regular network may be due to the direct imaging of the network, i.e., all of the observed contrast may not be due to diffraction contrast.

A good example of a relatively complicated diffraction situation is shown in Fig. 2a for the case of a small angle pure twist boundary, lying parallel to (001), with a twist angle, θ, of about 3.4°. Several Bragg reflections were excited in both the upper and lower crystals, and the square arrays of diffraction spots clustered around each major reciprocal lattice point in the selected area diffraction (SAD) pattern are readily explained on the basis of diffraction and double diffraction in the two crystals. The corresponding (000) bright field image, which was obtained by allowing all of the beams in the (000) cluster of spots into the aperture, is also shown. The contrast over most of the boundary is seen to be due to a complex superposition of effects arising from the strain field scattering of the square grid of intrinsic screw GBD's*, which run parallel to <110> directions, and which are known to make up the low angle boundary structure[1], and Moiré fringes caused by interference effects between the various (000) beams which entered the aperture. The Moiré interference effects may be seen most clearly in regions such as at B in Fig. 2a. Here the grain boundary is absent because of the presence of relatively large bubbles in the boundary which resulted from incomplete welding of the two crystals during specimen fabrication[1]. The pattern

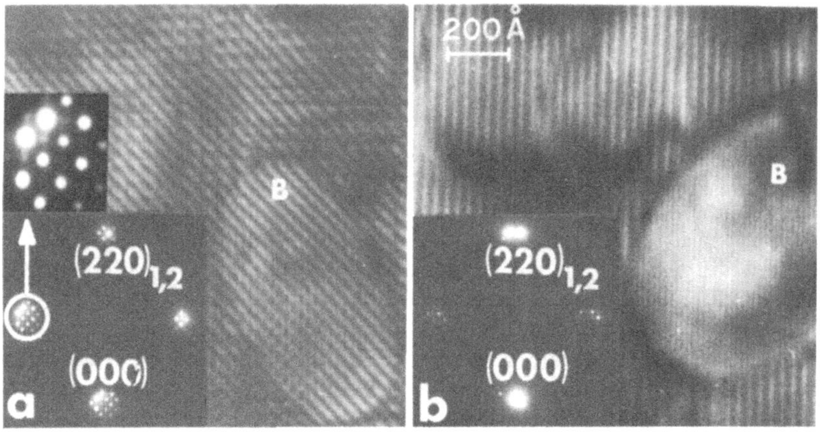

Fig. 2. (a) SAD pattern and bright field image of specimen containing small angle pure twist boundary lying parallel to (001) with θ ≃ 3.4°. The $(200)_{1,2}$ region of the SAD pattern which is encircled is shown enlarged. Bubble region at B.
(b) SAD pattern and $(220)_{1,2}$ dark field image of specimen containing (001) low angle twist boundary with θ ≃ 4.6°. Bubble region at B.

* All of the GBD network contrast seen in Fig. 2 (and also in Fig. 3) may not be due to diffraction contrast, since a contribution due to the direct imaging of the network may be present (see previous footnote).

in such a region is therefore a pure Moiré pattern due solely to the misorientation of the perfect crystalline regions located above and below the bubble. These bubbles, which were present in all of the boundaries examined, therefore gave us an extremely convenient means of separating out, and observing directly, any Moiré effects.

GBD Diffraction Contrast in Simplified Situations

In the present work we shall concentrate on situations which are simpler than those exemplified by Fig. 2a and which may be achieved experimentally and analyzed to obtain valuable information about the GBD's. We therefore restrict ourselves to the situation in Fig. 1 where no more than two strong beams exist in Crystal 1 and no more than four strong beams exist in Crystal 2. This corresponds to the case where the incident beam may excite only one strong diffracted beam (i.e., the G1 beam) in the upper crystal and then go on to diffract only one additional strong beam (i.e., the O1/G2 beam) in the lower crystal. In addition, the diffracted beam in the upper crystal (G1) may go on to excite only one strong doubly diffracted beam (i.e., the G1/G2 beam) in the lower crystal. (Here, O and G indicate "transmitted" and "diffracted" respectively, and 1 and 2 indicate Crystal 1 or 2. Hence, for example, O1/G2 indicates transmitted in Crystal 1 and diffracted in Crystal 2.) Under usual experimental conditions it is difficult to avoid the presence of a few weakly excited beams due either to primary diffraction or double diffraction[11]. However, the intensity of these beams can usually be reduced to the point where they can be ignored. We note that we often expect the diffraction vector of the G1/G2 beam in Crystal 2 to be the same as, or the negative of, the diffraction vector of the O1/G2 beam.

The following diffraction situations are therefore possible with the arrangement shown in Fig. 1:

(A) G1 beam strongly excited; O1/G2 and G1/G2 beams not excited. This situation corresponds, of course, to a simple two-beam case in Crystal 1.

(B) O1/G2 beam strongly excited; G1 and therefore, G1/G2 beams not excited. This is a simple two-beam case in Crystal 2.

(C) O1/G2 and G1 beams strongly excited; G1/G2 beam not excited. This is a three-beam situation in Crystal 2 in which two of the beams interact dynamically with each other but not with the third (see below).

(D) G1 and G1/G2 beams strongly excited; O1/G2 not excited.

This situation is similar to (C).

(E) G1, G1/G2 and 01/G2 beams strongly excited. This is a four-beam situation in Crystal 2 which consists of two independently diffracting systems.

In general, the GBD diffraction contrast in any one of the four beams exiting from the lower surface of the specimen may be obtained by employing a method which depends upon the fact that the four beams in Crystal 2 are dynamically coupled only in pairs* Thus, the 01/02 and 01/G2 beams are coupled separately, and the G1/02 and G1/G2 beams are coupled separately. No other coupling exists, because the reciprocal lattice vector which couples the 01 and G1 beams in Crystal 1 is <u>not</u> a reciprocal lattice vector in Crystal 2. The amplitudes of the 01 and G1 beams, as they exit from the upper crystal at z = T/2, may first be obtained by integrating the standard Howie-Whelan equations[12] for the two-beam dynamical case along a column in Crystal 1. The 01 beam at z = T/2 may then be regarded as an incident beam on Crystal 2, and the amplitudes of the 01/02 and 01/G2 beams, as they exit from the lower crystal at z = T, may be obtained by a further integration over a column in Crystal 2. Similarly, the G1 beam may be regarded as an incident beam on Crystal 2, and the G1/G2 and G1/02 beams may be found from still another integration.

Under many conditions when strong diffracted beams are excited in both Crystals 1 and 2, the various exiting beams are sufficiently divergent so that each may be enclosed separately by the objective aperture and examined for GBD contrast. Under these simple (and desirable) conditions Moiré interference effects are avoided, and the observed contrast should correspond to that obtained from the calculations described above.

Combined GBD Diffraction Contrast and Moiré Contrast in Simplified Situations

However, under certain conditions (see below), beams such as G1/02 and 01/G2 may be nearly coincident, and in such cases they cannot be examined independently but instead enter the objective aperture together. The resulting image then consists of a complicated superposition of the GBD strain field contrast and the Moiré pattern. Examples of such cases are given in Figs. 2b and 3. Figure 2b shows the case of a specimen containing a low angle twist boundary lying parallel to (001) in which strongly excited G1 and 01/G2 beams (corresponding to reciprocal lattice vec-

* By dynamically coupled, we mean that the two beams interact only with each other via Bragg diffraction.

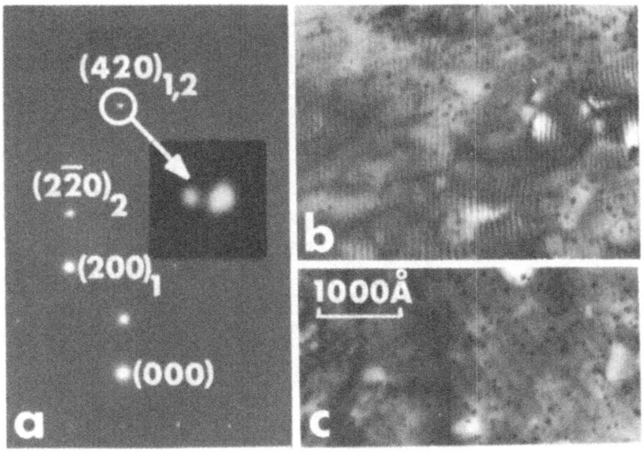

Fig. 3. (a) SAD pattern from specimen containing high angle twist boundary lying parallel to (001) with $\theta \approx 36.9^{\circ}$. (b) $(420)_{1,2}$ dark field image obtained by admitting both beams shown encircled and enlarged in (a) into objective aperture. (c) Bright field image.

tors $(220)_1$ and $(220)_2$ respectively) are nearly coincident. The $(220)_{1,2}$ dark field image obtained by admitting both beams into the aperture is seen to contain a Moiré pattern which is most clearly observed in the bubble region marked by B. In the grain boundary region the contrast is a result of the superimposed effects of the Moiré pattern and the contrast produced by the orthogonal network of intrinsic screw GBD's which run parallel to <110> directions in this boundary. In this case the Moiré pattern is overwhelmed by the GBD network contrast. Here, the spacing of the intrinsic screw GBD's is double that of the Moiré spacing, and only the set of GBD's in the network running parallel to the diffraction vector $\vec{g}$ is in strong contrast, since $\vec{g}\cdot\vec{b}$ is zero for the second orthogonal set ($\vec{b}$ = Burgers vector). A Moiré pattern obtained from a special high angle boundary is shown in Fig. 3. In this case the specimen contains a twist boundary lying parallel to (001) with $\theta \simeq 36.9°$. At this twist angle (420) planes in the two crystals are almost parallel, and the double diffraction spot obtained by exciting these two Bragg reflections (i.e., the G1/O2 and O1/G2 beams) is shown in Fig. 3a*.

The corresponding Moiré pattern in the dark field image formed from these two beams is seen in Fig. 3b. We note that in this case the Moiré contrast is considerably greater than the contrast from the network of intrinsic screw GBD's which is known to be present

* Note that several other reflections were strongly excited in Crystals 1 and 2 so that this situation does not correspond exactly to the simple situation shown in Fig. 1.

in this boundary*. These latter GBD's are clearly observed in the absence of Moiré effects in the bright field image, shown in Fig. 3c, which was obtained from the single strong (000) beam seen in Fig. 3a.

The diffraction contrast obtained in the case where two beams (designated now by 1 and 2) enter the objective aperture may be obtained by combining our previous solutions for beams 1 and 2 after adding in an appropriate phase factor which takes into account the fact that the beams are slightly divergent. The amplitude of the total wave field produced by combining the two waves may be written in the form,

$$\psi = \phi_1 \exp(2\pi i \vec{k}_1 \cdot \vec{r}) + \phi_2 \exp(2\pi i \vec{k}_2 \cdot \vec{r}) \ , \qquad (1)$$

where ϕ_i is the amplitude of wave i at $z = T$ as determined from the previous integration of the Howie-Whelan equations, k_i is the wave vector, and r is the position vector. (We note that the ϕ_i are generally complex amplitudes.) The total intensity is then

$$I = \psi\psi^* = |\phi_1|^2 + |\phi_2|^2 + 2|\phi_1||\phi_2| \cos\{2\pi(\vec{k}_1 - \vec{k}_2)\cdot\vec{r} + \delta\} \ , \qquad (2)$$

where δ is a constant phase factor. The first two terms are just the sum of the intensities of the two beams, while the third term is the added Moiré term which is seen to consist of a set of evenly spaced fringes of amplitude modulated by the function $|\phi_1||\phi_2|$. If we define the "Moiré wave vector" by $\Delta\vec{k} \equiv \vec{k}_1 - k_2$,[13] it is readily seen that the Moiré fringes run perpendicular to $\Delta\vec{k}$ and possess a spacing

$$d_M = \frac{1}{|\Delta\vec{k}|} \ . \qquad (3)$$

In the case of the low angle twist boundary, where the G1/02 and

---

* In previous work[2] we have demonstrated the existence of orthogonal networks of intrinsic screw GBD's with $\vec{b}$'s of the form a/10<310> in such twist boundaries which possess twist deviations from the critical twist angle $\theta = 36.9°$. The dislocations form in order to restore most of the grain boundary area to the low energy configuration corresponding to the exact $\theta = 36.9°$ structure. The latter structure is produced when the two crystals are misoriented to produce a relatively high density of coincidence sites in the boundary in the unrelaxed situation. Subsequent relaxations produce a grain boundary characterized by a two dimensional periodic structure of relatively short wavelength.

O1/G2 beams form the dark field Moiré pattern in Fig. 2b, $\Delta\vec{k}$ is obviously given by $\Delta\vec{k} = (\vec{k}_o+\vec{g}_1) - (\vec{k}_o+\vec{g}_2) = \vec{g}_1-\vec{g}_2$ where $\vec{g}_1$ and $\vec{g}_2$ are the diffraction vectors in the upper and lower crystals respectively, and $\vec{k}_o$ is the wave vector of the incident beam. The Moiré fringes should therefore run perpendicular to $\vec{g}_1-\vec{g}_2$ and should possess a spacing given by

$$d_M = \frac{1}{|\vec{g}_1-\vec{g}_2|} = \frac{d_{220}}{\theta} = \frac{a}{2\sqrt{2}\cdot\theta} \tag{4}$$

where d_{220} is the {220} interplanar spacing, and a is the lattice parameter. On the other hand, the intrinsic GBD spacing should be given by

$$d_D = \frac{|\vec{b}|}{\theta} = \frac{a}{\sqrt{2}\cdot\theta}\ , \tag{5}$$

since $\vec{b} = a/2\langle 110\rangle$. The geometries of the Moiré fringes and GBD's in Fig. 2b are seen to be consistent with these results.

DIFFRACTION CONTRAST DUE TO ISOLATED GBD

We now proceed to calculate (and compare with experiment in certain cases) the diffraction contrast expected from isolated GBD's for several simple situations.

One Crystal Strongly Excited

In many situations it is relatively easy to establish conditions so that only one crystal is strongly excited. This is particularly the case for high angle boundaries where the orientation of Crystal 2 differs widely from that of Crystal 1, and various low index directions in the two crystals lie in appreciably different directions. Examples are shown in Fig. 4 where GBD's have been imaged by diffraction contrast by establishing a two-beam condition in only one crystal. Under this particularly simple condition (either Case A or Case B in the previous section) we intuitively expect that we should be able to ignore any diffraction effects in the crystal which is far from the Bragg condition and treat the contrast as due solely to the excited crystal with the GBD embedded in its surface. In such a case, of course, there is no need to take into account any surface relaxation of the strain field, since the free surface is sufficiently far away so that such effects are negligible. It will be seen that these expectations are fully substantiated by the following calculations.

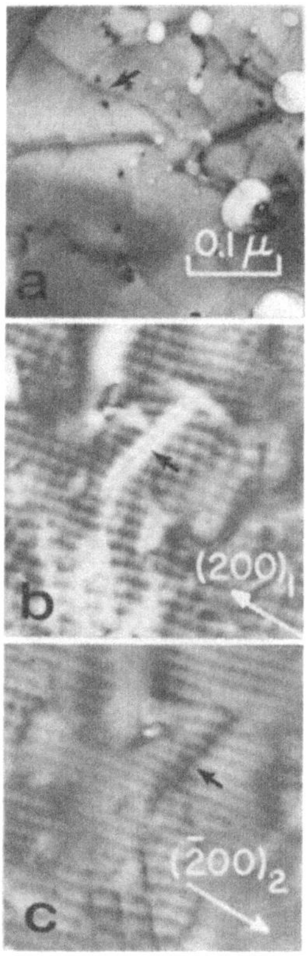

Fig. 4. (a) Screw GBD with $\vec{b}$ = a/10<310> (at arrow) in high angle twist boundary lying parallel to (001) with θ ≃ 36.9°. One crystal excited, $\vec{g}$ in plane of boundary. (b) Extrinsic edge GBD with $\vec{b}$ = a/2<001> (at arrow) in high angle twist boundary lying parallel to (001) with θ ≃ 36.9°. Background structure consists of orthogonal network of intrinsic screw GBD's. Detailed structure of extrinsic edge GBD described elsewhere[6]. Dark field image with one crystal excited with $\vec{g}$ as indicated. (c) Same as (b) except different $\vec{g}$ as indicated.

Case of $\vec{b}\perp\vec{n}$. Defining $\vec{n}$ as the unit vector normal to the boundary, this situation corresponds to the case where a non-zero value of $\vec{g}\cdot\vec{b}$ is the only source of contrast[14], since $\vec{g}$, $\vec{b}$, and the dislocation tangent vector, $\vec{u}$, all lie in the boundary and, therefore, the quantity $\vec{g}\cdot\vec{b}x\vec{u}$ = 0. We therefore consider the case of a screw GBD with $\vec{g}\cdot\vec{b} \neq 0$. The standard displacement field characteristic of a screw dislocation in an infinite medium was used. In order to produce a strong two-beam condition the value of the deviation parameter, s, was taken to be zero, and the "non-excited" situation was simulated by assigning to s the relatively large value of 0.1. Other details of the diffraction parameters which were used are given in the caption. The GBD scattering intensity profiles for the 01/02 (bright field) and G1/02 (dark field) beams when the upper crystal is strongly excited, and the profiles for the 01/02 (bright field) and 01/G2 (dark field) beams when the lower is strongly excited are shown in Fig. 5. As expected, these results, which show strong asymmetric GBD contrast, exhibit similarity between the bright field beams and complementarity between the dark field beams and are in agreement with the results expected on the basis of the previous calculations of Howie and Whelan[12] for screw dislocations very near the surfaces of a single crystal slab. We note that the sign of the contrast (i.e., above or below background) reverses with a change in the sign of the quantity $\vec{g}\cdot\vec{b}$. An experimentally observed example showing typical non-symmetric contrast of this type from a nearly screw GBD ($\vec{b}$ = a/10<310>) in a specimen containing a high angle twist boundary with one crystal strongly excited is shown in Fig. 4a.

<u>Case of $\vec{b}\|\vec{n}$</u>. In this case $\vec{b}$ is perpendicular to the boundary, and the dislocation is of edge type with $\vec{g}\cdot\vec{b} = 0$. A non-zero value of $\vec{g}\cdot\vec{b}x\vec{u}$ is then the only source of contrast, and we therefore choose $\vec{u}$ and $\vec{g}$ so that $\vec{g}$ has a strong component parallel to $\vec{b}x\vec{u}$. The GBD contrast in the various bright field and dark field beams was calculated as in the previous case, and the results are given in Fig. 6. Details are again given in the caption. The results show symmetric triple contrast of either the "white-black-white" (WBW) or "black-white-black" (BWB) type*. Again the results are in agreement with those expected on the basis of previous calculations[12] for the two-beam case in single crystals with $\vec{g}\cdot\vec{b} = 0$. Here, the bright field beams exhibit complementarity, and the dark field beams exhibit similarity in contrast to the previous case. In addition, the sign of the contrast reverses with a change in the sign of the quantity $m = 1/8(\vec{g}\cdot\vec{b}x\vec{u})$. Observed examples of this type of contrast are shown in Figs. 4b and 4c. In these cases an extrinsic edge GBD with effective Burgers vector $\vec{b}$ = a/2<001> was imaged in a high angle twist boundary. We see that the dark field contrast reversed due to the change in sign of m and the fact that Fig. 4b was obtained with the upper crystal excited and Fig. 4c with the lower crystal excited.

We conclude that the investigation of GBD's becomes quite simple when it is possible to excite only one crystal in a two-beam condition. Under these conditions the non-excited crystal may be ignored for all practical purposes, as expected. Burgers vector determinations may therefore be carried out in the usual way. Some applications of these techniques to the determination of the Burgers vectors of GBD's have been described elsewhere[1,2,6].

---

* This description is appropriate to a positive photographic print.

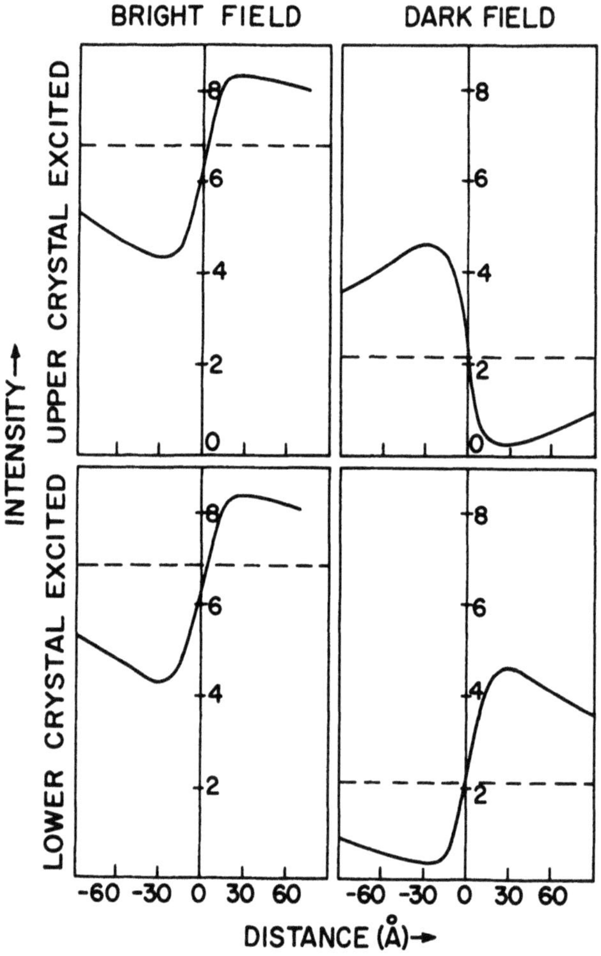

Fig. 5. Calculated diffraction contrast intensity profiles for screw GBD with one crystal strongly excited in two-beam situation (Case A or B, text). T = 800Å, $\vec{g}\cdot\vec{b} = +1$, $\vec{g}$ = (200), and extinction distance and absorption parameters appropriate to gold (ξ_g = 200Å, ξ_o = 93Å, ξ_g' = 2000Å, ξ_o' = 930Å). In each case strongly excited reflection has s = 0, and weakly excited reflections have s = 0.1. When $\vec{g}\cdot\vec{b}$ changes sign, the sign of the contrast of all profiles reverses. Background intensity far from dislocation indicated by dashed line.

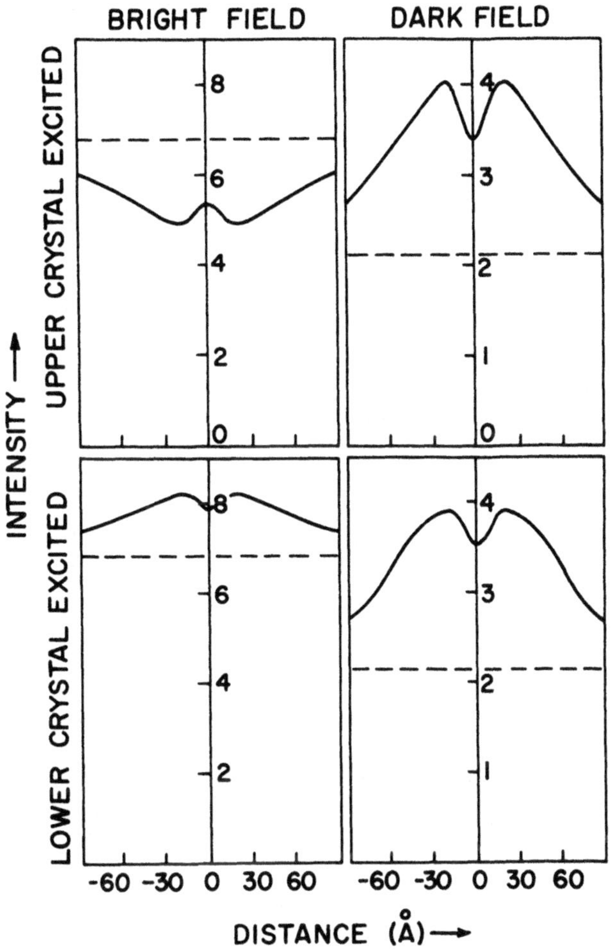

Fig. 6. Calculated diffraction contrast intensity profiles for edge GBD with one crystal strongly excited in two-beam situation (Case A or B, text). T = 800Å, $\vec{g}\cdot\vec{b} = 0$ ($\vec{g}$ is in boundary plane, and $\vec{b}$ is perpendicular to boundary plane). Extinction distance and absorption parameters appropriate to gold (ξ_g = 200Å, ξ_o = 93Å, ξ'_g = 2000Å, ξ'_o = 930Å). The quantity $m = 1/8(\vec{g}\cdot\vec{b}\times\vec{u}) = +0.102$. Sign convention chosen so that $(\vec{b}\times\vec{u})$ runs from the dislocation line away from the extra half plane. When m changes sign, the sign of the contrast of all profiles reverses.

Both Crystals Strongly Excited

We now consider Case C in the previous section where the G1 beam in Crystal 1 and the O1/G2 beam in Crystal 2 are both strongly excited. This condition may often be obtained in the case of low angle boundaries and under certain conditions in the case of high angle boundaries (see below). In order to simulate this situation we again took s = 0 for the G1 and O1/G2 beams, and s = 0.1 for the weak G1/G2 beam. The intensity profiles of the various beams were then calculated by carrying out successive integrations of the Howie-Whelan equations as described previously.

Case of $\vec{b}||\vec{n}$. In both crystals $\vec{g}$ was taken with a strong component parallel to $\vec{b}\times\vec{u}$ as in the one-crystal case described previously, and the calculated results are given in Fig. 7 (see caption for further details). Symmetric triple contrast was again obtained, as expected, since the contrast in each of the crystals is always of this general type (Fig. 6). In order to compare the calculated results with experiment, GBD's were examined under a number of three-beam conditions which were generally similar to those assumed in the calculations. The results are shown in Figs. 8 and 9. In Fig. 8 extrinsic edge GBD's with $\vec{b}||\vec{n}$ were imaged with $m_1>0$, $m_2>0$ using the O1/O2, G1/O2, and O1/G2 beams. The O1/O2 image, shown in Fig. 8a, exhibits exceedingly weak BWB contrast as predicted. Evidently, this is due to the fact that the final O1/O2 beam is the result of a bright field situation in the upper crystal (which tends to produce weak BWB contrast, Fig. 6) followed by a bright field situation in the lower crystal (which tends to produce opposite contrast of weak WBW type, Fig. 6). The G1/O2 beam image (Fig. 8b) exhibits stronger WBW contrast as predicted, and this is evidently due to the fact that the WBW contrast produced by the dark field situation in the upper crystal is stronger than the BWB contrast due to the bright field situation in the lower crystal (see Fig. 6). Finally, the WBW contrast found in the O1/G2 beam in Fig. 8c is also seen to be consistent with the calculated result. Another group of extrinsic edge GBD's imaged with $m_1<0$, $m_2>0$ using the O1/O2, G1/O2, and O1/G2 beams is shown in Fig. 9. Again, the experimental results are consistent with the predicted results given in Fig. 7 and can be understood qualitatively on the basis of the effects produced by the consecutive scattering in the upper and lower crystals.

Case of $\vec{b}\perp\vec{n}$. We have not investigated this case, but the contrast could easily be calculated by the present methods. Again, since the sign of the contrast produced by each crystal depends upon the sign of $\vec{g}\cdot\vec{b}$, whether the beam is bright field or dark field, and whether the crystal is upper or lower, it is possible to obtain a variety of results.

We conclude generally that the relatively more complicated

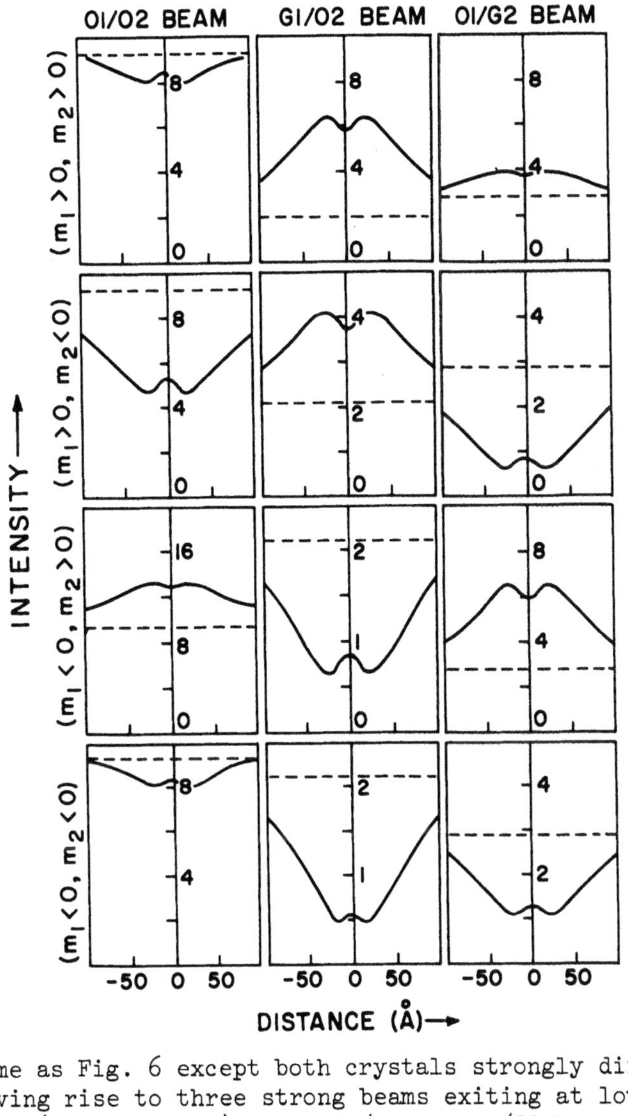

Fig. 7. Same as Fig. 6 except both crystals strongly diffracting giving rise to three strong beams exiting at lower surface (Case C, text). The O1/O2 and O1/G2 beams are coupled while the third is not. The quantity m = 1/8 $(\vec{g} \cdot \vec{b} \times \vec{u})$ has magnitude 0.102 in both crystals, but its sign varies as indicated.

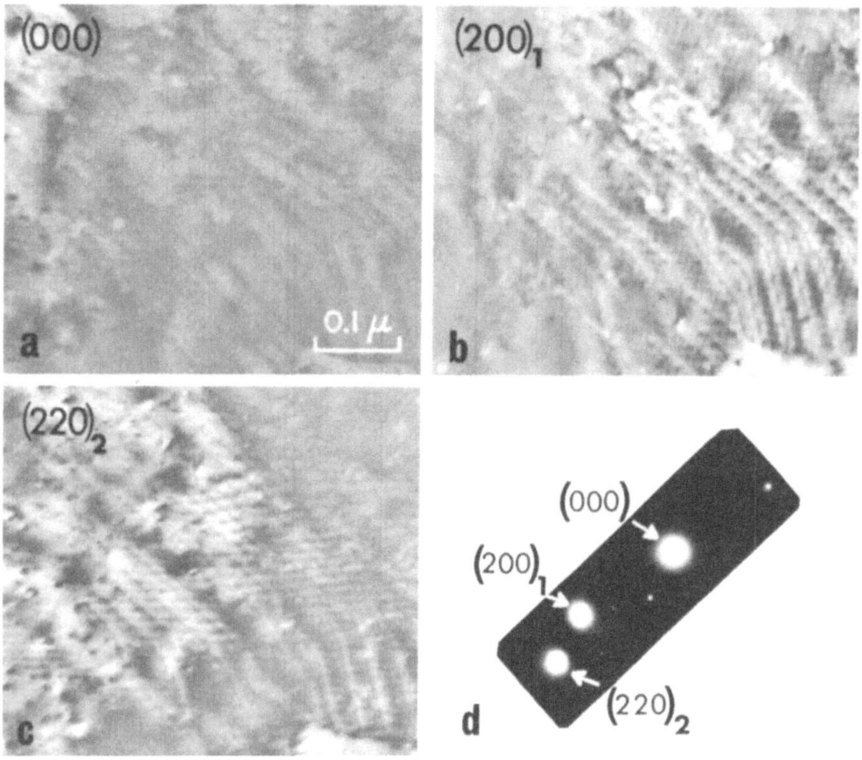

Fig. 8. Extrinsic edge GBD's with Burgers vectors perpendicular to boundary in high angle twist boundary lying parallel to (001) with $\theta \simeq 36.9°$. Background structure consists of orthogonal network of intrinsic screw GBD's. Detailed structure of extrinsic edge GBD's described elsewhere[6]. Three strong beams excited (Case C, text), m_1 and m_2 both positive. (a) O1/O2 image; (b) G1/O2 image; (c) O1/G2 image; (d) SAD pattern.

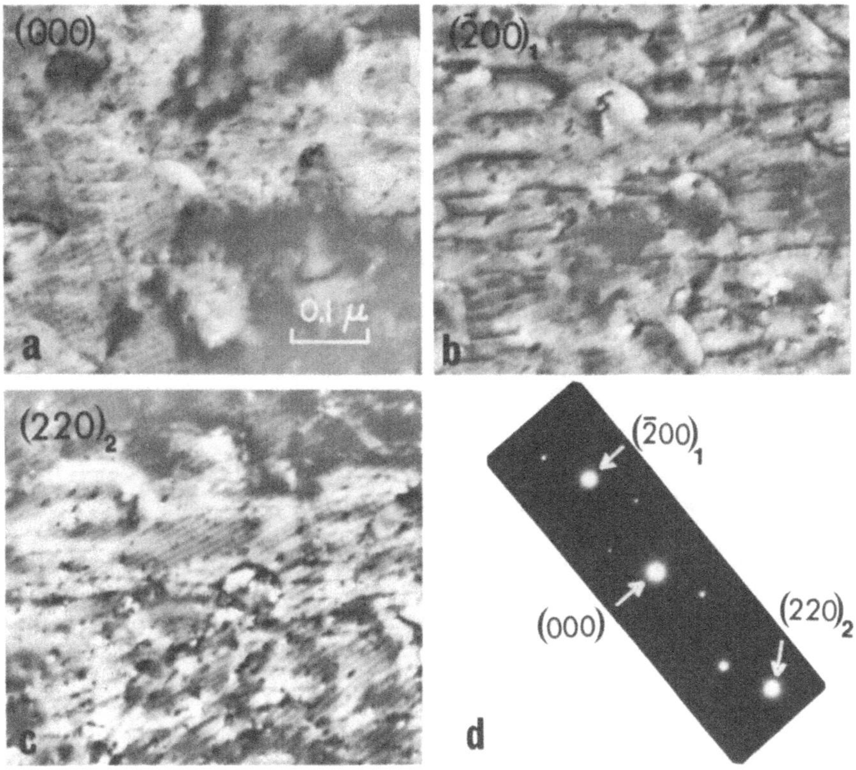

Fig. 9. Same as Fig. 8 except that $m_1 < 0$, and $m_2 > 0$.

observation of GBD's under three-beam conditions may be understood rather simply on the basis of the present analysis, and that it is obviously possible to obtain a variety of situations depending upon the details of the diffraction conditions in each crystal.

Effects of Moiré Interference

So far, we have considered only cases where the diffraction contrast in each beam could be examined independently, and we have not considered cases where more than one beam enters the objective aperture so that a Moiré pattern is produced*. Such cases are

*Such situations are generally difficult or impossible to avoid in the case of low angle boundaries where similar reflecting planes in both crystals make small angles with each other, and the doubly diffracted G1/G2 beam is almost parallel to the 01/02 beam. Of course, in such cases the situation is further complicated by the diffraction contrast from the background array of intrinsic GBD's which are present (see following section).

easily calculated by combining the beams as demonstrated previously. We shall not discuss such cases in detail, since the observation and analysis of isolated GBD's under such conditions are generally complicated, and a wide range of behavior may be expected. In addition, two-dimensional mapping of the image is required except in cases where the dislocation lies parallel to the Moiré fringes. It is worth pointing out that the maximum difficulty obviously occurs when the wavelength of the Moiré pattern is of the same magnitude as the width of the dislocation image. When the wavelength is either considerably longer or shorter, the locally smoothed GBD intensity profile obtained from Eq. (2) will be given essentially by

$$I = \psi\psi^* \simeq |\phi_1|^2 + |\phi_2|^2 + \text{constant}. \qquad (6)$$

CONTRAST DUE TO REGULAR ARRAYS OF INTRINSIC GBD'S

We now consider the contrast due to regular arrays of intrinsic GBD's. The displacement fields of such GBD's add up at a distance from the boundary greater than about the GBD spacing to produce a uniform misorientation between Crystals 1 and 2. The strains due to the closely spaced GBD's are then localized in a narrow region at the boundary of width $\sim d_D$, and long range strains of the type which exist around isolated GBD's are therefore absent.

For diffraction purposes, the bicrystal may therefore be approximated by the three regions shown in Fig. 10. Regions 1 and 2 consist of the major portions of Crystals 1 and 2 which are essentially strain free and are simply rotated with respect to each other, while the GBD Region contains the localized periodic strain field which is associated with the GBD's in both Crystals 1 and 2. As the network spacing decreases, the thickness of this latter region decreases. Obviously, the diffraction contrast due to the GBD array must then decrease, since the volumes of material which are highly strained are decreased and hence the electron scattering from these regions is decreased. When the spacing becomes considerably less than the extinction distance, we would expect to encounter difficulty in resolving the network by means of diffraction contrast, since the electron waves are then no longer very sensitive to the detailed atomic arrangement in the thin GBD Region.

The resolution of finely spaced GBD arrays therefore poses a difficult and important problem in the study of GBD behavior. Unfortunately, in many grain boundary studies in which fine structure was not detected in the boundary it has not been possible to

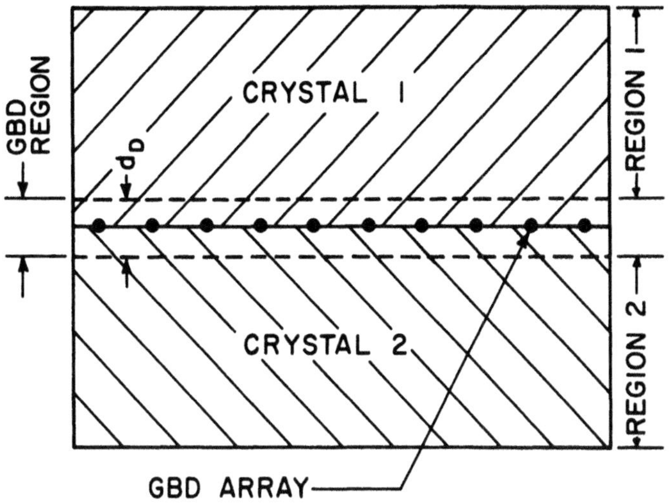

Fig. 10. Schematic drawing of bicrystal specimen showing crystal regions adjoining a regular array of intrinsic GBD's.

conclude that arrays of GBD's were not present because of the possibility that arrays with spacings below the limit of resolution actually were present[2,6].

Diffraction Contrast With One Crystal Strongly Excited

In this situation the diffraction contrast due to the intrinsic GBD array arises from the periodic strain field which is located in the portion of the GBD Region belonging to the excited crystal. Schober and Balluffi have recently imaged orthogonal networks of intrinsic screw GBD's of the type seen in Figs. 3c, 4b and 4c under conditions where one crystal was strongly excited and have published a series of electron micrographs (see Fig. 6 of reference 2) which clearly shows how the contrast and visibility of the individual GBD's decrease as the spacing decreases. So far, networks of this type with spacings as small as 40Å have been resolved by conventional microscopy[2*]. (We remark that the relatively small magnitude of the Burgers vector of these particular GBD's (1.29Å) tends to decrease the contrast and thereby increase

* We note, as we have done previously, that all of the observed contrast may not have been due to diffraction contrast, since a contribution due to the direct imaging of the network may have been present (see final section of the present paper).

the difficulty in resolving the network.) The possibility that higher resolution than this could be achieved by means of conventional diffraction contrast could be investigated to at least some extent by carrying out calculations using the same basic methods as those already described. The only input that would be required would be the displacement field of the array and preferably a computing system capable of mapping out the two-dimensional distribution of intensity in the various beams. So far, we have not yet performed any calculations of this type.

It is worth noting that the recently developed "weak beam" technique[15] could probably be used to improve matters. In this method the dislocations are imaged in dark field with a weakly excited reflection under conditions where another systematic reflection is strongly excited. In this situation the width of the dislocation strain field image is markedly reduced, and the ability to resolve closely spaced dislocations by diffraction contrast is therefore increased. So far, the method has been used to resolve closely spaced partial dislocations, but to our knowledge, the method has not yet been used to investigate dislocations in large regularly spaced arrays. Another potentially valuable method of decreasing the dislocation image width has been reported recently[16]. In this case high voltage microscopy is used, and the specimen is examined in bright field with a high order systematic reflection strongly excited.

Weatherly and Mok[17] have recently carried out calculations for the somewhat similar problem of the diffraction contrast expected from regular arrays of epitaxial dislocations in the case where only one crystal is strongly excited with one strongly diffracted beam. They estimated that dislocations of this type would just be visible when $d_D \simeq 0.33\xi_g$ where ξ_g is the extinction distance. This spacing is somewhat larger than the spacings actually resolved by Schober and Balluffi (above). The detailed explanation for this apparent discrepancy is not completely clear at present. However, it seems possible that contributions due to the direct imaging of the network may have helped the experimental situation appreciably (see previous footnote and also final section of the present paper). In addition, a number of relatively weak beams are generally excited under usual experimental conditions, and such conditions tend to improve the diffraction contrast resolution. More complicated "n-beam" calculations would therefore be of interest.

Diffraction Contrast With Both Crystals Strongly Excited

When both crystals are strongly excited the following two basically different situations may be distinguished:

(1) If the various beams which exit from the lower crystal (Fig. 1) can be examined separately in the objective aperture, the diffraction contrast in each beam will be the overall result of the sequential scattering which occurs in each crystal as discussed previously. In this case contrast due to the GBD array is built up in each crystal as a result of scattering in that portion of the GBD Region (Fig. 10) which is associated with that crystal. The method of calculating the contrast is therefore similar to that already described for the case where only one crystal is strongly excited.

(2) If two of the beams which exit from the lower crystal enter the objective aperture together, some attention must be given to the danger that parallel fringes in the resulting Moiré pattern may be assumed to be intrinsic GBD's in an array or _vice versa_. Taking a rotational Moiré pattern as an example, it is readily seen that simple methods exist for distinguishing between the two possibilities. According to Eqs. 4 and 5 Moiré and GBD spacings can only be equal when the very special condition $d_{hk\ell} = |\vec{b}|$ is satisfied. Furthermore, the single set of Moiré fringes always runs parallel to the average diffraction vector, whereas the orthogonal set of GBD images remains fixed relative to the crystal axes. A few tests using differing diffraction conditions should therefore suffice to distinguish Moiré fringes from intrinsic GBD contrast in situations where it is conceivable that they may be confused.

Further aspects of the relationship between simultaneously appearing Moiré patterns and intrinsic GBD diffraction contrast have been considered recently by Thölén[18]. Thölén calculated the combined dislocation and Moiré contrast under several diffraction conditions for a low angle twist boundary lying parallel to (001) of the type present in Fig. 2b. Instead of using the present procedure of carrying out sequential calculations in the two crystals and combining the beams, he calculated the total resulting contrast directly by integrating along a column extending throughout the entire specimen thickness and employing the full displacement field of the network of orthogonal screw GBD's. By systematically reducing the spacing of the network (i.e., reducing the thickness of the GBD Region) he showed that the intensity due directly to the dislocation scattering effects in the GBD region tends to fade away, and all that is left eventually is the pure Moiré pattern characteristic of Regions 1 and 2 rotated with respect to each other. Thölén concluded that when the network spacing becomes somewhat smaller than the extinction distance (i.e., $d_D \lesssim 0.3\xi_g$), the individual GBD's can no longer be resolved, since the electron waves are then no longer influenced by the GBD Region but instead are only affected by the total phase shift (due to the rotational misorientation of Regions 1 and 2) across the GBD Region.

Thölén's conclusion regarding the diffraction contrast is therefore seen to be quite similar to that of Weatherly and Mok[17] (above). However, again we have direct evidence that intrinsic GBD arrays of considerably smaller spacing than $d_D = 0.3\xi_g$ can be resolved under actual experimental conditions even in the presence of Moiré patterns. Consider first the example shown in Fig. 2b which corresponds closely to the calculation of Thölén for the case of a (001) twist boundary where $\vec{g}$ is parallel to one set of the orthogonal screw GBD's which should therefore be in contrast. Here, $\vec{g}\cdot\vec{b} = 2$, and d_D has the relatively small value of $0.16\xi_g$ (see Fig. 4 of Thölén's paper[18]). Even though d_D is considerably smaller than Thölén's limit of $0.3\xi_g$ the parallel GBD's are clearly imaged. In the case where the same orthogonal grid is imaged in dark field with $\vec{g} = (200)_{1,2}$, $\vec{g}\cdot\vec{b} = 1$ for each set of screw GBD's, and they should therefore both be in contrast. In addition, Moiré fringes should run parallel to $\vec{g} = (200)_{1,2}$ which lies at 45° with respect to each set of GBD's. Experimentally observed examples of the superimposed GBD and Moiré contrast obtained from boundaries of this type are shown in Figs. 11a, 11c, 11e and 11f for GBD spacings corresponding to $d_D = 0.6\xi_g$, $0.3\xi_g$, $0.15\xi_g$ and $0.08\xi_g$ respectively. An example of pure Moiré contrast is seen at the bubble region at B (lower right hand corner

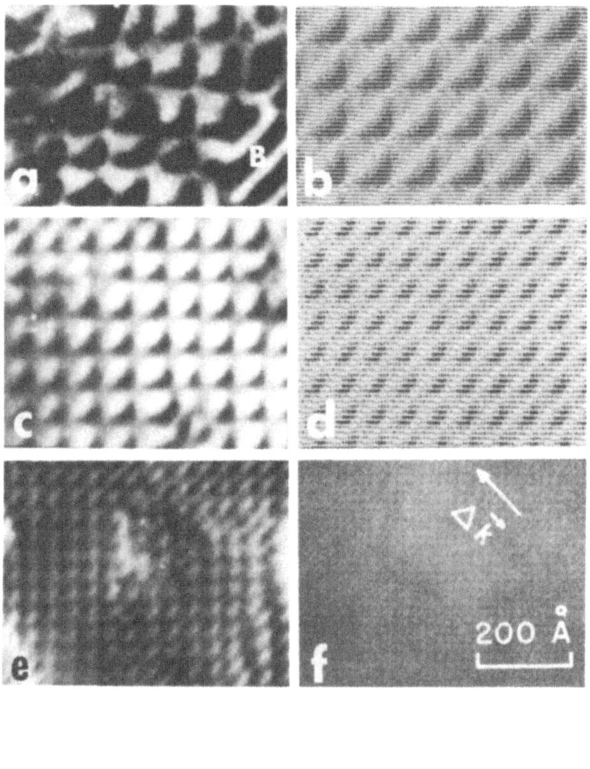

of Fig. 11a). Again, Thölén has calculated the total contrast for several cases which correspond almost exactly to those just described, and the calculated results (taken from Figs. 7c and 7d of Thölén's paper[18]) are shown in Figs. 11b and 11d and appear to be in at least qualitative agreement with experiment. However, the GBD network is still clearly visible in Figs. 11e and 11f, and it must be concluded that the GBD structure can still be readily observed when the spacing is as small as $d_D \simeq 0.08\xi_g$ ($\simeq$18Å) which is well below the Thölén limit of $d_D = 0.3\xi_g$. Again, it seems possible that contributions due to the direct imaging of the network may have helped the experimental situation appreciably. In addition, the excitation of a number of weak beams may have helped matters.

We conclude that, even though we expect the contrast of GBD networks to decrease steadily as the spacing decreases, it still appears possible to resolve the network GBD's under actual experimental conditions when their spacing is as small as $d_D \simeq 0.08\xi_g$ (Fig. 11f). (We note that the spacing of $\simeq$18Å in Fig. 11f was one of the finest spacings which we have been able to resolve using conventional microscopy.)

The GBD Network as a Diffraction Grating

In the course of our work another scattering phenomenon has been found which may be of considerable importance in the investigation of finely spaced intrinsic GBD arrays. Rather weak extra reflections have been found in the SAD patterns of specimens containing low angle twist boundaries which are evidently due to scattering from the two-dimensional periodic strain fields of the orthogonal grids of screw GBD's in the boundaries. In order to interpret this phenomenon we may regard the GBD region in Fig. 10 as essentially a two-dimensional diffraction grating. (The possibility of this phenomenon has apparently been suggested first by Spyridelis, et al.[19]). The diffraction pattern produced by

Fig. 11. Combined Moiré and GBD contrast from low angle twist boundaries lying parallel to (001). Orthogonal screw GBD's running parallel to <110>. Dark field images with $\vec{g} = (200)_{1,2}$. (a) Observed image of boundary with $d_D \simeq 0.6\xi_g$. Pure Moiré fringes at bubble at B; (b) Calculated (Thölén[18]) diffraction contrast image of boundary with $d_D = 0.5\xi_g$ [compare with (a)]. (c) Observed image of boundary with $d_D \simeq 0.3\xi_g$. (d) Calculated (Thölén[18]) diffraction contrast image of boundary with $d_D = 0.25\xi_g$ [compare with (c)]. (e) Observed image of boundary with $d_D = 0.15\xi_g$. (f) Observed image of boundary with $d_D = 0.08\xi_g$. $\Delta\vec{k}$ lies perpendicular to Moire fringes.

scattering from such a GBD grid (due to a beam incident upon it in a normal direction) should consist of square arrays of spots centered about the FCC reflections in the same orientation as the square GBD grid. Furthermore, the spacing, p, of the spots within these square arrays should be just the reciprocal of the spacing of the orthogonal GBD array, i.e.,

$$p = \frac{1}{d_D} \quad . \tag{7}$$

An example of these extra diffraction spots is shown in Fig. 12a. Also shown is a $(200)_{1,2}$ dark field image of the boundary which

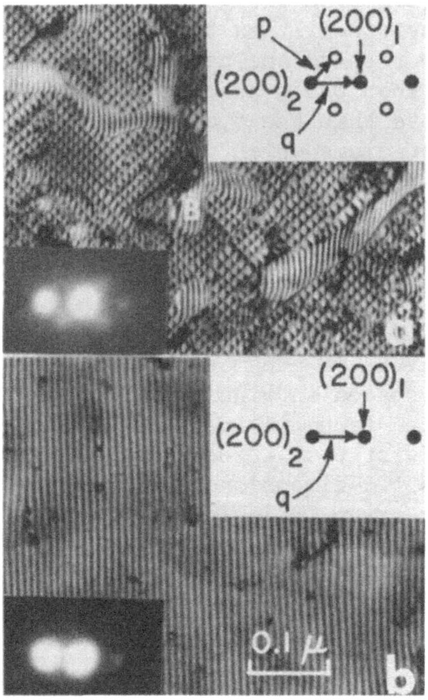

Fig. 12. Extra SAD diffraction spots due to scattering by orthogonal grid of screw GBD's in low angle twist boundary lying parallel to (001). (a) $(200)_{1,2}$ dark field image of boundary showing GBD network. Pure Moiré fringes seen at bubble regions such as B. $(200)_{1,2}$ region of SAD pattern at lower left shows $(200)_{1,2}$ spots and extra weaker spots due to GBD scattering. Schematic diagram of SAD pattern at upper right. (b) Same as (a) except that grain boundary is absent. GBD network and extra SAD spots have disappeared.

shows the orthogonal grid of intrinsic screw GBD's having $\vec{b}$ = a/2<110> which constitutes the boundary. Moiré fringes formed from the $(200)_1$ and $(200)_2$ beams which entered the objective aperture are clearly visible in the bubble regions such as at B. At the upper right is a schematic diagram of the SAD pattern in which the $(200)_1$ diffracted beam (which was incident upon the grain boundary) and the $(200)_2$ diffracted beam are represented by filled circles, and the weaker spots caused by diffraction from the GBD grating are indicated by open circles. These latter spots are arranged on a square lattice which has the same orientation as the GBD grid and which is centered around the (200) region of the SAD pattern. The spacing, p, from Eq. 7 is given by

$$p = \frac{\theta}{|\vec{b}|} = \frac{\sqrt{2}\theta}{a} , \tag{8}$$

while the distance, q, is given by

$$q = \frac{\theta}{d_{200}} = \frac{2\theta}{a} = \sqrt{2}\, p \quad , \tag{9}$$

The geometry of the SAD pattern in Fig. 12a is therefore in complete agreement with the predicted geometry. (We note that similar extra SAD spots due to diffraction from GBD's may also be seen clearly in Fig. 2a.)

In order to obtain a final confirmation regarding the origin of these extra diffraction spots, SAD patterns were obtained from isolated patches of the specimens where the Crystals 1 and 2 had not been properly welded together to form the grain boundary. In these regions the grain boundary, and the GBD's, were therefore absent, and no extra spots were detected in the corresponding SAD patterns as expected. An example of this result is shown in Fig. 12b which shows both the SAD pattern and the $(200)_{1,2}$ dark field image from a non-welded region of the same specimen which was used to obtain Fig. 12a. It may be seen that both the GBD's and the GBD diffraction spots are missing, and that the dark field image consists of a Moiré pattern which is identical to that seen in the bubble regions of Fig. 12a.

Further examples and discussion of extra diffraction spots from GBD's have been given elsewhere[20]. So far, the phenomenon has been found for arrays of screw GBD's and also edge GBD's in a variety of twist and tilt boundaries[20].

As we have seen, the SAD spot spacing, p, increases with a decrease in dislocation spacing, and this behavior is therefore helpful in the detection of closely spaced GBD arrays by this method. However, the width of the GBD Region decreases with a decrease in the GBD spacing, and this effect should tend to decrease the intensity of the scattering from the GBD grating. It therefore remains to be seen whether the direct observation of the extra diffraction spots will be of use in the study of extremely fine GBD arrays. Calculations of the scattering factors of such arrays are being carried out at present[21], and further experimental work is in progress.

Finally, as we have already mentioned earlier in the paper, diffraction of the present type from the GBD grids may be helpful in resolving the grids by direct imaging. As is well known if several of the closely spaced diffracted beams, such as those shown in Fig. 2c, are allowed to enter the objective aperture together, they will build up fringe patterns which tend to synthesize a direct image of the GBD grid. As we have already pointed out, the intensity from this effect may contribute to the total image intensity (along with the usual diffraction contrast intensity) and may therefore be of assistance in resolving closely spaced arrays. Calculations of this effect are in progress[21].

SUMMARY AND CONCLUSIONS

(1) The electron microscope contrast produced by grain boundary dislocations (GBD's) in thin-film bicrystal gold specimens was studied both experimentally and theoretically. In each case the boundary containing the GBD's was parallel to the thin-film surfaces and was observed at normal incidence. GBD's of both the intrinsic and extrinsic type were considered. (The former occur in regular networks and are part of the equilibrium structure of the boundary, whereas, the latter are extra GBD's which are generally distributed in a disorganized fashion.)

(2) In the most general case a number of Bragg reflections was excited in both the upper and lower crystals, and the total observed contrast then arose from a complex superposition of dislocation strain field scattering effects in the various electron beams and possible Moiré patterns which were produced when more than one beam was allowed to enter the objective aperture.

(3) Attention was therefore focused on simpler situations which could be readily achieved experimentally and analyzed to obtain information about the GBD's. These corresponded to cases where no more than one strong beam was diffracted in the upper crystal and no more than two strong beams were diffracted in the lower crystal. A general technique was developed for calculating the diffraction contrast due to both isolated extrinsic GBD's and intrinsic GBD arrays under these diffraction conditions both in the

presence and absence of Moiré interference effects. Detailed calculations of the diffraction contrast due to both isolated screw and edge extrinsic GBD's were carried out. The results in several cases were found to be in good apparent agreement with experimental observations.

(4) The problem of resolving individual intrinsic GBD's in closely spaced arrays by diffraction contrast was considered. In this case resolution is expected to become difficult when the spacing becomes considerably less than the extinction distance, ξ_g, since the electron waves are then no longer very sensitive to the detailed atomic arrangement in the thin grain boundary region which is highly strained by the GBD's. It was demonstrated experimentally, however, that intrinsic GBD spacings as small as $0.08\xi_g$ could actually be resolved using conventional microscopy. It was suggested that contributions due to the direct imaging of the array may have improved the experimental situation significantly [see (5) below]. "N-beam" effects due to weakly excited reflections may also have helped. The use of recently developed special techniques for reducing the diffraction contrast image widths of dislocations was suggested as a future aid in the problem of obtaining higher resolution.

(5) Evidence that a regular grid of intrinsic GBD's acts as a two-dimensional diffraction grating was found. Extra diffraction spots due to this phenomenon were observed in diffraction patterns. Since the spacing of these extra spots increases with a decrease in the intrinsic GBD grating spacing, the observation of these extra diffraction spots could be a useful means of detecting and measuring extremely fine intrinsic GBD networks. It was also pointed out that these extra spots could build up fringe patterns which would tend to synthesize a direct image of the GBD grid. The intensity from this effect could then contribute to the total image intensity of the grid (along with the usual diffraction contrast intensity) and would therefore be of assistance in resolving closely spaced grids [see (4) above].

ACKNOWLEDGEMENT

The writers would like to thank Professor S. L. Sass for useful discussions.

REFERENCES

1. T. Schober and R. W. Balluffi, Phil. Mag., 20, 511-518, (1969).

2. T. Schober and R. W. Balluffi, Phil. Mag., 21, 109-123, (1970).

3. T. Schober, Phil. Mag., 22, 1063-1068, (1970).

4. T. Schober and R. W. Balluffi, Phys. Stat. Sol. (b), 44, 103-114, (1971).

5. T. Schober and R. W. Balluffi, Phys. Stat. Sol. (b), 44, 115-126, (1971).

6. T. Schober and R. W. Balluffi, Phil. Mag., 24, 165-180, (1971).

7. T. Schober and R. W. Balluffi, Phil. Mag., 24, 469-474, (1971).

8. R. W. Balluffi, Y. Komem and T. Schober, Report No. 1606, Materials Science Center, Cornell University, Ithaca, New York, July 1971 (to be published in Proceedings of the International Conference on Structure and Properties of Grain Boundaries and Interfaces, IBM Thomas J. Watson Research Center, Yorktown Heights, New York, August 23-25, 1971).

9. G. A. Bassett, J. W. Menter and D. W. Pashley, Proc. Roy. Soc., A246, 345-368, (1958).

10. P. B. Hirsch, A. Howie, R. B. Nicholson, D. W. Pashley and M. J. Whelan, Electron Microscopy of Thin Crystals, 1st ed., Butterworths, London (1965), p. 345.

11. P. B. Hirsch, A. Howie, R. B. Nicholson, D. W. Pashley and M. J. Whelan, Electron Microscopy of Thin Crystals, 1st ed., Butterworths, London, (1965), p. 148.

12. A. Howie and M. J. Whelan, Proc. Roy. Soc., A267, 206-230, (1962).

13. See: S. Amelinckx, The Direct Observation of Dislocations, Academic Press, New York, (1964), p. 414 et seq.

14. P. B. Hirsch, A. Howie, R. B. Nicholson, D. W. Pashley and M. J. Whelan, Electron Microscopy of Thin Crystals, 1st ed., Butterworths, London, (1965), p. 247 et seq.

15. D. J. H. Cockayne, I. L. F. Ray and M. J. Whelan, Phil. Mag., 20, 1265-1270, (1969).

16. R. Osiecki, L. C. DeJonghe, W. L. Bell and G. Thomas, Report UCRL-20700, Lawrence Radiation Laboratory, University of California, Berkeley, California, June 1971.

17. G. C. Weatherly and T. D. Mok, Department of Metallurgy and Materials Science, University of Toronto (private communication). To be published in Proceedings of the International Conference on Structure and Properties of Grain Boundaries and Interfaces, IBM Thomas J. Watson Research Center, Yorktown Heights, New York, August 23-25, 1971.

18. A. R. Thölén, Phys. Stat. Sol. (a), 2, 537-550, (1970).

19. J. Spyridelis, P. Delavignette and S. Amelinckx, Mat. Res. Bull., 2, 615-620, (1967).

20. R. W. Balluffi, S. L. Sass and T. Schober, Report No. 1707, Materials Science Center, Cornell University, Ithaca, New York, January, 1972.

21. D. Y. Guan, Y. Komem, and S. L. Sass, Department of Materials Science and Engineering, Cornell University, Ithaca, New York (private communication).

SOME PROPERTIES OF THE DISCLINATION STRUCTURE OF GRAIN BOUNDARIES

J. C. M. Li

Department of Mechanical & Aerospace Sciences

University of Rochester, Rochester, N.Y. 14627

ABSTRACT

The interaction between a simple tilt boundary made of wedge disclination dipoles and a single edge dislocation is studied. It is found that in general edge dislocations can be generated or absorbed at the wedge disclination dipoles. The dilatation field of such a boundary is calculated and from which the interaction with impurity atoms which cause dilatational distortions of the lattice is studied. The resulting periodic distribution of impurity atoms along the boundary is suggested to play a role in intergranular embrittlement and grain boundary sliding.

INTRODUCTION

In a previous communication,[1] a disclination model was proposed for high angle grain boundaries. The essential feature of the model is that a grain boundary consists of alternating regions of low energy boundaries of different misfit angles, some higher and others lower than the average angle of the boundary. These angles could correspond to angles of energy cusps, coincidence orientations, twinning orientations, or simply perfect lattice. The local strains caused by the neighboring regions of different angles are those of disclinations. The strain energy of these disclinations contributes to the energy of the boundary in addition to the energies of the regions themselves. The model is capable of describing the energy-angle relations for high angle boundaries.

Since disclinations are defects created by rotational misfits, they are the proper defects to be incorporated into a grain boundary. Although it is possible to use dislocations instead of disclinations, it is not as convenient to produce a lattice rotation by dislocations than by disclinations. For example, there are a variety of Burgers vectors to be selected from and the resultant dislocation structure depends on the selection. But there is only one rotation axis for the disclinations, namely, the common direction of the two grains.

The essential idea of a periodic structure of low energy regions is similar to that of Bishop and Chalmers,[2,3] except that boundary coincidence was their concern rather than energy. They do, however, recognize the possible existence of long range stresses extending to distances comparable to the period in the structure and possible long-range interactions with impurity atoms. These ideas are best treated quantitatively by the disclination model. Of course the fit-misfit structure of high angle boundaries was proposed very early by Mott[4] and extended later by Brandon.[5-6]

In view of the foregoing, it seems worthwhile to study some properties of the disclination boundaries. Two properties are presented here: The capability of the disclination boundary to emit or to absorb dislocations and the interaction with impurity atoms which cause only dilational distortions of the lattice.

GRAIN BOUNDARY AS SOURCE AND SINK OF DISLOCATIONS

It was suggested by many people that grain boundaries are sources and sinks of dislocations. Exactly how these sources and sinks operate is not known, although various possibilities have been suggested. As pointed out by DeWit,[7] a wedge disclination can act, at least geometrically, as a source or sink of edge dislocations. Thus it is expected that a high angle grain boundary made of wedge disclination dipoles is capable of emitting or absorbing edge dislocations, at least geometrically. The question is whether it is also possible energetically.

Interaction between a Wedge Disclination and an Edge Dislocation

For a wedge disclination of rotation ω situated on the axis (z axis) of a cylinder of radius R, the stress field is given by Huang and Mura:[8]

$$\sigma_{xx} = \frac{1}{2}\,\ell n\,\frac{R^2}{x^2+y^2} - \frac{y^2}{x^2+y^2} \tag{1}$$

$$\sigma_{yy} = \frac{1}{2}\,\ell n\,\frac{R^2}{x^2+y^2} - \frac{x^2}{x^2+y^2} \tag{2}$$

$$\sigma_{zz} = \nu(\ell n\,\frac{R^2}{x^2+y^2} - 1) \tag{3}$$

$$\sigma_{xy} = \frac{xy}{x^2+y^2} \tag{4}$$

in units of $\mu\omega/2\pi(1-\nu)$. The stress components are at a point (x,y,z) sufficiently far removed from the core of the disclination.

Interacting with this stress field, a parallel edge dislocation with a Burgers vector b in the x direction has the following potential energy at (x,y) relative to that at any point (x,0):

$$\Delta E = \frac{\mu b\omega y}{4\pi(1-\nu)}\,\ell n\,\frac{R^2}{x^2+y^2}\ . \tag{5}$$

Eq. (5) is plotted as contours in Fig. 1. The points at which the energy is extremum are $(0, \pm R/e)$ and the energy is $\pm(2/e)\mu b\omega R/4\pi(1-\nu)$ with e being the base of natural logarithm. It is seen that an edge dislocation can be emitted or absorbed at the tip of a wedge disclination without any elastic resistance except for some short range interaction near the core which can be considered to be chemical in nature.

However, for the wedge disclination to be able to emit a dislocation, the reduction in energy of the wedge disclination must be sufficient to provide for the energy of the dislocation. Since the energy of the

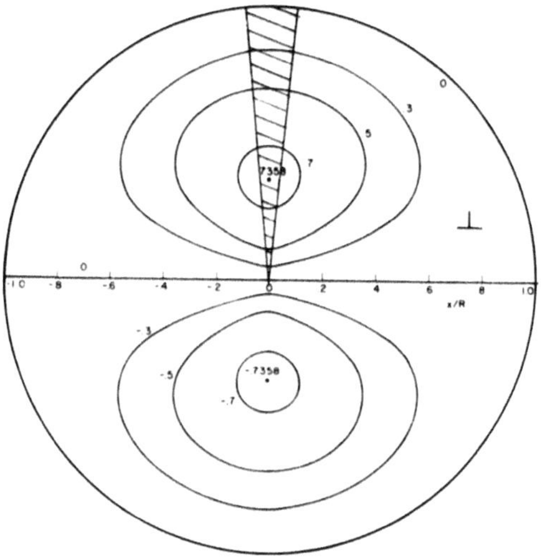

FIG 1
INTERACTION BETWEEN AN EDGE DISLOCATION AND A WEDGE DISCLINATION[UNIT OF ENERGY: $\mu b\omega R/4\pi(1-\nu)$]

wedge disclination as given by Huang and Mura is $\mu\omega^2R^2/16\pi(1-\nu)$ per unit length, it is equivalent to $\mu N^2b^2/16\pi(1-\nu)$ by using the dislocation model where N is the total number of dislocations. Hence the reduction in energy after a positive edge dislocation is emitted is $\mu Nb^2/8\pi(1-\nu)$. This is to be compared with the energy of the dislocation. Thus the emission process is possible if

$$\omega R > 2b\ (\ell n\ \frac{R}{r_o} + C) \qquad (6)$$

where r_o is the core radius and C is the core energy per unit length of the dislocation in units of $\mu b^2/4\pi(1-\nu)$. On the other hand, the process of absorption is possible when the inequality sign is reversed. Note that it is impossible to emit a negative edge dislocation but it is always possible to absorb one.

The foregoing consideration is for the case in which the dislocation takes a gliding path along the xz plane. If the dislocation is allowed to climb, an easier path is along the yz plane. Along this path a further reduction of energy is possible so that the condition for emission becomes

$$\omega R > \frac{2be}{4+e}\ (\ell n\ \frac{R}{r_o} + C)\ . \qquad (7)$$

This condition applies to either emitting a positive edge dislocation to the -y side or a negative one to the +y side. Note that in Fig. 1 the shaded region is actually perfect for a perfect wedge disclination. When the inequality sign is reversed, it gives the condition for absorption along the yz plane except that the process has to be thermally activated to overcome the energy barrier. This is true for either a positive edge dislocation entering from the -y side or a negative one entering from the +y side. Note that it is always possible to absorb a negative edge dislocation from the -y side (without thermal activation) or a positive one from the +y side, but it is impossible to emit them.

Interaction between an Edge Dislocation and a Wedge Disclination Dipole

The stress field of a wedge disclination dipole can be obtained by using Eqs. (1-4) twice, once for each wedge disclination. The stress field is also the same as that of a finite edge dislocation wall.[9] Let the positive wedge disclination ($+\omega$) be situated at $y = -L$, $x = 0$ and the negative one ($-\omega$) at $y = L$, $x = 0$. Interacting with the stress field of such a dipole, an edge dislocation parallel to the z axis and having a Burgers vector b in the x direction has a potential energy at (x,y) relative to that at (0,0) as follows:

$$\Delta E = (q-1)\ln[p^2+(q-1)^2] - (q+1)\ln[p^2+(q+1)^2] \qquad (8)$$

in units of $\mu b\omega L/4\pi(1-\nu)$ per unit length, where $p = x/L$ and $q = y/L$. Eq. (8) is plotted in Fig. 2. It is seen that there is no elastic barrier if the dislocation is emitted (or a negative one is absorbed) from anywhere inside the disclination dipole.

The energy of a wedge disclination dipole has been given previously.[1] In terms of a dislocation model, it is

$$E = \frac{\mu N^2 b^2}{4\pi(1-\nu)} \ln \frac{R}{2L} + 2\gamma L \qquad (9)$$

where N is the total number of edge dislocations in the dipole and $2\omega L = Nb$. Assuming constant γL, the reduction

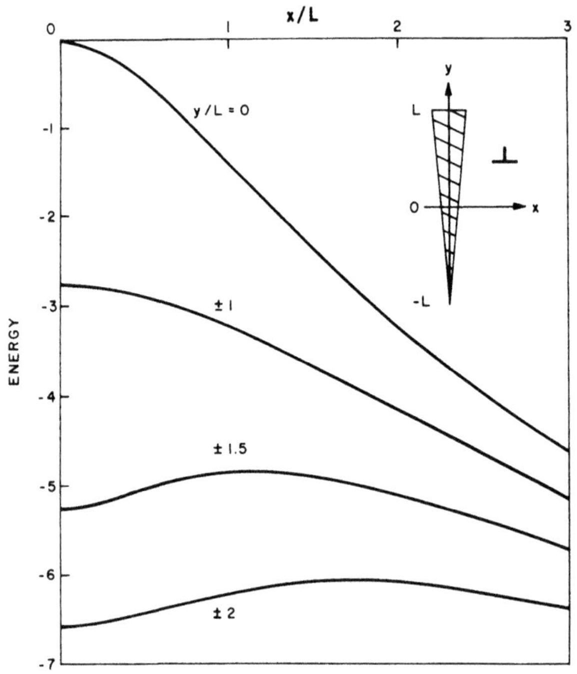

FIG. 2. INTERACTION BETWEEN AN EDGE DISLOCATION AND A WEDGE DISCLINATION DIPOLE [UNIT OF ENERGY: $\mu b\omega L/4\pi(1-\nu)$]

in energy after emitting a dislocation is

$$\Delta E = \frac{\mu\omega b L}{\pi(1-\nu)} \ell n \frac{R}{2L} . \tag{10}$$

For R >> L, a similar amount of energy is released when a dislocation is emitted and move to x = R/2. Hence the condition at which the emission process is possible is

$$\omega L > \frac{b}{8} (\ell n \frac{R}{r_o} + C)/\ell n (\frac{R}{2L}) . \tag{11}$$

When the inequality sign is reversed, it is the condition for the absorption process. Note that it is impossible to emit a negative edge dislocation but it is always possible to absorb one.

Instead of the gliding path, a climbing edge dislocation can be emitted along x = 0. For R >> L and by allowing the dislocation to move to y = R/2, the condition for the emission process is

$$\omega L > \frac{b}{4} \left(\ln \frac{R}{r_o} + C\right) / \ln \left(\frac{eR^2}{8L^2}\right) \tag{12}$$

slightly easier than the gliding case. This condition applies to a positive edge dislocation emitted from either end of the dipole along the +y or -y directions. When the inequality sign is reversed, it is a condition for absorption of the positive edge dislocation into either end of the dipole. Note again that it is impossible to emit a negative edge dislocation from either end of the dipole but it is always possible to absorb one.

Interaction between an Edge Dislocation and a Wall of Disclination Dipoles

A typical disclination model for a symmetric tilt boundary is a wall of disclination dipoles.[1] Let the plane of the dipoles be the yz plane and the spacing between dipoles be H. Let all the wedge disclinations be parallel to the z axis and the separation within a dipole be 2L. Then the stress component of interest is

$$\sigma_{xy} = \frac{\mu\omega x}{2\pi(1-\nu)} \sum_{n=-\infty}^{\infty} \left[\frac{y+L+nH}{x^2+(y+L+nH)^2} - \frac{y-L+nH}{x^2+(y-L+nH)^2} \right] \tag{13}$$

$$= \frac{\mu\omega\alpha}{4\pi(1-\nu)} \left[\frac{\sin(\beta+\lambda)}{\cosh\alpha-\cos(\beta+\lambda)} - \frac{\sin(\beta-\lambda)}{\cosh\alpha-\cos(\beta-\lambda)} \right] \tag{14}$$

where $\alpha = 2\pi x/H$, $\beta = 2\pi y/H$, $\lambda = 2\pi L/H$. The situation along the planes $y = \pm L$, namely, the planes of the wedge disclinations, is shown in Fig. 3 for several ratios of L/H. It is seen that for all values of L/H less than 1/4, the emission of a dislocation from any wedge disclination is assisted by the stress field of the wall. Similarly for all values of L/H between 1/4 and 1/2, the emission of a negative edge dislocation or the absorption of a positive one at any wedge disclination is assisted by the stress field.

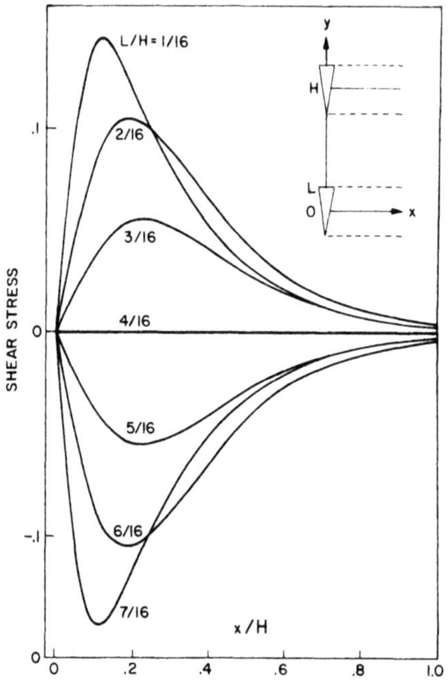

FIG. 3. THE SHEAR STRESS, σ_{xy}, ALONG $y = \pm L$ OF A WALL OF WEDGE DISCLINATION DIPOLES [UNIT OF STRESS: $\mu\omega/2(1-\nu)$]

THE INTERACTION OF GRAIN BOUNDARY WITH MOBILE IMPURITIES

It is known that grain boundaries play an important role in the mechanical behavior of materials such as grain boundary sliding, intergranular embrittlement, and grain size strengthening. It is also known that these effects are modified, sometimes to a large extent, by the existence and redistribution of impurities. Exactly how the impurity atoms get involved with the grain boundary is still not clear probably because of the difficulty involved in dealing with impurity interactions quantitatively.

The present model proposes a structure of a high angle grain boundary which interacts, at least elastically, with impurities. Such interaction can be treated quantitatively. An attempt is made here for a symmetric tilt boundary interacting with mobile impurities which produce only dilatational distortions of the lattice.

The Dilatational Strain Field of a Wedge Disclination

The dilatational strain field of a wedge disclination can be obtained from the results of Huang and Mura and is given by

$$\Delta V/V = \frac{\omega(1-2\nu)}{4\pi(1-\nu)}\left(2\ell n\,\frac{R}{r} - 1\right) \tag{15}$$

at a distance r sufficiently far removed from the core of the disclination. The dilatation is zero at $r = R/\sqrt{e}$ which describes a cylinder inside which the dilatation is positive and outside which it is negative. The total dilatation is zero. The radial variation is a monotonously decreasing function as shown in Fig. 4 and, therefore, impurity atoms which expand the lattice tend to move radially toward the core and those which shrink the lattice tend to move in the opposite direction as shown in Fig. 4.

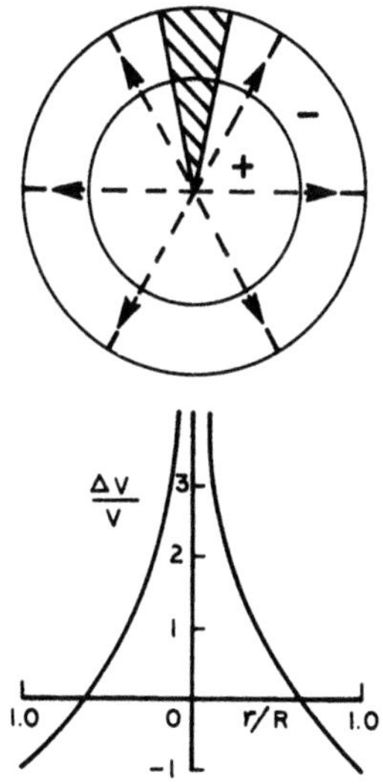

FIG. 4. DILATATIONAL STRAIN FIELD OF A WEDGE DISCLINATION [UNIT: $(1-2\nu)\omega/4\pi(1-\nu)$]

The Dilatational Strain Field of a Wedge Disclination Dipole

The dilatational strain field of a wedge disclination dipole can be obtained by using Eq. (15) twice, one for each disclination. Let the positive wedge disclination be at (0,-L) and that of the negative one be at (0,L). The dilatational strain field is given by

$$\Delta V/V = \frac{\omega(1-2\nu)}{4\pi(1-\nu)} \ln \frac{x^2+(y-L)^2}{x^2+(y+L)^2} \quad (16)$$

which is the same as that of a finite edge dislocation wall.[9] The contours are circles as shown in Fig. 5 and the dotted circles are the flow lines of impurity atoms. Depending on the size of atoms, they flow from one disclination to the other.

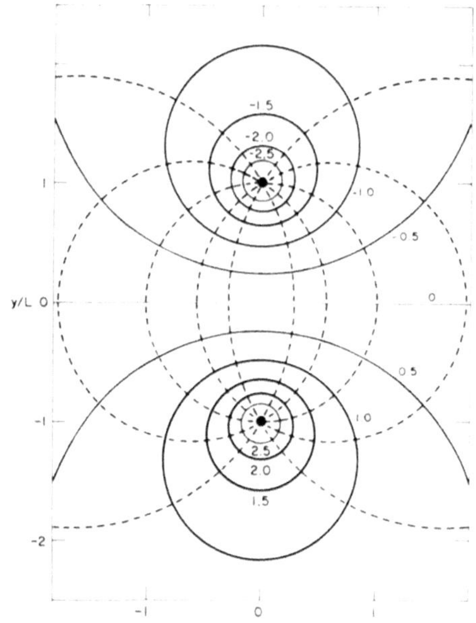

FIG. 5. DILATATIONAL STRAIN FIELD OF A WEDGE DISCLINATION DIPOLE
[UNIT: $(1-2\nu)\omega/2\pi(1-\nu)$]

The Dilatational Strain Field of a Wall of Disclination Dipoles

For a wall of disclination dipoles, let the positive wedge disclinations be located at (0,-L), (0,-L±H), (0,-L±2H), etc., and the negative ones be at (0,L), (0,L±H), (0,L±2H), etc., all parallel to the z axis. Then the dilatational strain field is, from Eq. (16):

$$\Delta V/V = \frac{\omega(1-2\nu)}{4\pi(1-\nu)} \sum_{n=-\infty}^{\infty} \ln \frac{x^2+(y-L+nH)^2}{x^2+(y+L+nH)^2} \tag{17}$$

$$= \frac{\omega(1-2\nu)}{4\pi(1-\nu)} \ln \frac{\cosh\alpha - \cos(\beta-\lambda)}{\cosh\alpha - \cos(\beta+\lambda)} \tag{18}$$

where $\alpha = 2\pi x/H$, $\beta = 2\pi y/H$, and $\lambda = 2\pi L/H$. The contours are plotted in Fig. 6 for L/H = 1/4. These contours can be expressed also by, $\lambda = \pi/2$:

$$\cosh\alpha = \coth\xi \sin\beta \tag{19}$$

where ξ is the dilatation in units of $\omega(1-2\nu)/2\pi(1-\nu)$.

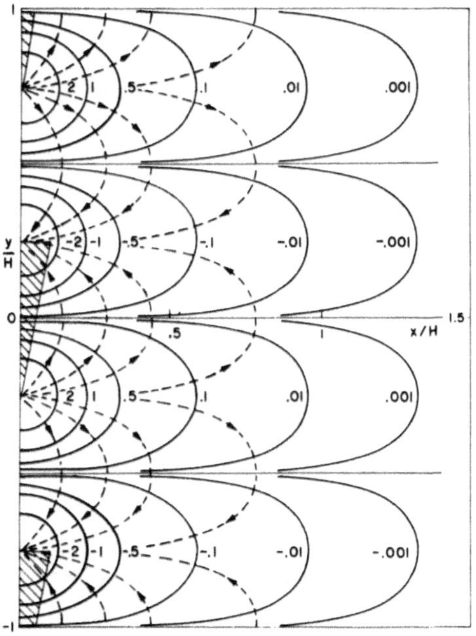

FIG. 6. DILATATIONAL STRAIN FIELD OF A WALL OF WEDGE DISCLINATION DIPOLES [UNIT: $\omega(1-2\nu)/4\pi(1-\nu)$]

The trajectories of the general contours of Eq. (18) are the flow paths of impurity atoms

$$\cos \beta = \cos \lambda \cosh \alpha + C \sinh \alpha \tag{20}$$

where C is any constant. The case of L/H = 1/4 or $\cos \lambda = 0$ is shown by dotted lines in Fig. 6.

It is seen that, depending on the spacing H, large scale solute movements may take place during annealing. As a result of such movements, some regions of the grain boundary may be loaded with impurities and others may be free from them. Such impurity distributions should play a role in intergranular embrittlement and grain boundary sliding.

REFERENCES

1. J. C. M. Li, "Disclination Model of High Angle Grain Boundaries," Proc. Int. Conf. on the Structure and Properties of Grain Boundaries and Interfaces, IBM Watson Research Center, Aug. 23-25, 1971.

2. G. H. Bishop and B. Chalmers, Scripta Met. 2, 133 (1968).

3. K. T. Aust and B. Chalmers, Met. Trans. 1, 1095 (1970).

4. N. F. Mott, Proc. Phys. Soc. Lond. 69, 391 (1948).

5. D. G. Brandon, B. Ralph, S. Ranganathan, and M. S. Wald, Acta Met. 12, 813 (1964).

6. D. G. Brandon, Acta Met. 14, 1479 (1966).

7. R. de Wit, "Relation between Dislocations and Disclinations," to be published.

8. Wen Huang and T. Mura, J. Appl. Phys. 41, 5175 (1970).

9. J. C. M. Li, Acta Met. 8, 563 (1960).

COINCIDENCE AND NEAR-COINCIDENCE GRAIN BOUNDARIES IN HCP METALS

G. A. Bruggeman*, G. H. Bishop* and W. H. Hartt†

*Army Materials and Mechanics Res Ctr, Watertown, Mass.

†Florida Atlantic University, Boca Raton, Florida

ABSTRACT

Coincidence concepts, commonly used in treating the structures of grain boundaries in cubic metals, are used to develop coincidence structural-unit models for grain boundaries in hcp metals. In doing this the concept of coincidence is extended to one of *near-coincidence*, in which the shared atoms at the boundary need not occupy exact coincidence sites but instead occupy compromise positions between two nearly coincidence lattice sites. This extension allows one to define near-coincidence-site lattices (near-CSL's) from which are derived short-period near-coincidence structural units. Exact and near-CSL's in the common hcp metals are tabulated for rotations about [0001], $<10\bar{1}0>$, and $<11\bar{2}0>$, and short-period coincidence structural units are presented. Geometric models of a variety of symmetric and asymmetric $<10\bar{1}0>$ tilt boundaries are developed as examples to show how a relatively few coincidence structural units are combined in the structures of both simple and complex boundaries. Although hcp boundaries are treated here, the approach is quite general and can be applied to any crystal structure.

INTRODUCTION

The concept of the coincidence-site lattice (CSL) has been found to be a useful geometric tool for predicting the existance of certain special misorientations at which special grain boundaries occur. These coincidence misorientations, first referred to in a limited way by Hargreaves and Hills (1), were later recognized by Kronberg and Wilson (2) in connection with the development of secondary recrystallization textures. Boundaries at or near these special misorientations have been shown to have a higher-than-average mobility in the presence of impurities (3-5) and a lower-than-average energy (6-8). Initially it was proposed that the special properties of coincidence boundaries are due to the high density of coincidence sites in the boundary when it lays along close-packed planes in the coincidence lattice (9-10). Subsequently it was suggested that it is not a high density of coincidence sites per se, but rather a related crystallographic property, namely the periodic repetition of short structural units, that is the important feature of coincidence boundaries (11-14). The short-range elastic stress field that results from such a uniform periodic structure should yield the special properties possessed by coincidence boundaries, i.e. a lower surface energy and higher mobility, the latter due to a reduced interaction with impurity atoms.

It has been proposed that the structural units characteristic of these special short-period coincidence boundaries will also be found as part of the structures of high angle boundaries deviating substantially from the ideal coincidence misorientations (11-14). The structures of these off-ideal coincidence boundaries can be treated as mixtures of the short structural units characteristic of two or more ideal coincidence boundaries, or as the superposition of an array of secondary dislocations onto the nearest ideal coincidence boundary (11, 12, 15 - 18). These alternative views of off-ideal coincidence boundary structures have been shown to be equivalent in the coincidence-ledge-dislocation (CLD) description of grain boundaries by Bishop and Chalmers (11, 12). These more general grain boundaries will also possess a periodic structure, although of a much longer period than the ideal coincidence boundaries, in which the short structural units derived from the ideal coincidence boundaries are combined in a systematic and repetitive manner. For example Figure 1 shows an off-ideal coincidence <100> fcc symmetric tilt boundary at $\theta = 50.0°$ (measured between <110> directions). This boundary lies between the ideal coincidence boundaries occurring at $\theta = 53.1°$ and $\theta = 36.9°$ and is made up of six units (ledges two atoms long) of the 53.1° boundary and one unit (ledges three atoms long) of the 36.9° boundary.

The coincidence structural-unit approach to grain boundary structures has been applied in detail only to symmetric tilt boundaries in cubic metals. However its extension to more complex boundaries in cubic metals, and further to boundaries in non-cubic metals can be accomplished with certain modifications. Each coincidence misorientation is capable of contributing several possible structural units besides those appearing in the symmetric tilt boundaries. It can be shown that units characteristic of symmetric

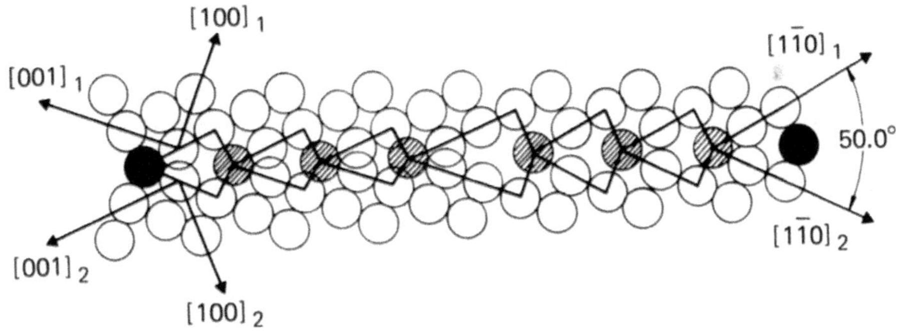

Figure 1. Hard sphere model of a 50.0° ⟨100⟩ fcc non-ideal symmetric tilt coincidence boundary. One period is shown. Only the boundary sites at the end of the repeat unit, indicated by the filled circles, are exact coincidence sites. The boundary sites indicated by the cross-hatched circles are near-coincidence sites. Six units characteristic of the 53.1° ⟨100⟩ ideal coincidence boundary (ledge length 2·a/2·⟨110⟩) and one unit of the 36.9° ⟨100⟩ ideal coincidence boundary (ledge length 3·a/2·⟨110⟩) make up the larger repeat unit.

tilt, twist, and mixed tilt and twist boundaries can be derived from the same coincidence-site lattice, enabling boundaries of many misorientations and inclinations to be described. Thus many (perhaps most) grain boundaries are made up of combinations of a relatively small number of elementary structural units, where the structural units may be derived from the coincidence-site lattices and are simply short segments of coincidence boundaries*. This is the simplest and most general statement of the coincidence structural-unit model for high angle grain boundaries.

In this paper the coincidence structural-unit model will be applied to the description of grain boundaries in hcp metals. Coincidence-site lattices in hcp crystals will be investigated, and qualitatively relaxed coincidence structural-unit models for symmetric and asymmetric tilt grain boundaries will be developed following the approach taken in the CLD treatment. In doing this it will be seen that the concept of the CSL can be generalized to include what will be termed near-coincidence-site lattices (near-CSL's).

* In certain misorientation ranges structural units characteristic of the lattice are part of the boundary structure, e.g. the "1" units in <100> fcc symmetric tilt boundaries for $\theta > 53.1°$ (11). Therefore, the $\theta = 0$ boundary (i.e. a single crystal) is considered a coincidence boundary (from the $\Sigma=1$ CSL) for the purpose of this statement.

COINCIDENCE VERSUS NEAR-COINCIDENCE

In dealing with off-ideal coincidence boundaries in cubic metals by the CLD treatment (11, 12) it was noted that atom sharing could be maintained at the ends of each short structural unit in the boundary, at the expense of some lattice strain, even though lattice coincidence was destroyed or severely altered. This is accomplished by some boundary atoms occupying compromise positions in the boundary between two nearly-coincident lattice sites. In the boundary in Figure 1 only the boundary atoms at the ends of the longer repeat pattern (the solid circles) are exactly coincident. The atoms at the ends of the interior ledges which are shared by the two adjoining crystals (the shaded circles) occupy compromise, or near-coincidence sites (termed boundary coincidence sites in the CLD treatment). By having shared atoms in these near-coincidence sites, normal atomic coordination and spacing are maintained as closely as possible at the boundary and boundary coherency is maximized. Thus in order for combinations of structural units to exist in an off-ideal coincidence boundary, the stringent geometric requirements of exact coincidence must be relaxed to permit near-coincidence to occur in the boundary between the exact coincidence sites.

In this paper the concept of near-coincidence will be extended to include situations in which there is no repetition (no series) of exact coincidence sites in the boundary, but in which a series of near-coincidence sites can be found. In such a boundary a misfit strain will tend to accumulate along the boundary. This strain can be relieved, however, by periodically introducing a dislocation, or structural unit of different ledge length. Such a boundary, which contains a series of near-coincidence instead of exact coincidence sites, will be termed a near-coincidence boundary.

The introduction of the concept of near-coincidence removes a dilemma which one encounters in trying to apply coincidence concepts to non-cubic crystals, namely that symmetric coincidence boundaries exist for which there is no corresponding exact CSL. That is, there are misorientations at which a series of exact coincidence sites in the boundary is geometrically possible in certain symmetric inclinations, but at which no lattice coincidence away from the boundary can be present. In fact in non-cubic crystals, exact CSL's cannot be found for rotations about most axes, except at special values of the lattice parameters. In hcp metals, except for rotations about [0001], exact CSL's are obtained only when $(c/a)^2$ is a rational number. As a result coincidence models as developed thus far for cubic metals are special cases.

When a coincidence boundary exists without the presence of

an exact CSL, it is found that at this same misorientation, there is an alternate symmetric boundary inclination, at approximately 90° to the first inclination. The second inclination deviates slightly from a coincidence boundary, i.e. a slight change in misorientation can cause some or all of the shared atoms at the boundary to come into exact coincidence. These two symmetric coincidence boundaries at approximately 90° to each other but at slightly different crystal misorientations can be used to construct a near-CSL in the same manner that two such coincidence boundaries occuring at precisely the same misorientation define an exact CSL. The boundaries occurring along planes of asymmetry in the near-CSL contain no exact coincidence sites, regardless of the misorientation, but rather a series of near-coincidence sites and are therefore the near-coincidence boundaries referred to above.

It will be shown that in hcp metals (and presumably in other non-cubic metals) these near-CSL's can be used to elucidate the existence of short-period structural units. Thus they provide a convenient three-dimensional frame of reference for relating the different structural units which must be introduced as the boundary changes inclination at constant misorientation.

The coincidence structural-unit model is inherent in the thought first expressed by Hargreaves and Hills (1) that the minimum energy configuration adopted by a grain boundary would be the one in which the atoms are displaced as little as possible from their normal lattice positions. The model proposes that this can most easily be accomplished by the joining of structural units at a series of shared atoms or near-coincidence sites along the boundary. The structure of the units would reflect the atomic configurations on the planes being joined at the boundary. The concept of the CSL is used to define these units and the misorientations and inclinations at which they occur.

Some questions have been raised concerning whether atoms are exactly shared along the boundary plane (13, 19). This presently unsettled point does not alter the basic concepts of the coincidence structural-unit model, only the details of the atomic configurations within particular structural units. Consequently for the purposes of this paper, coincidence (either exact or near-coincidence) will be assumed to exist in the boundary where indicated by the type of geometric treatment used in the CLD model, and possible structural units will be presented accordingly. In support of this approach, it can be shown (20) that the structural units identified with some of the boundaries presented in the CLD model have been corroborated by computer calculations (21) of the same kind that raised the question of atom sharing initially.

HCP COINCIDENCE RELATIONS

Since $(c/a)^2$ is an irrational number in all the cases of interest, there is no simple analytical method for determining the axis/angle relationships associated with near-coincidence. The discussion which follows is based on computer calculations in which, for each axis and angle of rotation, an atom by atom search for near coincidence was performed. The discussion will be restricted to rotations about the three directions of lowest indices, i.e. [0001], $\langle 10\bar{1}0\rangle$, and $\langle 11\bar{2}0\rangle$, and will consider only those cases in which the unit cell of the near-CSL is less than twenty atomic diameters on a side.

In cubic crystals, for a given crystal orientation there are 24 uniquely different ways of representing the usual set of cubic axes. Consequently for each CSL in the cubic system, there are 24 different rotations that can bring about the same crystal misorientation, or 24 equivalent axis/angle pairs which will generate the same CSL. The 23 additional axis/angle pairs can be obtained from the first pair by the suitable operation on the initial rotation matrix by the 24 rotation matrices relating the 24 sets of axes to one another (22). In hexagonal-close-packed crystals there are 6 different ways of representing the set of hexagonal axes. Consequently there are 6 axis/angle pairs that produce equivalent rotations of the crystal and therefore 6 axis/angle pairs for each near-CSL. Elements of the appropriate rotation matrices contain terms relating to $(c/a)^2$ which cause the various equivalent rotation axes to be non-rational whenever $(c/a)^2$ is non-rational. For this reason no attempt will be made to present any of the equivalent rotation axes and rotation angles besides those involving the three major directions.

THE COINCIDENCE RATIO AND ROTATIONS ABOUT [0001]

The crystallography of a grain boundary is described in terms of five parameters,* three to describe the crystallographic misorientation of the two crystals, i.e. the common rotation axis and the angle of rotation of one crystal relative to the other, and two to describe the inclination of the boundary relative to the crystallographic axes of one of the crystals. A coincidence-site lattice is generated whenever a particular rotation about a common axis brings a fraction of the lattice points of one crystal into exact registry with lattice points of the other. The reciprocal of this fraction of common lattice is denoted by Σ, the coincidence ratio.

* If translational relaxation of one crystal relative to the second is considered, then there are eight degrees of freedom (13).

Causing the boundary plane to pass through the array of coincidence lattice points such that the coincidence sites lie in the plane of the boundary produces a coincidence boundary. If this boundary plane is normal to the axis of rotation, the boundary is a pure twist boundary. If the boundary plane is parallel to the rotation axis and passes along a plane of symmetry of the CSL, the boundary is usually a symmetric tilt boundary.† If the boundary plane is parallel to the rotation axis but passes along a plane of asymmetry in the CSL, the boundary is an asymmetric tilt boundary.

It is well to distinquish between the coincidence ratio in the plane of the CSL normal to the axis of rotation and the coincidence ratio in the volume of the CSL. The former will be referred to as the planar coincidence ratio Σ_p, and the latter as the volume coincidence ratio or simply as the coincidence ratio Σ. In cubic crystals the two-dimensional array of atoms on any {hkl} plane can be described in terms of a rectangular unit cell lying in the plane, having an axial ratio R. Various specific rotations about <hkl>, the normal to the plane, will bring about coincidence in the {hkl} plane with values of $\Sigma_p = n(x^2 + R^2y^2)$ where x and y are integers and n is a rational number that causes nx and nR^2y both to be integers (23). The planar unit cell in the resulting coincidence lattice turns out also to be rectangular with an axial ratio of nR.* The shorter edge of the rectangular unit cell of the CSL is the lattice vector $x\bar{i} + yR\bar{j}$, where $\bar{i}$ and $\bar{j}$ are unit vectors that point in the direction of the rectangular axes in the {hkl} plane. The above expression for Σ_p is given by the ratio of the areas of these two unit cells in the {hkl} plane. The true value of Σ_p differs from the one calculated from the above equation whenever the number of atoms per planar unit cell in the crystal differs from the number of coincidence sites per planar cell in the CSL. When this difference is appropriately accounted for, Σ_p is always an odd integer.

The volume coincidence ratio Σ, however, depends upon the coincidence that is established throughout the entire stacking sequence of {hkl} planes. Take for example the three-layered stacking sequence of {111} planes that generates the face-centered cubic

† For a given misorientation the number of possible symmetric tilt boundaries will equal the degree of rotational symmetry about the tilt axis exhibited by the crystal, e.g. for rotations about a two-fold rotation axis there are two possible symmetric tilt boundaries.

* The unit cell of the coincidence-site lattice will always display the same degree or a higher degree of rotational symmetry about the rotation axis than does the crystal itself.

crystal structure. For a given angle of rotation one may find that the same fraction of coincidence sites are found in each layer of the three-layered sequence. This is the condition produced by a 38.2° rotation about <111> which results in a planar coincidence ratio of $\Sigma_p = 7$ on each and every {111} plane of the three-layered stacking sequence. The volume coincidence ratio Σ is therefore equal to the planar coincidence ratio in this case, i.e. $\Sigma = 7$. On the other hand, for some angles of rotation one may find that coincidence sites exist on only one layer of the three-layered sequence. This is the condition produced by a 21.8° rotation about <111> which results in a planar coincidence ratio of $\Sigma_p = 7$ on every third plane of the three-layered sequence. The volume coincidence ratio is therefore three times as large as the planar coincidence ratio, i.e. $\Sigma = 3\Sigma_p = 21$ in this case.

The hcp structure is generated by the two-layered stacking of the close-packed (0001) planes analogous to the three-layered stacking of the close-packed {111} planes that give the fcc crystal structure. It is clear that any angle of rotation about <111> that produces an exact CSL in fcc crystals will also produce an exact CSL in hcp crystals for the same rotation about [0001]. The Σ_p values in the (0001) plane will be the same as those in the {111} plane of fcc; however the volume coincidence ratio Σ may be different. Whenever a particular rotation causes coincidence on every layer of the stacking sequence, e.g. the 38.2° rotation above, the volume coincidence ratio and the planar coincidence ratio will be the same in both fcc and hcp crystals. Whenever a rotation causes coincidence in only one layer of an m-layered sequence, e.g. the 21.8° rotation above, the volume coincidence ratio is m times the planar coincidence ratio. Therefore the rotation that produced the CSL with $\Sigma = 21$ in fcc produces a CSL with $\Sigma = 14$ in hcp; or any rotation that produces a CSL with $\Sigma = 3\Sigma_p$ in fcc produces a CSL with $\Sigma = 2\Sigma_p$ in hcp where Σ_p is always an odd integer. Thus the possibility of even values of Σ exist in hcp crystals, an occurrence not encountered in cubic crystals. This will be seen to be rather common when rotations about $\langle 10\bar{1}0\rangle$ and $\langle 11\bar{2}0\rangle$ are considered.

Table I presents the coincidence-site lattices formed by rotations about [0001] in hcp crystals. Values of Σ up to 37 and the corresponding angles of rotation about [0001] are listed. Possible grain boundary structural units derived from these and other CSL's will be presented later.

ROTATIONS ABOUT $\langle 10\bar{1}0\rangle$

In hcp crystals directions of the form <hki0> are normal to {hki0} planes. Thus the $\langle 10\bar{1}0\rangle$ direction is normal to the $\{10\bar{1}0\}$

TABLE I. COINCIDENCE-SITE LATTICES FORMED BY ROTATIONS ABOUT $[0001]_{hcp}$ OR $<111>_{fcc}$

Σ_{hcp}	θ*	Σ_{fcc}
2	60.00	3
7	38.21	7
13	27.80	13
14	21.79	21
19	46.83	19
26	32.20	39
31	17.90	31
37	50.57	37

*Also 120°±θ

plane. Looking along the $<10\bar{1}0>$ direction, the hcp structure can be described in terms of a four-layered stacking of $\{10\bar{1}0\}$ planes in which the interplanar spacing is not uniform. This is shown in Figure 2 which also shows a rectangular planar unit cell defined on the $\{10\bar{1}0\}$ plane having an axial ratio $R = c/a$. Rotations about $<10\bar{1}0>$ will generate CSL's with a planar coincidence ratio Σ_p on the $\{10\bar{1}0\}$ plane given by the same generating function used previously in the cubic system, namely $\Sigma_p = n(x^2 + R^2y)$.

Figure 3 illustrates a $<10\bar{1}0>$ CSL generated by a rotation of 78.5° in an hcp crystal having an ideal c/a ratio, where $(c/a)^2 = 8/3$. The background shows the four layers in the $\{10\bar{1}0\}$ stacking sequence denoted by four different symbols.(see Figure 2). The array of large filled circles are the coincidence sites in the plane of the small circle-like symbols. This array of coincidence sites is represented by the rectangular unit cell OABC. The complete array of coincidence sites is shown in only this one layer; coincidence sites in the other three layers are shown by the large open circles only in the unit cell. A second rectangular array of lattice points can be identified on the $\{10\bar{1}0\}$ plane that is geometrically identical to the array of large filled circles except that it is misoriented by 78.5° with respect to the first array. This second array is represented in Figure 3 only by its unit cell OA'B'C', where

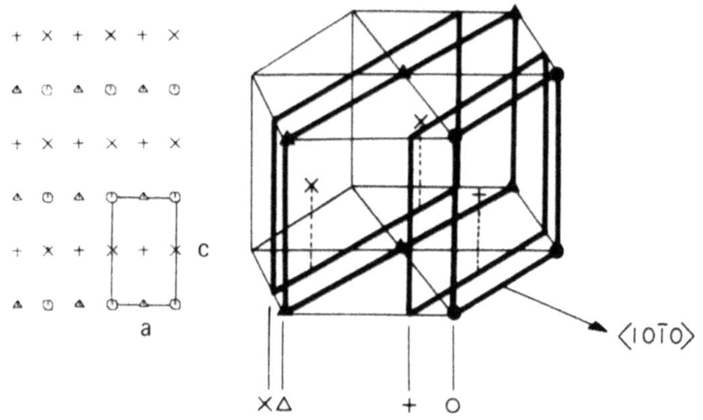

Figure 2. The four-layered stacking sequence of $\{10\bar{1}0\}$ planes in a hexagonal-close-packed crystal. The planar unit cell is shown on the left in a plan view along the $\langle 10\bar{1}0 \rangle$ axis.

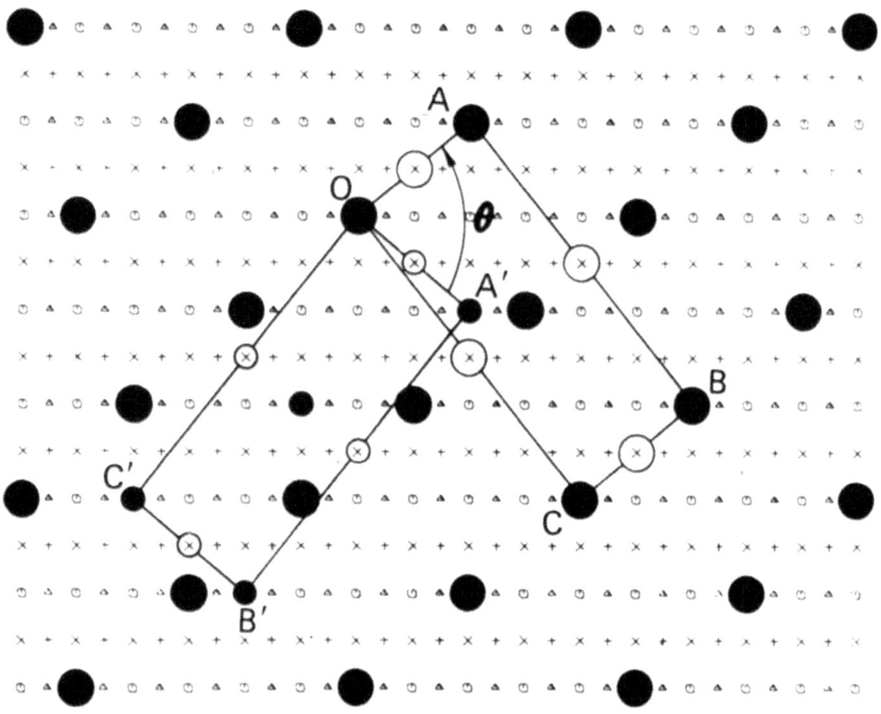

Figure 3. The $\Sigma = 10$ coincidence site lattice formed by a 78.46° rotation about $\langle 10\bar{1}0 \rangle$ in a hcp crystal having an ideal c/a ratio. This rotation brings the planar unit cell OA′B′C′ into exact coincidence with the planar unit cell OABC. The filled circles all lie on the same layer.

the small filled circles lie in the plane of the large filled circles. The rotation of 78.5° about <10$\bar{1}$0> brings these two arrays (and their representative unit cells) into exact superposition. The rotation of unit cells as shown in Figure 3 and in subsequent figures is understood to represent the rotation of interpenetrating crystal lattices to bring about coincidence of three-dimensional arrays of lattice points to define a coincidence-site lattice (CSL).

Since $R^2 = (c/a)^2$ is a rational number in the case illustrated in Figure 3, the CSL is exact. In general when $(c/a)^2$ is a rational number, Σ_p will take on exactly integral values and exact CSL's will be generated. When $(c/a)^2$ is irrational, Σ_p calculated from the generating function deviates slightly from integral values and only near-CSL's are generated. The latter is the case for all the common hcp metals at room temperature. In either case rotations about <10$\bar{1}$0> will yield odd values of Σ_p. The volume coincidence ratio will take on both odd and even values, however, depending on the coincidence established throughout the stacking sequence. Whenever the same planar coincidence ratio is produced on all four layers of the sequence, an odd value of Σ results. Figure 4 shows an example of this case in which $\Sigma = 7$ for a rotation of 64.7° about <10$\bar{1}$0> in titanium. Whenever the same planar coincidence ratio is produced on only two of the four layers (the remaining two layers are always found to contain no coincidence sites) an even value of Σ results, equal to $2\Sigma_p$. Figure 3 shows an example of this latter case in which $\Sigma = 10$ for a rotation of 78.5° about <10$\bar{1}$0> in a crystal having an ideal c/a ratio. A near-CSL of this same type exists for a rotation of 78.6° about <10$\bar{1}$0> in magnesium which has close to the ideal c/a ratio. These are the only two types of situations that have been encountered in the searches for near-coincidence, namely coincidence on all four layers or coincidence on only two of the layers.

Invariably, whenever Σ is an odd number, the unit cell in the CSL contains coincidence sites located as in Figure 4, in which each layer contributes one site to the total number in the unit cell. In the plan view projected onto the {10$\bar{1}$0} plane, sites from one layer are located at the corners of the cell, and the sites from the other layers project onto the centers of the edges and the face-center. On the other hand, whenever Σ is an even number, the coincidence cell is centered on the {10$\bar{1}$0} face as in Figure 3 and the sites from the other contributing layer project onto the centers of the edges of the unit cell. The coincidence cell thus contains coincidence sites on the top and bottom {10$\bar{1}$0} planes midway along the face diagonal. This centering is significant since it allows the asymmetric tilt boundary passing along this diagonal of the unit cell of the CSL to have a period half as long as what it would have been had Σ been an odd number and the

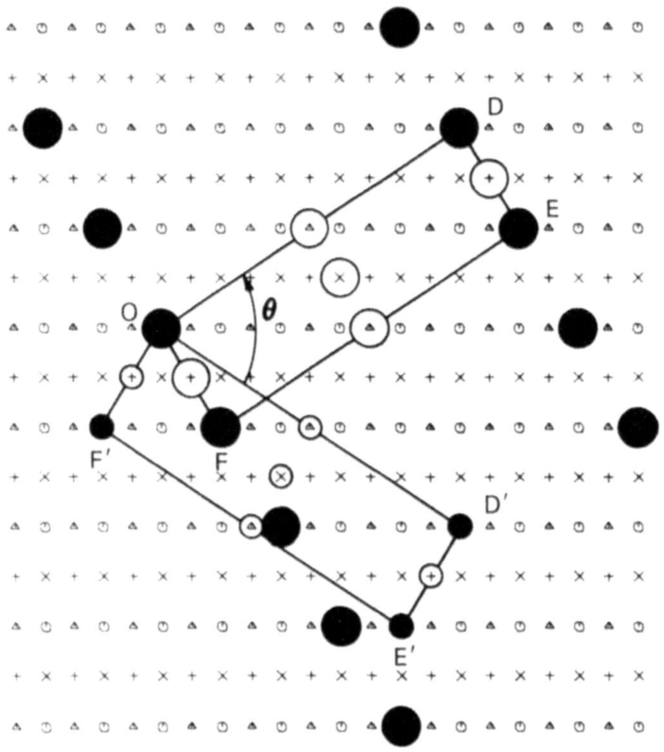

Figure 4. The $\Sigma = 7$ near-coincidence-site lattice formed by a $64.7^{o}\,{}^{+0.1}_{-0.3}$ rotation about ⟨10$\bar{1}$0⟩ in titanium. This rotation brings the planar unit cell OD'E'F' into near-coincidence with the planar unit cell ODEF. The filled circles all lie on the same layer.

unit cell not so centered. In such CSL's the asymmetric structural unit will be shorter than the symmetric structural unit. As a result many of the shorter structural units for <10$\bar{1}$0> tilt boundaries will be asymmetric.

Several of the features which distinquish a near-CSL from an exact CSL can be illustrated by reference to Figures 3 and 4. In both cases the spacing of lattice points is such that A can be made to coincide with A', C with C', D with D', and F with F'. In the exact CSL in Figure 3, the angles AOC and A'OC' are both 90° so that the edges OA and OA', OC and OC', and the lattice points A and A', and C and C' can be brought into coincidence simultaneously. In the near-CSL in Figure 4, however, the angles DOF and D'OF' are not 90°, but rather 90.20° and 89.80° respectively.*

* In order to more clearly distinquish near-coincidence from exact coincidence, angles are given to an extra significant figure in the next few paragraphs and the departure from exact coincidence in Figure 4 has been exaggerated.

As a result, a rotation of 64.42° will bring OF and OF', and F and F' into exact coincidence, or a rotation of 64.83° will bring OD and OD', and D and D' into exact coincidence, but not both simultaneously. Between these two angles none of the points will coincide exactly. The points E and E' never coincide because OE $\neq$ OE' in length, so that asymmetric boundaries associated with a near-CSL are always near-coincidence boundaries.

Thus there is a range of misorientations over which exact or near-coincidence can be achieved, with the degree of registry at different points in the near-CSL being better or worse depending on the precise angle of rotation within this misorientation range. In the tabulations of near-CSL's in Tables II and III, this range of misorientations is indicated by a median angle, at which the misregistry of D with D' is equal to that of F with F', and also a positive and negative angular variance which define the two misorientations at which the edges of the cell come into exact coincidence.

Because there cannot be the same degree of coincidence along all directions at the same θ in a near-CSL, the misregistry this introduces must be relaxed. As an example assume the misorientation is 64.42° in titanium (Figure 4). If the boundary lies along the direction OF/OF' (OF and OF' coincide at this misorientation) it forms a symmetric coincidence boundary made up solely of one type of structural unit. If it bisects the directions OD/OD' which do not coincide, it is slightly off from a second symmetric coincidence boundary. This deviation would be accomodated by introducing units of different ledge length characteristic of an adjacent short-period coincidence boundary, or equivalently, by introducing secondary dislocations. If the misorientations were 64.83°, the situation would be reversed. At any misorientation between these two angles both symmetric inclinations would be off-coincidence boundaries into which structural units characteristic of the appropriate adjacent short-period coincidence boundary were periodically introduced to accomodate the misregistry.

Next consider asymmetric inclinations. The directions OE and OE' coincide at 64.79°. At this misorientation there is still, however, the misregistry resulting from the unequal lengths of OE and OE'. Since this misregistry is parallel to the boundary plane it can be compensated by the introduction of an array of misfit dislocations, analogous to those in semi-coherent interphase interfaces, with Burgers vectors parallel to the boundary plane. Deviations from the optimum misorientation of 64.79° increases the misregistry by adding a component normal to the boundary plane. This would require the periodic introduction of an array of dislocations with a new Burgers vector normal to boundary plane. In many cases both components of the misregistry could be accommodated simultaneously by the periodic introduction of other types of structural units which have the equivalent effect of the simultaneous introduction of both misfit and secondary dislocations.

Table II presents values of Σ and the corresponding angles of rotation about <10$\bar{1}$0> for near-CSL's with Σ < 25 in the common hcp metals. For the purposes of this table, two lattice points were said to be near-coincident when, after the lattice rotation, they were displaced from one another by no more than 5% of an atom diameter. A more extensive listing of near-CSL's, giving the lattice vectors that define the edges of the coincidence unit cell of each near-CSL, may be found in Reference 24. Inasmuch as <10$\bar{1}$0> is an axis of two-fold rotational symmetry, the same CSL can be obtained by two different angles of rotation about <10$\bar{1}$0>; the sum of these two rotation angles is 180°. In Table II only the rotation angle that is less than 90° is reported.

It should be noted in Table II that a given near-CSL is never achieved in every hcp metal. For example, the near-CSL with Σ = 7 is obtained only in those metals in which c/a < ideal; the near-CSL with Σ = 9 is obtained only in those metals in which c/a > ideal. Consequently each hcp metal must be treated separately; one c/a ratio cannot be taken as typical of all the hcp metals. Furthermore, when the same near-CSL is obtained for several hcp metals, the angle of rotation at which it occurs varys systematically with the c/a ratio. The quality of the near-CSL will vary from one c/a ratio to another, i.e. for one c/a ratio the near-CSL will be more nearly exact than for the others. The smaller the angular variance associated with the angle of rotation, the more nearly exact is the near-CSL. In the limit of exact coincidence no angular variance is required since the same rotation angle brings about exact registry in all directions in the CSL. Finally it should be noted that the same Σ value can be obtained with two or more different CSL's which are produced by different angles of rotation. These different CSL's can exist in different metals (c.f. Σ = 13: Zn vs. Zr, Ti, Be) or in the same metal (c.f. Σ = 11: Mg).

ROTATIONS ABOUT <11$\bar{2}$0>

The hcp crystal structure can be described in terms of the two-layered stacking of {11$\bar{2}$0} planes in which, as was the case with the {10$\bar{1}$0} stacking sequence, the interplanar spacing is not uniform. On the {11$\bar{2}$0} plane a rectangular planar unit cell can be defined having an axial ratio $R = c/a\sqrt{3}$. The asymmetric array of lattice points on {11$\bar{2}$0} reflects the fact that <11$\bar{2}$0> is an axis of one-fold rotational symmetry. This asymmetry is usually transmitted to the near-CSL's that result from specific rotations about this axis. Furthermore because of the asymmetry, the near-CSL formed by a rotation of θ degrees is usually different from the CSL formed by a rotation of $(180 - \theta)$ degrees; both rotations will generate a near-CSL but the near-CSL's and/or the coincidence ratios may be different. This point is illustrated in Figures 5a and 5b which show CSL's formed in a crystal having an ideal c/a ratio (hence exact CSL's) for rotations of 70.53° and 109.47° respectively. The lower symmetry of these CSL's as compared with those encountered with rotations about $\langle 10\bar{1}0 \rangle$ is obvious.

Table III presents values of Σ and the corresponding angles of rotation about $\langle 11\bar{2}0 \rangle$ for several near-CSL's in the common hcp metals. A more extensive listing, giving the lattice vectors that define the edges of the unit cells is to be found in Reference 24. These near-CSL's can form the basis for defining possible grain boundary structural units for crystal misorientations resulting from rotations about $\langle 11\bar{2}0 \rangle$ or axes close to $\langle 11\bar{2}0 \rangle$. None of the actual examples of possible structural units that are used in the ensuing discussion however, come from CSL's found in this table. That discussion is limited to $\langle 10\bar{1}0 \rangle$ tilt boundaries, although the extension to $\langle 11\bar{2}0 \rangle$ and other boundaries is straight forward.

GRAIN BOUNDARY STRUCTURAL UNITS

As presented earlier in the statement of the coincidence structural-unit model of high angle grain boundaries, the structural units that are combined to give a more-or-less general grain boundary structure are actually short segments of coincidence boundaries. Such a boundary in three different inclinations is shown in Figure 6. This is a 78.46° $\langle 10\bar{1}0 \rangle$ tilt boundary in a crystal with an ideal c/a ratio. The boundary is passing along planes of the $\Sigma = 10$ CSL shown in Figure 3, following its two possible symmetric inclinations and one of the possible asymmetric inclinations. The structural units identified with the boundary in each of these inclinations are indicated. Here and in future discussions these structural units will be defined in terms of the lattice vectors which, with the rotation axis, define the crystallographic plane on each side of the coincidence boundary. These lattice vectors will be denoted in terms of their components in the c and the a directions (the vectors c[0001] and $a/3\langle 11\bar{2}0 \rangle$). Thus the structural units making up the horizontal portion of the boundary in Figure 6 will be denoted 1c,2a/1c,2a or simply 1,2/1,2. The vertical portion of the boundary contains the structural unit denoted by 3c,4a/3c,4a or simply 3,4/3,4. The asymmetric portion of the boundary is made up of asymmetric structural units with the designation 1c,3a/2c,1a or simply 1,3/2,1.

TABLE II. NEAR-COINCIDENCE-SITE LATTICES FORMED BY ROTATIONS ABOUT <10$\bar{1}$0>

Σ	Θ – Misorientation*						
	Cd c/a: 1.8858	Zn 1.8564	Ideal 1.6330	Mg 1.6237	Zr 1.5933	Ti 1.5873	Be 1.5680
7					$64.8^{+0.2}_{-0.6}$	$64.7^{+0.1}_{-0.3}$	$64.4^{+0.7}_{-0.2}$
9	$56.5^{+0.1}_{-0.6}$	$56.0^{+0.6}_{-0.1}$					
10			78.46	$78.6^{+0.2}_{-0.5}$	$79.0^{+0.9}_{-1.9}$		
11a			62.96	$62.8^{+0.5}_{-0.1}$			
11b				$35.3^{+0.7}_{-1.1}$	$35.2^{+0.1}_{-0.4}$	$35.1^{+0.1}_{-0.1}$	$35.0^{+0.4}_{-0.2}$
13a		$58.0^{+0.2}_{-1.4}$					
13b					$76.4^{+0.7}_{-0.2}$	$76.5^{+0.4}_{-0.1}$	$76.9^{+0.2}_{-0.7}$
14			44.42	$44.4^{+0.2}_{-0.2}$	$44.5^{+0.9}_{-1.1}$		
15a	$30.0^{+0.2}_{-0.3}$	$29.9^{+0.2}_{-0.2}$					
15b	$85.9^{+0.7}_{-0.2}$	$86.3^{+0.3}_{-0.6}$					
16							$75.1^{+1.1}_{-0.2}$

TABLE II. (Continued)

Σ	Cd	Zn	Ideal	Mg	Zr	Ti	Be
17		$40.1^{+0.6}_{-0.6}$					
19a	$87.2^{+0.1}_{-0.6}$						
19b					$87.1^{+0.3}_{-0.5}$	$87.0^{+0.2}_{-0.2}$	$86.9^{+0.6}_{-0.4}$
21a	$64.8^{+0.1}_{-0.5}$						
21b							$76.3^{+0.0}_{-0.1}$
22							$24.8^{+0.5}_{-0.8}$
23a	$77.6^{+0.3}_{-0.6}$	$77.3^{+0.6}_{-0.3}$					
23b					$55.4^{+0.5}_{-0.2}$	$55.5^{+0.3}_{-0.1}$	$55.8^{+0.2}_{-0.6}$
23c					$77.7^{+0.1}_{-0.6}$	$77.8^{+0.2}_{-0.9}$	
25a	$63.7^{+0.6}_{-0.2}$	$64.1^{+0.2}_{-0.6}$					
25b			23.07	$23.0^{+0.2}_{-0.1}$			

*Also 180°−Θ

TABLE III. NEAR-COINCIDENCE-SITE LATTICES FORMED BY ROTATIONS ABOUT <11$\bar{2}$0>

Σ	Θ - Misorientation						
	Cd	Zn	Ideal	Mg	Zr	Ti	Be
9a			70.53	$70.5^{+0.3}_{-0.3}$			
9b			109.47	$109.5^{+0.3}_{-0.3}$			
9c				$56.3^{+0.0}_{-0.1}$			
10							$95.8^{+0.0}_{-0.1}$
11a	$84.5^{+0.6}_{-0.1}$	$83.9^{+2.1}_{-0.4}$					
11b					$95.5^{+0.1}_{-0.7}$	$95.4^{+0.0}_{-0.4}$	$94.9^{+0.8}_{-0.2}$
12a	$85.3^{+0.0}_{-0.2}$						
12b					$94.8^{+0.1}_{-0.0}$		
13a	$122.4^{+0.5}_{-0.1}$						
13b	$85.9^{+0.1}_{-0.8}$	$85.3^{+0.7}_{-0.2}$					
13c				$94.9^{+0.2}_{-1.2}$	$94.1^{+0.7}_{-0.1}$		
13d					$57.6^{+0.2}_{-0.5}$	$57.5^{+0.1}_{-0.3}$	$57.2^{+0.6}_{-0.2}$

TABLE III. (Continued)

Σ	Cd	Zn	Ideal	Mg	Zr	Ti	Be
15a	$87.0^{+0.2}_{-1.9}$	$86.3^{+0.0}_{-0.3}$					
15b				$93.9^{+0.0}_{-0.2}$	$93.1^{+1.7}_{-0.2}$	$92.9^{+2.1}_{-0.2}$	
16a					$71.7^{+0.2}_{-0.0}$		
16b					$108.3^{+0.0}_{-0.2}$		
17a	$49.9^{+0.1}_{-0.6}$	$49.5^{+0.6}_{-0.2}$					
17b			93.37	$93.1^{+0.6}_{-0.1}$			
17c					$130.2^{+0.4}_{-0.2}$		
17d					$139.8^{+0.4}_{-0.2}$	$139.8^{+0.2}_{-0.1}$	$140.0^{+0.2}_{-0.4}$
18a		$123.8^{+0.1}_{-0.2}$					
18b				$123.7^{+0.1}_{-0.0}$			
18c						$82.7^{+2.3}_{-0.5}$	$83.2^{+1.1}_{-0.3}$
18d						$97.3^{+0.5}_{-2.3}$	$96.8^{+0.3}_{-1.1}$

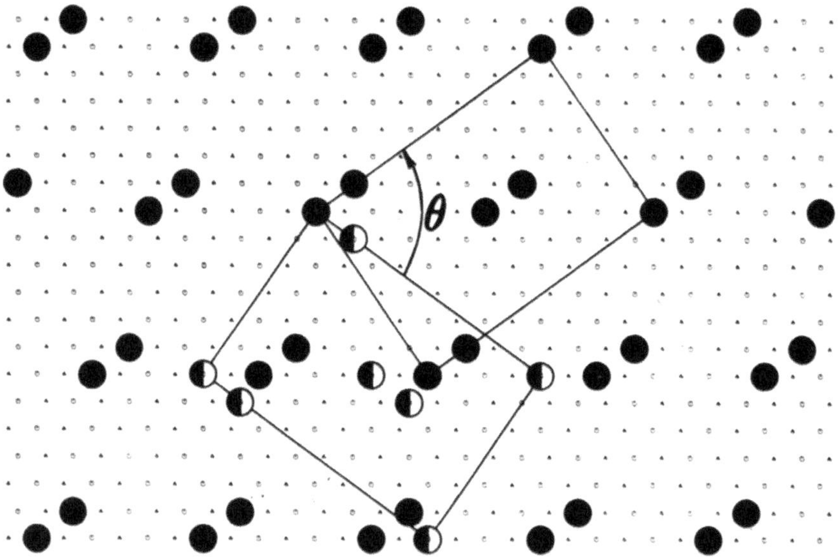

Figure 5a. The $\Sigma = 9$ coincidence-site lattice formed by a rotation of 70.53° about $\langle 11\bar{2}0 \rangle$ in a hexagonal-close-packed crystal having an ideal c/a ratio. Only coincidence sites in one layer have been indicated.

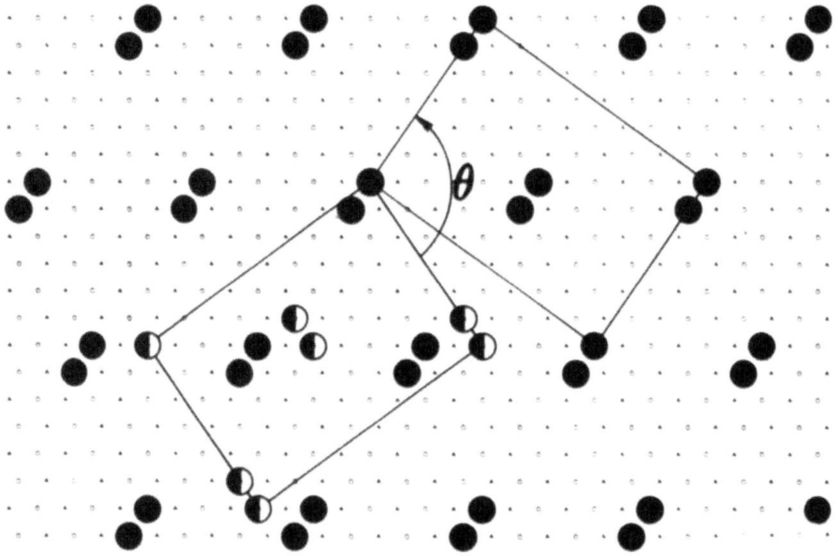

Figure 5b. The $\Sigma = 9$ coincidence-site lattice formed by a rotation of 109.47° about $\langle 11\bar{2}0 \rangle$ in a hexagonal-close-packed crystal having an ideal c/a ratio. Only coincidence sites in one layer have been indicated..

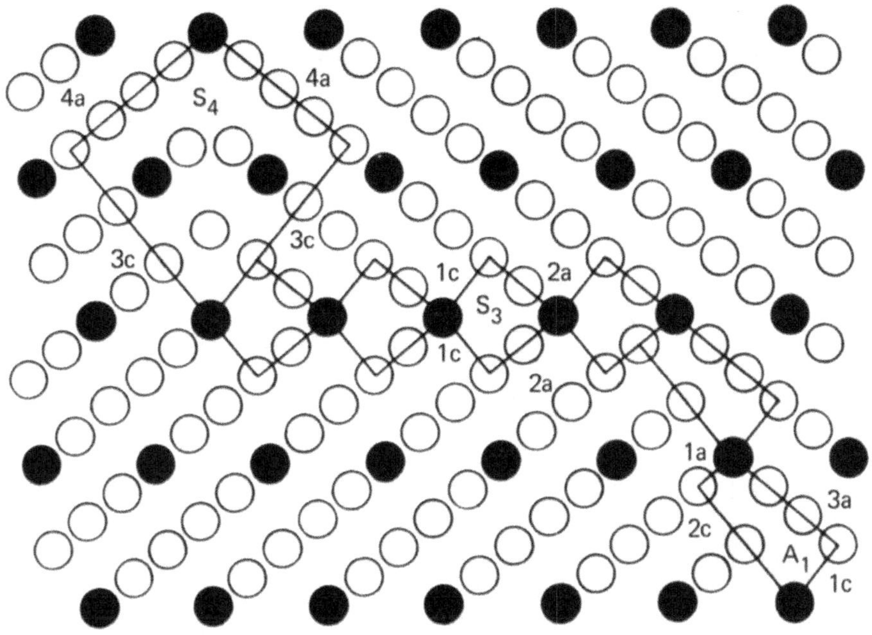

Figure 6a. A coincidence boundary passing along densely packed planes of the $\Sigma = 10$ coincidence-site lattice formed by a rotation of 78.46° about $\langle 10\bar{1}0\rangle$ in a hexagonal-close-packed crystal having an ideal c/a ratio. Three boundary inclinations are shown characterized by three different structural units. These are designated S_3, S_4, and A_1.

Figure 6b. The boundary of Figure 6a defining the boundary inclination α relative to the symmetric boundary. By convention α will be measured relative to that symmetric boundary which bisects the acute angle between the basal planes.

Possible symmetric structural units for $<10\bar{1}0>$ tilt boundaries are presented in Table IV. The angles of rotation associated with these structural units are given for each of the common hcp metals along with the values of the near-CSL's from which that particular unit was derived or in which the unit may be found. It will be noted that the boundary misorientation at which each of the structural units occurs is different for each hcp metal.

TABLE IV. STRUCTURAL UNITS FOR SYMMETRIC <1010̄> TILT BOUNDARIES

Structure mc,na/mc,na	Boundary Plane		θ - Misorientation/Σ - Coincidence Ratio						
			Cd	Zn	Ideal	Mg	Zr	Ti	Be
1,5/1,5	$\{1\ 1\ \bar{2}\ 10\}$	θ:	41.3	40.7	36.2	36.0	35.3	35.2	34.8
		Σ:	40	17	83	11b	11b	11b	11b
1,4/1,4	$\{11\bar{2}8\}$	θ:	50.5	49.8	44.4	44.2	43.4	43.3	42.8
		Σ:	44	34	14	14	14	37	60
1,3/1,3	$\{11\bar{2}6\}$	θ:	64.3	63.5	57.1	56.8	55.9	55.8	55.2
		Σ:	21a	29	35	31	27	23b	23b
2,5/2,5	$\{11\bar{2}5\}$	θ:	74.1	73.2	66.3	66.0	65.0	64.8	64.2
		Σ:	55	62	107	64	7	7	7
1,2/1,2	$\{11\bar{2}4\}$	θ:	86.6	85.7	78.5	78.1	77.1	76.9	76.2
		Σ:	34	26	10	10	36	13b	21b
2,3/2,3	$\{11\bar{2}3\}$	θ:	103.0	102.1	94.9	94.5	93.4	93.2	92.5
		Σ:	23a	23a	59	26	19b	19b	19b
1,1/1,1	$\{11\bar{2}2\}$	θ:	124.1	123.4	117.0	116.7	115.8	115.6	114.9
		Σ:	9	9	11a	11a	7	7	7
3,2/3,2	$\{33\bar{6}4\}$	θ:	141.1	140.5	135.6	135.4	134.6	134.4	133.9
		Σ:	54	17	14	14	14	53	39
2,1/2,1	$\{11\bar{2}1\}$	θ:	150.3	149.9	146.0	145.8	145.2	145.0	144.6
		Σ:	15a	15a	35	11b	11b	11b	11b
3,1/3,1	$\{33\bar{6}2\}$	θ:	160.0	159.6	156.9	156.8	156.4	156.3	156.0
		Σ:	34	31	25b	25b	47	47	22
m,n/m,n	$\{1\ 1\ \bar{2}\ 2\frac{n}{m}\}$								

An example of the 1,3/1,3 unit in titanium is shown in Figure 7 where, at a misorientation of 55.8°, it makes up a <10$\bar{1}$0> ideal coincidence tilt boundary. An example of the 2,1/2,1 unit in titanium is shown in Figure 8 where the misorientation is 145.0° (or equivalently 35.0°). The same structural unit may be found in more than one near-CSL, e.g. the 1,2/1,2 unit in zirconium appears in the Σ = 10, 13, 23, and 36 near-CSL's. Whenever this happens the misorientation ranges over which those near-CSL's occur, as reported in Table II, will be seen to overlap. The Σ value selected for inclusion in Table IV is that for the near-CSL which generally is most nearly exact. The fact that very large Σ's appear in the table reflect the fact that at certain c/a ratios only cells with a very high axial ratio fulfilled the acceptance criteria for near-coincidence. Along certain directions in these near-CSL's, however, the repeat distance is short and hence the density of near-coincidence sites in boundaries following these directions would be relatively high in spite of the large Σ value.

Table V presents possible asymmetric structural units for <10$\bar{1}$0> tilt boundaries. The boundary misorientations θ and inclinations α at which these structural units appear are given for the common hcp metals as well as the Σ values of the near-CSL's from

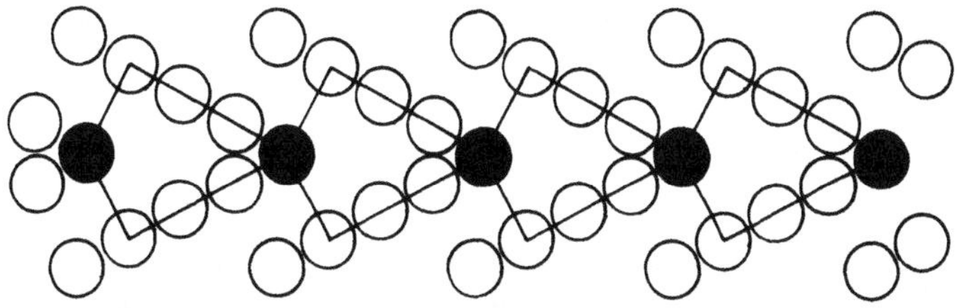

Figure 7. A symmetric ⟨10$\bar{1}$0⟩ tilt boundary characterized by the 1c, 3a/1c, 3a structural unit. This boundary occurs at a misorientation of 55.8° in titanium.

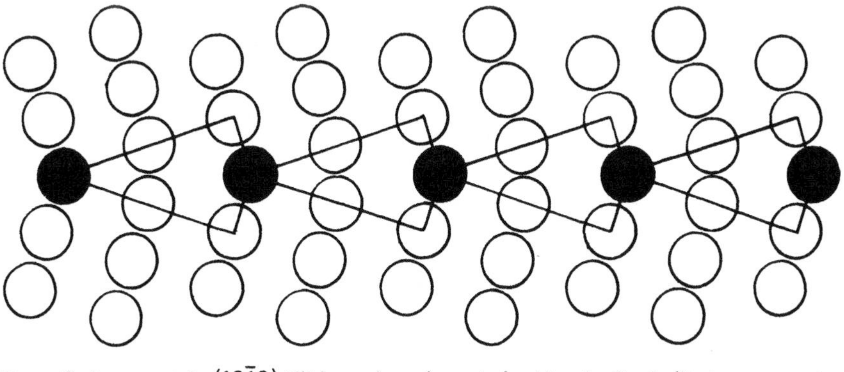

Figure 8. A symmetric ⟨10$\bar{1}$0⟩ tilt boundary characterized by the 2c, 1a/2c,1a structural unit. This boundary occurs at a misorientation of 145.0° in titanium.

TABLE V. SHORT-PERIOD ASYMMETRIC STRUCTURAL UNITS FOR <10$\bar{1}$0> TILT BOUNDARIES

Structure mc,na/*mc*,*n*a	Σ	α – Inclination, Θ – Misorientation						
		Cd	Zn	Ideal	Mg	Zr	Ti	Be
0,5/3,0	30			45.0,90.0	*			
1,5/3,2	26			24.9,85.9	24.8,85.7	24.8,85.0	24.8,84.8	*
1,5/3,2	34			42.9,49.7	42.8,49.7	42.5,49.6	42.4,49.6	*
2,5/3,3	18	*	77.5,81.7					
2,5/3,3	42	*	49.1,25.1					
1,6/3,3	30	*	22.2,78.9					
1,6/3,3	42	*	39.4,44.5					
1,3/2,1	10			67.8,78.5	67.8,78.7	67.7,79.5	67.7,79.6	67.6,80.1
1,3/2,1	14			50.8,44.4	50.7,44.5	50.3,44.6	50.2,44.6	50.0,44.7
0,5/3,1	10			39.2,78.5	39.2,78.4	*	*	*
0,4/2,1	16	37.6,75.2	37.5,74.9					
2,4/3,2	16			*	*	75.6,74.2	75.6,74.3	75.6,74.9
2,4/3,2	32			*	*	52.9,28.8	52.8,28.8	52.5,28.9
0,6/3,2	36	35.3,70.5	35.1,70.2					

TABLE V. (Continued)

Structure	Σ	Cd	Zn	Ideal	Mg	Zr	Ti	Be
1,4/2,3	10			*	12.6,69.4	12.5,68.4	12.5,68.3	12.4,67.7
1,4/2,3	22			*	34.7,25.2	34.2,25.0	34.1,25.0	33.8,24.9
0,2/1,1	4	*	30.8,61.7	29.3,58.5	29.2,58.4	*	*	*
1,5/2,4	4	*	*	10.6,57.3	10.5,57.1	*	*	*
1,5/2,4	28	*	*	28.7,21.1	28.5,21.1	*	*	*
1,6/2,5	14	9.8,54.5	9.7,53.8					
1,6/2,5	34	27.2,19.6	26.9,19.4					
2,4/3,0	24	*	66.4,47.1					
0,5/2,4	20			*	*	*	19.2,38.4	19.0,38.1
0,6/2,5	24			16.6,33.1	16.5,33.0	16.3,32.5	16.2,32.4	16.0,32.1
3,0/3,1	6	85.0,10.0	84.9,10.2	84.2,11.5	84.2,11.6	84.1,11.8	84.1,11.9	84.0,12.0

*For this table the acceptance criterion required coincidence within 10% of an atomic diameter. Units indicated by an asterisk fell just outside this acceptance limit

which these asymmetric units were derived. The angle α is measured, as in Figure 6b, from the symmetric inclination. For near-CSL's formed by rotations about <10$\bar{1}$0> there are two symmetric inclinations, one inclined at an angle of 90° from the other. The values of α quoted in the table are measured from that symmetric boundary that bisects the acute angle between the basal planes. The table was arbitrarily restricted to asymmetric units that are shorter than the longest symmetric unit given in Table IV. The listing, as a result, includes only units that are derived from centered CSL's (note the even Σ's) and that are therefore one-half the length of the diagonal of the unit cell of the near-CSL. Note that there are two entries, one in italics for a number of the structural units. The two entries denote two related but distinct asymmetric structural units that appear as diagonals of different near-CSL's at different misorientations. These two units are related to one another by a rotation of one side of the unit relative to the other by 180° about the normal to the boundary plane. This relation is illustrated for the second asymmetric unit in the list (the 1,5/3,2 unit) in Figure 9. All asymmetric units from near-CSL's are near-coincidence units.

The next section will show how the <10$\bar{1}$0> structural unit presented in Tables IV and V may be combined to give the structures of a variety of grain boundaries. The principles demonstrated are easily applied to combinations of structural units associated with [0001] and <11$\bar{2}$0> tilt boundaries also, although examples of these will not be shown. Tables VI and VII present some of the possible

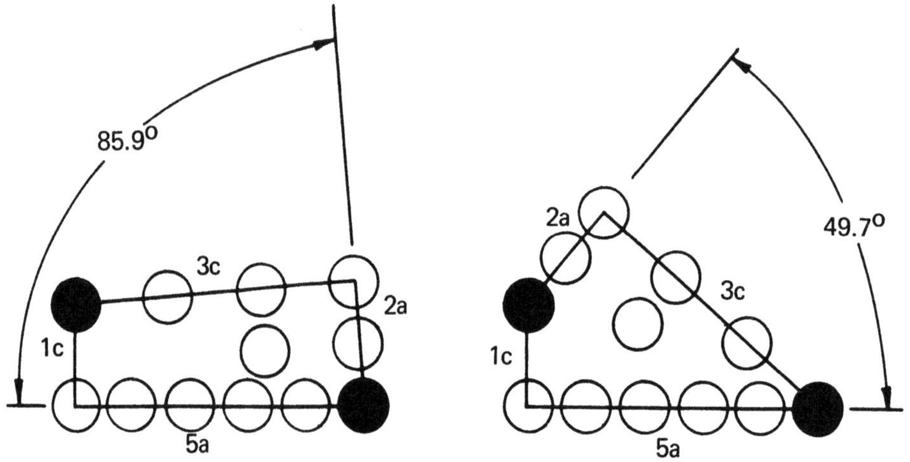

Figure 9. Two asymmetric structural units both designated 1c, 5a/3c, 2a. The crystallographic planes in the two crystals that are joined at the boundary are the same in both cases. The units differ only by a 180° rotation of the upper half of the unit relative to the bottom half about the normal to the boundary plane (the plane containing the two filled circles). c/a = ideal.

structural units from which symmetric [0001] and $\langle 11\bar{2}0\rangle$ tilt boundaries may be constructed.

The same notation is used in Tables VI and VII for two slightly different structural units, one referred to as a symmetric unit, the other as a pseudo-symmetric unit. We define a pseudo-symmetric boundary as one in which only a fraction of the lattice points of one crystal are reflected across the boundary plane. These boundaries will result whenever the degree of rotational symmetry about the rotation axis is higher in the CSL than it is in the crystal itself. This situation has been encountered in the CSL's and near-CSL's formed by rotations about [0001] and $\langle 11\bar{2}0\rangle$ respectively. The pseudo-symmetric [0001] tilt boundaries are symmetric in only one of the two layers of the (0001) stacking sequence*, the pseudo-symmetric $\langle 11\bar{2}0\rangle$ tilt boundaries have only one-half of the lattice points reflected across the boundary plane, namely those sites lying on alternate (0001) planes. In both instances, the symmetric and pseudo-symmetric boundaries are related to one another by a translational displacement of one crystal relative to the other. At this time this difference is being noted with an eye toward possible effects that it may have upon boundary properties. However, no suggestions will be made here as to what such effects may be.

Tables IV - VII are not intended to be an exhaustive listing of hcp structural units. They partially cover only three axes of rotation (a more extensive listing of structural units for rotations about [0001] $\langle 10\bar{1}0\rangle$ and $\langle 11\bar{2}0\rangle$, and the related coincidence lattices may be found in Reference 24). Rather, the tables and discussion thereof are intended to demonstrate the principles involved in deriving possible structural units. These principles should be common to all grain boundaries.

GRAIN BOUNDARY MODELING

Utilizing the structural units presented in the previous discussion, the structures of a wide variety of high-angle grain boundaries can be described. While the structural units are characteristic of a particular angular rotation about a specific rotation axis, it will be shown that boundaries of many different misorientations can be successfully modeled. Although examples will be limited to a few $\langle 10\bar{1}0\rangle$ tilt boundaries, combinations of structural units of the sort found in Tables IV-VII can be used to describe even the most complex of boundaries.

---

* Notice that pseudo-symmetric $\langle 111\rangle$ tilt boundaries can exist in cubic metals in which the boundary is symmetric in only one of the three layers.

TABLE VI. STRUCTURAL UNITS FOR SYMMETRIC [0001] TILT BOUNDARIES

Structure* $m\,a_1, n\,a_2/m\,a_1, n\,a_2$	Boundary Plane	Θ - Misorientation†		Σ_{hcp}	
		pseudo-symmetric	symmetric	pseudo-symmetric	symmetric
1,5/1,5	$\{15\bar{6}0\}$	17.90	42.10	31	62
1,4/1,4	$\{14\bar{5}0\}$	21.79	38.21	14	7
1,3/1,3	$\{13\bar{4}0\}$	27.80	32.20	13	26
1,2/1,2	$\{12\bar{3}0\}$	38.21	21.79	7	14
2,3/2,3	$\{23\bar{5}0\}$	46.83	13.17	19	38
3,4/3,4	$\{34\bar{7}0\}$	50.57	9.43	37	74
1,1/1,1	$\{11\bar{2}0\}$	60.00	0.00	2	1
m,n/m,n	$\{m\,n\,(\overline{m+n})\,0\}$				

*

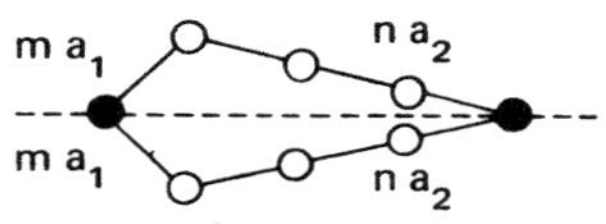

†Also 120°±Θ

TABLE VII. STRUCTURAL UNITS FOR SYMMETRIC AND PSEUDO-SYMMETRIC <11$\bar{2}$0> TILT BOUNDARIES

Structure* mc,na'/mc,na'	Boundary Plane			Θ - Misorientation Cd	Zn	Ideal	Mg	Zr	Ti	Be
1,3/1,3	{10$\bar{1}$6}	S†	Θ:	39.9	39.3	34.9	34.7	34.1	34.0	33.6
			Σ:	34	27	178	46	46	46	36
		P†	Θ:	140.1	140.7	145.1	145.3	145.9	146.0	146.4
			Σ:	34	27	89	23	23	46	36
1,2/1,2	{10$\bar{1}$4}	S	Θ:	57.1	56.4	50.5	50.2	49.4	49.2	48.7
			Σ:	26	36	44	44	34	34	24
		P	Θ:	122.9	123.6	129.5	129.8	130.6	130.8	131.3
			Σ:	13	18	22	22	17	34	24
1,1/1,1	{10$\bar{1}$2}	S	Θ:	94.9	94.0	86.6	86.3	85.2	85.0	84.3
			Σ:	24	30	34	30	24	22	20
		P	Θ:	85.1	86.0	93.4	93.7	94.8	95.0	95.7
			Σ:	12	15	17	15	12	11	10
2,1/2,1	{10$\bar{1}$1}	S	Θ:	130.7	130.0	124.1	123.8	122.9	122.8	122.2
			Σ:	46	34	82	18	26	26	26
		P	O:	49.3	50.0	55.9	56.2	57.1	57.2	57.8
			Σ:	23	17	41	9	13	13	13
3,1/3,1	{30$\bar{3}$2}	S	Θ:	146.0	145.4	141.1	140.8	140.2	140.0	139.6
			Σ:	23	23	27	27	17	17	17
		P	Θ:	34.0	34.6	38.9	39.2	39.8	40.0	40.4
			Σ:	46	23	27	27	34	34	17
m,n/m,n	{1 0 $\bar{1}$ $2\frac{n}{m}$}									

* $a' = \frac{a}{2}$<10$\bar{1}$0>

† S - Symmetric, P - Pseudo-symmetric

The purpose of grain boundary modeling, as with other geometric approaches to crystal and defect structures, is to provide a basis for a general conceptual understanding allowing further extension to more complex and detailed treatments. Many of the difficulties encountered in trying to extend dislocation models of low-angle boundaries into the high-angle range (e.g. the definition of Burgers vectors, core structure, and core overlap) are avoided in the coincidence structural-unit model. Although this type of model has not yet been developed to a highly quantitative level, it has proven to be a useful tool in defining starting structures for detailed computer calculations of grain boundary structures and energies (19, 21). In the case of <100> tilt boundaries in fcc metals, comparison of the results of computer modeling studies with the coincidence structural-unit models given in the CLD model (11, 12) show that the geometric models give quite a good first approximation to the computer relaxed structures (21). In addition coincidence structural-unit models have been used to predict the inclination of grain boundary facets in zinc with excellent agreement with experiment (25, 26).

In forming grain boundaries from combinations of structural units, there appear to be three requirements imposed upon the units: (1) they should be short-period units, i.e. units derived from short-period coincidence boundaries, (2) the units being mixed should be associated with nearly the same misorientation, and (3) they should not be far from the same boundary inclination. The first requirement arises from the observation made by Bishop and Chalmers (11, 12) following Read and Shockley (15), and supported by Hassen et al (21) in their computer calculations on symmetric fcc tilt boundaries, that boundaries with longer periods, and hence longer-range elastic stress fields, have higher energies due to the larger elastic contribution to the boundary energy. Since the elastic stress fields of the structural units of more general boundaries will also most likely be related to the length of the unit, it has seemed most reasonable to attempt to minimize the period of the units in developing a grain boundary structure. The second requirement is imposed because of the excessive lattice distortion that must accompany the forced combinations of units with slightly differing misorientations. The third requirement simply states that it is unreasonable to expect that grain boundaries will develop facets on an atomic scale in which the apex angle of the facet will be less than say 90°. All of the boundary models presented are well within this last criterion.

A convenient way of anticipating favorable combinations of structural units is illustrated in Figure 10. Here, for each near-CSL within the misorientation range represented, the inclination α of a particular structural unit derived from that near-CSL is plotted versus the misorientation θ. This will be referred to as an *inclination/misorientation diagram* (IMD). Loberg, Norden and Smith (27) have recently presented misorientation-inclination data in a somewhat different but related plot.

Figure 10 was constructed to depict potential structural units and combinations of structural units in an hcp crystal with an ideal c/a ratio in the misorientation range $55° < \theta < 80°$. The size of each point is inversely proportional (approximately) to the length of the structural unit, i.e. large points correspond to short-period units. All the points shown in the IMD have not been presented in Tables IV and V. The three structural units shown earlier in Figure 6 are plotted on the IMD at $\theta = 78.46°$ and are designated S_3, S_4, and A_1 for the two symmetric units and the asymmetric unit, respectively. Notice that the asymmetric unit A_1 is shorter than the symmetric unit S_4. This is characteristic of a centered CSL with an even Σ. Because these units are derived from an exact CSL they are plotted at precisely the same misorientation angle. Other asymmetric units derived from this same CSL are also plotted at the same misorientation.

All structural units derived from the same exact CSL are plotted at the same misorientation angle and hence appear on the same vertical line in the IMD, such as the lines designated $\Sigma = 35$, 11, 107, and 10. Since structural units derived from the same near-CSL occur at slightly different misorientation angles, they will appear on a line that is slightly off from vertical, such as the lines designated $\Sigma = 31$, 39, and 15 in Figure 10. For a diagram of this type based upon an ideal c/a ratio, both vertical and off-vertical lines will appear, corresponding to exact CSL's and near-CSL's respectively. For similar diagrams in which $(c/a)^2$ is not rational and exact CSL's do not exist, only the off-vertical lines will be found. The closer the line containing structural units from a near-CSL comes to being exactly vertical, the more nearly exact is the near-CSL. More than one line emanating from a single symmetric unit signifies that the unit is associated with more than one near-CSL as was pointed out earlier (c. f. the $\Sigma = 11$ and 15 lines on Figure 10 which show the 1,1/1,1 unit associated with both the $\Sigma = 11$ exact CSL and the $\Sigma = 15$ near-CSL).

Boundaries containing combinations of two different structural units will lie at inclinations and misorientations midway between those characteristic of the two component units. As was shown by Bishop and Chalmers for cubic <100> tilt boundaries, combinations of two symmetric units produce a symmetric boundary with a misorientation between that of the two symmetric units. Exactly the same result occurs in hcp boundaries. As shown on the IMD in Figure 10, combinations of the 1,3/1,3 unit (designated S_1) and the 1,2/1,2 unit (designated S_3) can produce the 2,5/2,5 boundary $(\ldots S_1S_3\ldots)$, the 3,8/3,8 boundary $(\ldots S_1S_1S_3\ldots)$, or the 4,11/4,11

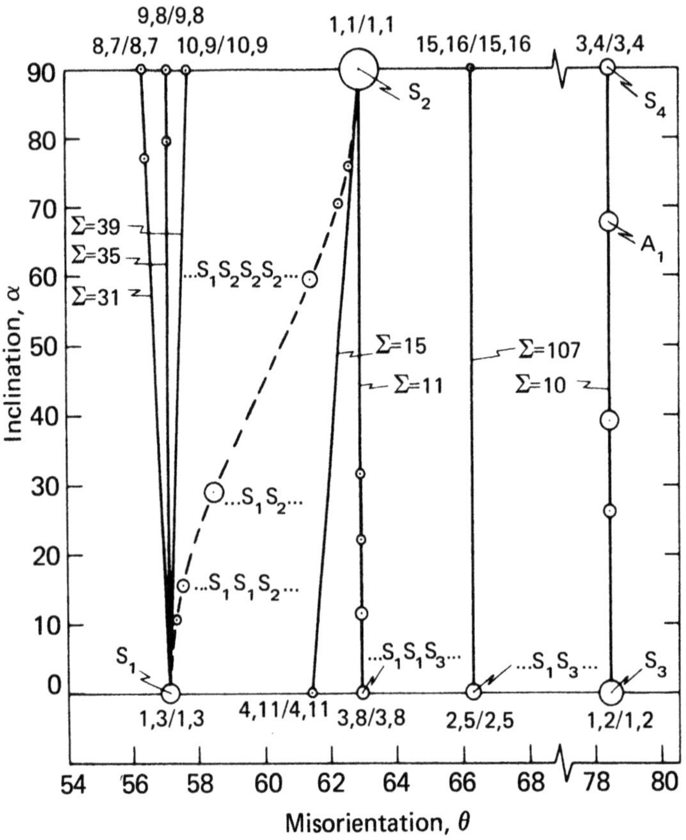

Figure 10. An inclination/misorientation diagram of ⟨10$\bar{1}$0⟩ tilt boundary in a hexagonal-close-packed crystal having an ideal c/a ratio. Vertical lines represent exact coincidence-site lattices; near-vertical lines represent near-coincidence-site lattices. The sizes of the circles representing the various possible structural units and composite boundary structures are inversely proportional (approximately) to the length of period of the units, i.e., large circles correspond to short-period units. Several possible structural unit combinations are illustrated.

boundary, all of which are symmetric boundaries. Furthermore combinations of two symmetric units inclined 90° with respect to one another will produce boundaries at asymmetric inclinations. In Figure 10 combinations of the S_1 unit with the 1,1/1,1 unit (S_2) produce boundaries lying along the dashed line joining these two component units. Some of the combinations of S_1 and S_2 that give these asymmetric boundaries are indicated. It should be understood that in each of the above composite boundaries, near-coincidence will occur in the boundary at the ends of each structural unit or segment.

Two points should be clarified with regard to the positioning of the structural units from Tables IV and V on an IMD such as that in Figure 10. One should notice that the order in which the symmetric units are presented in Table IV corresponds to a gradual opening of the angle between the basal planes and hence a gradual increase in θ. This is exactly the order in which these units will appear with increasing θ along the line $\alpha = 0°$ on the IMD. Because of the two-fold symmetry of the $<10\bar{1}0>$ axis, these same units will also make up boundaries formed by rotations of $(180° - \theta)$. The units when associated with these latter rotations, however, appear along the line $\alpha = 90°$ in the IMD and are then of course in the reverse order with increasing θ. In this manner a self-consistency is established between the relative positions of the various near-CSL's on the diagram. This explains why the 1,1/1,1 unit, which is listed in Table IV at a misorientation of 117.0°. is plotted in Figure 10 at $\theta = 63.0°$ and $\alpha = 90°$. Similarly for the asymmetric units in Table V, they will be positioned on an IMD such as that in Figure 10 at the coordinates (α, θ) as listed and also at the coordinates $(90° - \alpha, 180° - \theta)$.

The second point requiring clarification is that the inclination α measured from a symmetric inclination can be either positive or negative. For example the asymmetric portion of the boundary in Figure 6 can bend either upward or downward relative to the adjoining symmetric portion. The implication which this fact holds for Figure 10 is that the entire IMD can be reflected across the line $\alpha = 0°$ so that structural units corresponding to a positive value of α can exist in combination with units corresponding to a negative value of α. The consequences of this will become apparent shortly.

Observing the conventions mentioned above, an IMD has been constructed for $<10\bar{1}0>$ tilt boundaries in titanium in the misorientation range $52° < \theta < 78°$. This diagram has been presented in Figure 11 and includes an expanded section near 68° in which units corresponding to negative values of α appear. This IMD will be used to illustrate the formation of a variety of boundaries from the structural units that appear on the diagram.

All of the near-vertical lines are associated with near-CSL's. Particular reference will be made to the $\Sigma = 7$ and the $\Sigma = 10$ near-CSL's, so that these two near-CSL's are shown together in Figure 12. The distortions of the unit cells have been exaggerated to emphasize the fact that coincidence is inexact. Furthermore it should be noted that the greater distortion associated with the $\Sigma = 10$ near-CSL is reflected in the IMD by the $\Sigma = 10$ line being more off-vertical than the $\Sigma = 7$ line.*

* Actually the $\Sigma = 10$ near-CSL did not meet the previously established acceptance criteria for near-CSL's and therefore is not listed in Table II. Nonetheless structural units derived from even this poor quality near-CSL are viable components of $<10\bar{1}0>$ tilt boundaries in titanium.

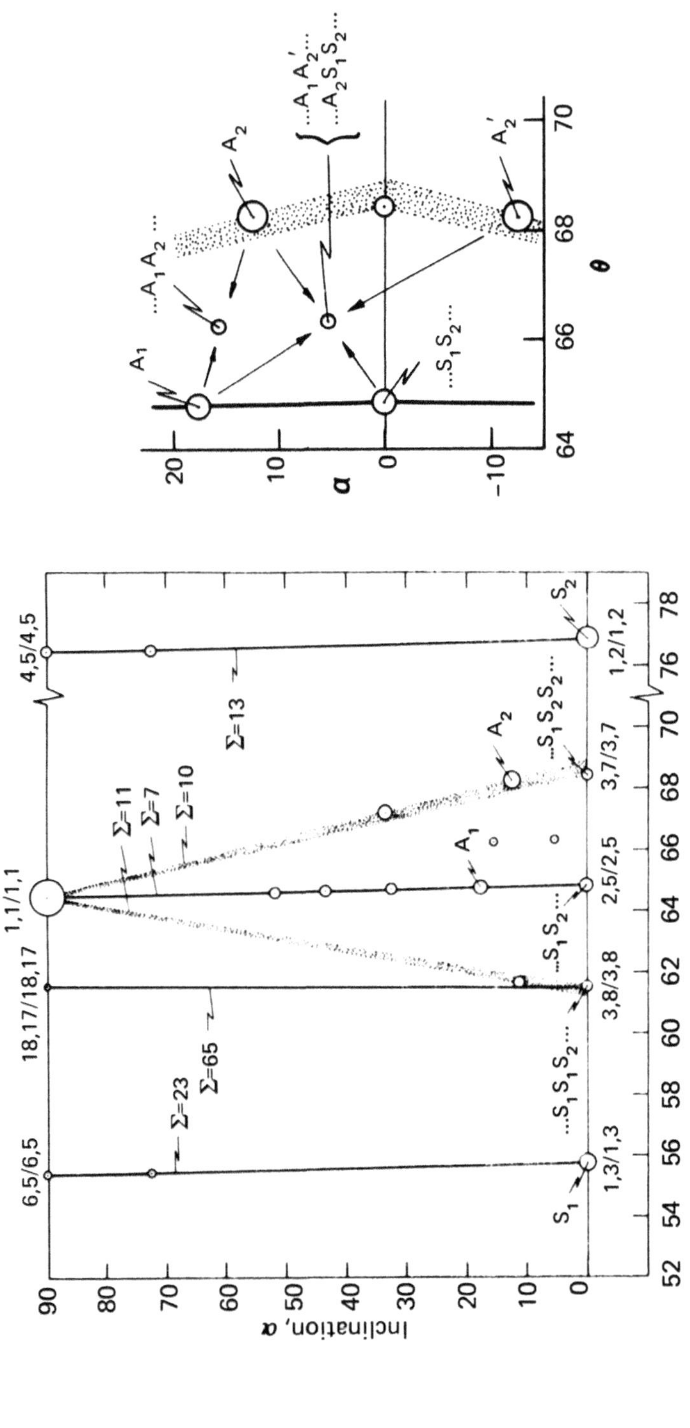

Figure 11. An inclination/misorientation diagram of $\langle 10\bar{1}0\rangle$ tilt boundaries in α- titanium. The near-vertical lines represent near-coincidence-site lattices; the sizes of the circles are inversely proportional to the period of the boundary or structural unit. The inset figure shows the IMD reflected across the $\alpha = 0^{o}$ line to enable the presentation of units with negative values of α. Possible structural unit combinations are illustrated.

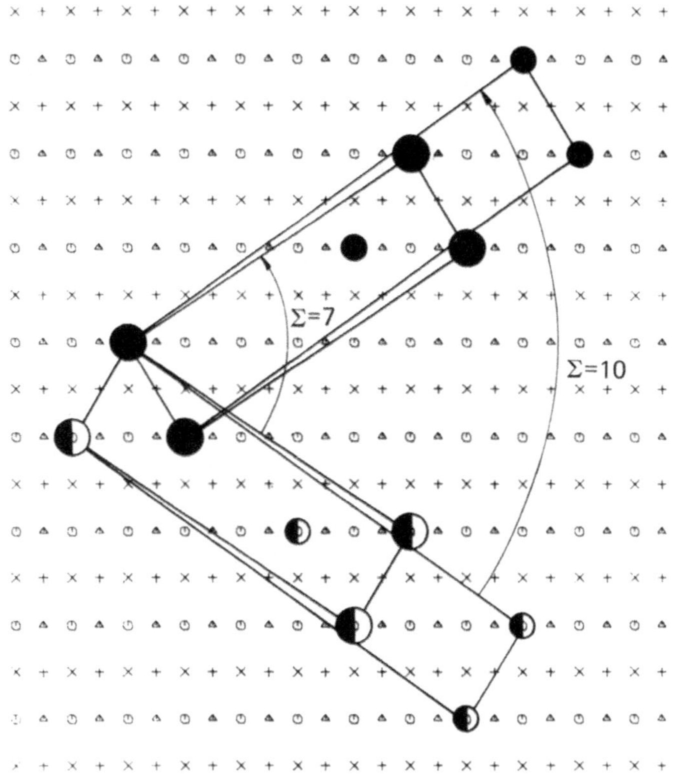

Figure 12. The $\Sigma = 7$ and the $\Sigma = 10$ near-coincidence lattices in titanium formed by rotations about $\langle 10\bar{1}0\rangle$. The distortions of the unit cells have been exaggerated to emphasize the concept of near-coincidence. The greater distortion associated with the $\Sigma = 10$ near-CSL is reflected in the greater departure from verticality of the $\Sigma = 10$ line on Figure 11. Only near-coincidence sites in one layer are shown.

Based upon the units presented in this inclination/misorientation diagram, several structural-unit models of $<10\bar{1}0>$ tilt boundaries will be presented. These will involve a variety of combinations of structural units in an attempt to illustrate the almost limitless number of combinations that are possible. Depending upon the proportions in which the various structural units are combined, nearly every boundary misorientation and inclination could conceivable be described.

1. *Boundaries containing one symmetric unit:* Figures 7 and 8 are representative of this boundary. Figure 7 shows the boundary designated S_1 on the IMD of Figure 11. In principle, this is the only type of boundary that can exhibit exact boundary coincidence in most hcp metals.

2. *Boundaries containing mixtures of symmetric units:* Figure 1 showed a boundary of this type in a cubic crystal; Figure 13 shows a boundary of this type in titanium. The filled circles are exactly coincident whereas the shaded circles are only near-coincident. This is a combination of the 1,3/1,3 unit designated S_1 on the IMD and the 1,2/1,2 unit designated S_2. It forms the 3,7/3,7 boundary on the IMD via the combination $...S_1S_2S_2...$ Other combinations of these two units lead to the 2,5/2,5 boundary and the 3,8/3,8 boundary as indicated in Figure 11, as well as a myriad of others, all of which lie on a line joining the two component units on the IMD.

3. *Boundaries containing combinations of the same asymmetric unit in its two different inclinations:* Figure 14 shows a boundary of this type made up of alternating 1,4/2,3 units, one being the unit designated A_2 on the IMD, the other the unit designated A_2'. Notice that the combination $...A_2A_2'...$ lies midway between A2 and A_2' and that this combination is equivalent to the 3,7/3,7 symmetric boundary. Notice too that it is an alternate structure for the boundary shown in Figure 13. Alternating asymmetric units of

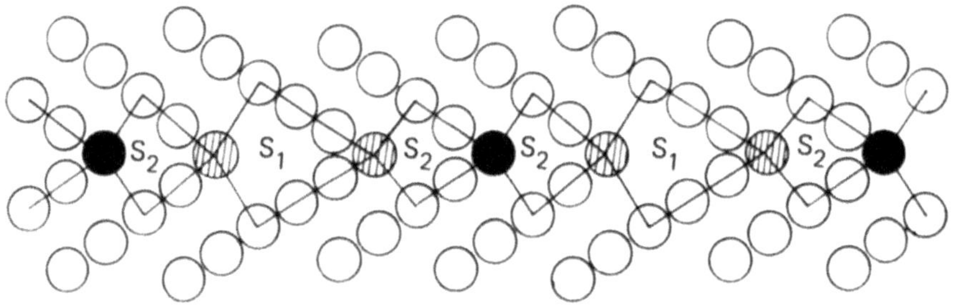

Figure 13. A symmetric $\langle 10\bar{1}0\rangle$ tilt boundary in titanium at a misorientation of 68.4°. It combines units of the 55.8° boundary (S_1) with units of the 76.9° boundary (S_2) in the sequence ... $S_1S_2S_2$...

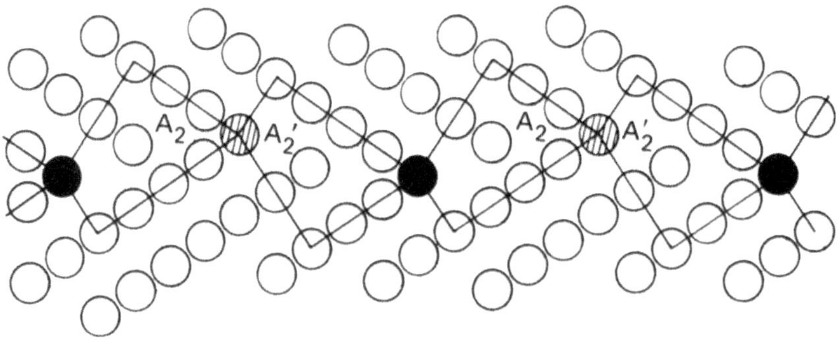

Figure 14. A $\langle 10\bar{1}0\rangle$ tilt boundary in titanium at a misorientation of 68.4°. It combines two asymmetric structure units (the 1c, 4a/2c, 3a units of Figure 11) in the sequence ... A_2A_2' ... Alternating asymmetric units result in an overall symmetric inclination for the boundary.

the same type always result in the net boundary inclination being symmetric and hence capable of containing exact boundary coincidence sites. In this case the filled circles are exactly coincident whereas the shaded circles are near-coincident. The alternate structures in Figure 13 and 14 suggest the possibility of a grain boundary phase transformation (see paper by E. Hart, this volume).

If instead of alternating units at every opportunity as is shown in Figure 14, many asymmetric units in one inclination were joined before the units of the alternate inclination were introduced, the boundary would develop macroscopic facets consisting of these asymmetric segments. This is illustrated in Figure 15. This is exactly what happens in the faceted grain boundaries observed by the authors in zinc (25,26). Notice that only near-coincidence exists along each asymmetric leg of such a boundary. Consequently, as mentioned earlier, any mismatch in the boundary tends to build up as the length of the asymmetric leg increases. This may explain why macroscopic faceting of grain boundaries is not often observed although microscopic faceting (alternating single asymmetric units) may be quite common.

4. *Boundaries containing mixtures of two or more different asymmetric units:* This type of boundary is illustrated in Figure 16 in which the 1,6/3,4 unit from the $\Sigma = 7$ near-CSL alternates with the 1,4/2,3 unit from the $\Sigma = 10$ near-CSL. The former is the unit designated A_1 on the IMD, the latter is the A_2' unit. The IMD shows that this combination leads to the asymmetric 3,9/4,8 boundary via the sequence $...A_1A_2'...$ Also shown on the IMD is that the $...A_1A_2...$ combination leads to the 2,10/5,7 asymmetric boundary. These boundaries and all boundaries of this type will contain only near-coincidence sites.

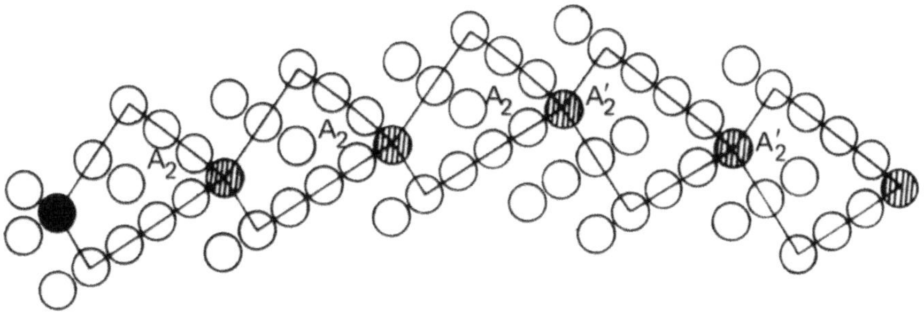

Figure 15. A ⟨10$\bar{1}$0⟩ tilt boundary in titanium that combines the same asymmetric structural units as in Figure 14, i.e. the 1c, 4a/2c, 3a asymmetric units. Each asymmetric leg of the boundary can contain several units of the same type as shown here. Notice, however, that the lattice strains in the vicinity of near-coincident atoms in the boundary increase with the consecutive number of A_2 units. This strain is recovered as A_2' units are introduced. There is no net strain when the number of A_2 units and A_2' units are equal

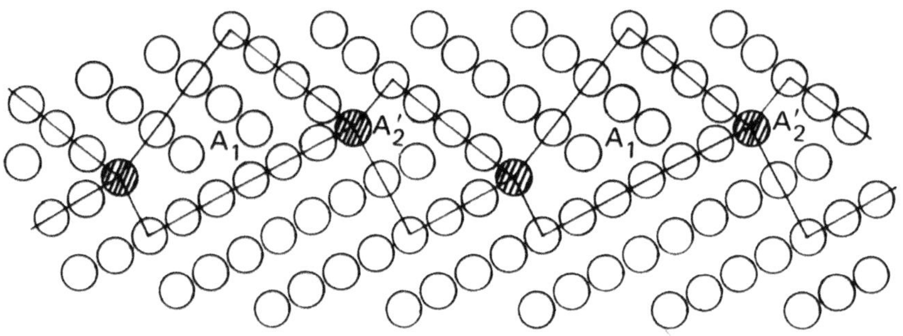

Figure 16. An asymmetric ⟨10$\bar{1}$0⟩ tilt boundary in titanium at a misorientation of 66.3°. It combines two asymmetric structural units (the 1c, 6a/3c, 4a unit at 64.8° and the 1c, 4a/2c/3a unit at 68.3°) in the sequence ... $A_1 A_2'$ ...

5. *Boundaries containing mixtures of symmetric and asymmetric units:* Figure 17 shows a mixture of the units S_1, S_2, and A_2 from Figure 11 in the combination ...$S_1 S_2 A_2$... As seen in the expanded view of Figure 11, the result is the asymmetric 3,9/4,8 boundary which lies along the line joining A_2 with the ...$S_1 S_2$... boundary. Again because this is an asymmetric tilt boundary, it will contain only near-coincidence sites. Further notice that this is an alternate description of the boundary appearing in Figure 16. The coincidence structural-unit model, being geometric in nature, permits one to develop these alternate descriptions but not to judge which of these alternate descriptions might in fact exist.

The above models have been presented as typical of simple <10$\bar{1}$0> tilt boundaries made up of combinations of <10$\bar{1}$0> structural units. It is felt that more complex boundaries can be modeled in much the same manner although their construction would of course be more complicated. One very obvious possibility suggested by the coincidence structural-unit model is that very few boundaries will be flat on an atomic scale; the existance of micro-facets could be a rather common feature of most grain boundaries. This

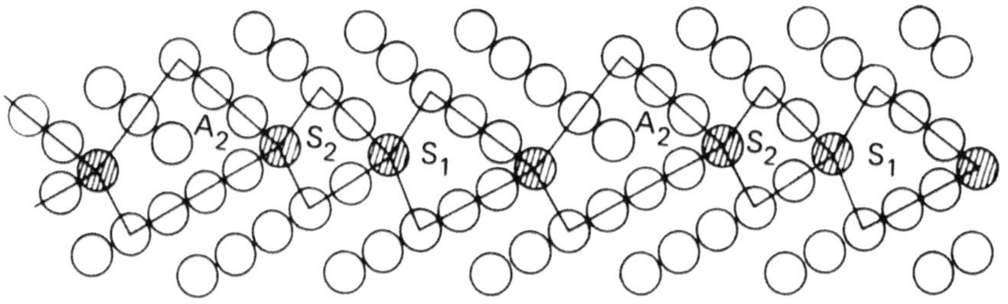

Figure 17. An asymmetric ⟨10$\bar{1}$0⟩ tilt boundary in titanium at a misorientation of 66.3°... It combines units of the 55.8° symmetric boundary, S_1, units of the 76.9° symmetric boundary S_2, and units of the asymmetric 68.3° boundary, A_2 (the 1c, 4a/2c, 3a unit) in the sequence ... $S_1 S_2 A_2$ ...

suggests that computer modeling studies of grain boundary structures should consider such micro-faceted structures as starting points. Thus far only planar boundaries have been considered in any detail. Starting with planar structures could prejudice the results of such studies toward planar relaxed structures, and a deeper energy minimum associated with a micro-faceted structure could be overlooked.

Finally it should be pointed out that the coincidence structural-unit approach taken here with hcp metals is quite general. The use of CSL's (either exact or near-CSL's) to define the structural units (either exact or near-coincidence units) and the construction of an IMD to disclose favorable combinations of units that form the structures of symmetric and asymmetric grain boundaries is applicable to all crystal systems. From a general standpoint the concept of exact lattice coincidence finds applicability primarily in the cubic system, whereas near-coincidence concepts are of much broader usefulness when considering crystal classes of lower symmetry. Near-coincidence is not necessarily limited to non-cubic crystals, however; near-coincidence boundaries in cubic crystals are also possible.

SUMMARY

The coincidence structural-unit model for grain boundaries in hexagonal-close-packed metals has been discussed. The structural units are shown to be related to near-coincidence-site lattices in the hcp crystal, which exist in spite of the impossibility of achieving exact lattice coincidence. The manner in which possible structural units are derived from the near-CSL's is presented and the use of the inclination/misorientation diagram for predicting favorable combinations of structural units is demonstrated. Extensive tables presenting near-CSL's and possible structural units associated with rotations about [0001], $\langle 10\bar{1}0 \rangle$, and $\langle 11\bar{2}0 \rangle$ are included making possible the construction of grain boundary models for a very large number of boundary misorientations and inclinations.

REFERENCES

1. F. Hargreaves and R. J. Hills, J. Inst. Met. 41, 257 (1929).
2. M. L. Kronberg and F. H. Wilson, Trans. AIME 185, 501 (1949).
3. K. T. Aust and J. W. Rutter, in Recovery and Recrystallization of Metals, ed. by L. Himmel, Interscience (1963), p.131.
4. J. W. Rutter and K. T. Aust, Acta Met. 13, 181 (1965).
5. B. B. Rath and Hsun Hu, Trans. TMS-AIME 236, 1193 (1966).
6. K. T. Aust and J. W. Rutter, Trans. TMS-AIME 218, 1023 (1960).
7. K. T. Aust, Trans. TMS-AIME 221, 758 (1961).

8. K.T.Aust, in Surfaces and Interfaces I, Proceedings of the 13th Army Materials Research Conference, ed. by Burke et al., Syracuse Univ. Press (1967), p.435.
9. D. G. Brandon, B. Ralph, S. Ranganathan, and M. S. Wald, Acta Met. 12, 813 (1964).
10. D. G. Brandon, Acta Met. 14, 1479 (1966).
11. G. H. Bishop and B. Chalmers, Scripta Met. 2, 133 (1968).
12. G. H. Bishop and B. Chalmers, Phil. Mag. 24, 515 (1971).
13. B. Chalmers and H. Gleiter, Phil. Mag. 23, 1541 (1971).
14. H. Gleiter, Phys. Stat. Sol. (b) 45, 9 (1971).
15. W. T. Read and W. Shockley, Phys. Rev. 78, 275 (1950).
16. W. Bollmann, Crystal Defects and Crystal Interfaces, Springer-Verlag, New York, Heidelberg and Berlin (1970), p.209.
17. T. Schober and R. W. Balluffi, Phil. Mag. 21, 109 (1970).
18. T. Schober and R. W. Balluffi, Phys. Stat. Sol. (b) 44, 103 (1971).
19. M. Weins, H. Glieter, and B. Chalmers, J. Appl. Phys. 42, 2639 (1971).
20. G. H. Bishop and G. A. Bruggeman, J. of Mat. Sci. (in press).
21. G. C. Hasson, J. B. Guillot, B. Baroux and C. Goux, Phys. Stat. Sol. (a) 2, 551 (1970).
22. P. H. Pumphrey and K. M. Bowkett, Scripta Met. 5, 365 (1971).
23. S. Ranganathan, Acta Cryst. 21, 197 (1966).
24. G. A. Bruggeman, G. H. Bishop and W. H. Hartt, "Coincidence Structural-Unit Models for Grain Boundaries" AMMRC TR 72-7, February 1972.
25. G. H. Bishop, W. H. Hartt and G. A. Bruggeman, Acta Met. 19, 37 (1971).
26. W. H. Hartt, G. H. Bishop, and G. A. Bruggeman, J. of Metals 21, 124 A (1969).
27. B. Loberg, H. Norden, and D. A. Smith, Phil. Mag. 24, 897 (1971).

COMPUTER SIMULATION OF ASYMMETRIC GRAIN BOUNDARIES AND THEIR INTERACTION WITH VACANCIES AND CARBON IMPURITY ATOMS

R.E. Dahl, Jr., J.R. Beeler, Jr., and R.D. Bourquin

WADCO Corporation, North Carolina State University, and Battelle-Northwest

ABSTRACT

Computer simulation has proven to be a powerful tool for determination of atomistic structure in complex zones in metals. This method has been applied to study asymmetric tilt boundaries in γ-iron and their interaction with vacancies and carbon atoms. A program, "GRAINS", which employs classical equations of motion was used for these experiments. A first neighbor function described the interaction between atom pairs. Volume forces were imposed by the computational cell boundaries. Observations and conclusions of particular interest were: The determination of values for migration and formation energies of vacancies in the vicinity of a grain boundary so vacancy transport and diffusion mechanisms in the boundaries can be estimated more accurately. These values were found to differ appreciably from those in ordered regions. The zone of interaction extends into the grains along close-packed lines from misfit regions much farther than would be anticipated from measurements of grain boundary width. Local ordering in the misfit region of a grain boundary caused by a very few impurity atoms which demonstrated how trace concentrations of impurities could significantly affect the strength of a metal.

INTRODUCTION

The atomistic structure of grain boundary regions is of considerable interest as mechanisms are developed for the phenomena and properties of polycrystalline metals. Structural definition had been extremely difficult to deduce with sufficient accuracy and detail prior to the advent of high resolution electron microscopy and computer simulation. In the recent past, a number of studies have been reported [1-7] in which there is remarkable agreement between structures postulated theoretically, determined through computer simulation and observed through microscopy.

The work reported here is an extension of structural studies utilizing computer simulation to determine the effects of grain boundaries on point defects. The defects chosen for this venture were the simple monovacancy and interstitial impurity atom. The host material was fcc (γ) iron so that the conclusions would be at least qualitatively applicable to stainless steel. Results sought were an insight into the interactions and perturbations which could affect diffusional processes.

The term "computer experiment" is perhaps unfamiliar to many readers and, therefore, it is somewhat obligatory to provide an explanation of just what a computer experiment is. A computer experiment has been described by Beeler [8] as "a computational method in which physical processes are simulated according to a given set of physical mechanisms". In this study, the computations were performed in a high-speed, digital computer in which a crystalline array of atoms and the forces which they exert upon each other, are calculated according to numerical analysis equations representing interatomic potential functions, which were in this case, the physical mechanisms. The studies are empirical in nature since the potential functions are derived from experimental measurements rather than from theoretical models. Properties and measurements upon which the potential functions are based must be carefully chosen to be as closely related to the objectives of the study as possible. A potential function based upon the elastic constants is the appropriate choice to achieve the goals of this study rather than one which would probably permit more accurate determination of some of the physical properties of the metals. Derivation of an exact potential function which would describe all properties of a metal, is not practical because of the extreme complexity of the forces in a lattice, particularly one in which some disorder is present.

During the course of a computer experiment study, the investigator should try to obtain agreement between his calculated results and experimentally determined parameters of interest. Having established confidence in the method and the potential function through agreement with other results, the experimenter can proceed by extending the study into areas of original research. This is, of course, no different than the procedure in any laboratory experiment in which method, theories, and apparatus are tested and calibrated by reproducing standards or accepted measurements before exploratory investigations are made.

Computer experimentation has, since its inception, been largely directed toward investigations of various phenomena occurring in metals. The methods are particularly well suited to this field since the atomistic structure can be studied in detail and perturbed at will, in discrete lattice analyses. The microscopic structure and processes which determine macroscopic behavior can then be studied in detail. Thus, computer experimentation affords a method for determining basic mechanisms and obtaining a fundamental understanding of metals which may not be possible with the conventional laboratory techniques.

COMPUTER SIMULATION

The computer experimentation has been used predominantly to describe lattice defect phenomena in metals. The computer experiment method is a useful technique for (1) estimating the arrangement of atoms in the "core" region of a lattice defect where the displacements from perfect lattice positions are too large to be accurately described by elasticity theory, and (2) exploring the consequences of assuming a given mechanism for atom movement processes in a defected crystal.

Typically, these computer experiments were performed in three stages. First, a perfect crystallite was constructed with a specific orientation appropriate to the symmetry of the defect to be studied. Next, a first approximation to the perturbed lattice configuration for the defect concerned was instituted, either on the basis of displacement field libraries from previous computer experiments or using elasticity theory. Finally, the atoms in the crystallite were allowed to relax from their initial approximate positions to positions which gave the lowest potential energy state. This relaxation process was done in accordance with the atomic interaction function assumed and the laws of Newtonian mechanics. Constraints on the crystallite of movable, relaxing atoms were imposed at its boundary to preserve the inherent characteristics of the enveloping perfect lattice. The characteristics preserved were the lattice constant, elastic constants, compressibility and volume cohesive forces.

The GRAINS computer code [9] was developed to conduct the experiments described in this discussion.

The lattice model used in GRAINS consists of a central computational cell (crystallite) of movable atoms enveloped by a mantle of immovable atoms on four of the six sides. The crystallites formed by this code are rectangular parallepipeds. The defect one wishes to study is introduced by setting up an approximate displacement field for this defect in the lattice. This displacement field is introduced both in the computational cell and in the mantle. At the same time, atoms are either added or deleted in the core of the defect to institute any required change in the atom population associated with the defect. Thereafter, the positions of the mantle atoms remain fixed throughout the simulation. All atoms in the computational cell are free to move about in accordance with the interatomic forces they feel. This movement leads to the equilibrium configuration.

Atoms are moved using the dynamical computer simulation method developed by Gibson et al [10]. The 3N Newtonian equation of motion for N atoms are numerically integrated simultaneously via the central difference approximation with a pairwise interaction model. The potential energy of the system is the sum of the energies of the atoms which result from their positions relative to each other. The net force on each atom is the vector sum of the forces resulting from the pairwise interactions in which it participates.

The boundary of a computational cell must simulate the essentially infinite extent of a crystal beyond the region which encompasses the stress and strain fields of the defects studied. The boundary must also preserve the crystallinity of the material studied by exerting volume forces. Two types of boundaries are used in each of the cells constructed by GRAINS (Fig. 1). A mantle of fixed atoms encompasses the cells on the four X-Z and Y-Z boundary planes. The X-Y faces of the computational cells are boundaries in which atomic movement is possible in a restricted manner.

In most instances the computational cell was composed of an odd multiple of the crystallographic repeat distance so that symmetry existed about a center X-Y plane. Point defects were located in or very near this plane of symmetry. Extended defects such as edge dislocations and grain boundaries are constructed so they extended into the computational cell normal to the X-Y faces.

The success of a computer experiment depends primarily upon the accuracy of the functions used to describe the interaction between atoms as a function of their separation distance. The functions which are included in this case were derived empirically from the elastic constants of the materials (iron and carbon) rather than from first principles. Thus they permit solution of complex many-body problems by summation of pairwise interactions. The use of short-range potentials was permissible because the cell boundary constraint simulates the volume effects in a lattice which maintain crystallinity. Thus the local structural components of energy due to ion-ion interaction can be nonequilibrium in

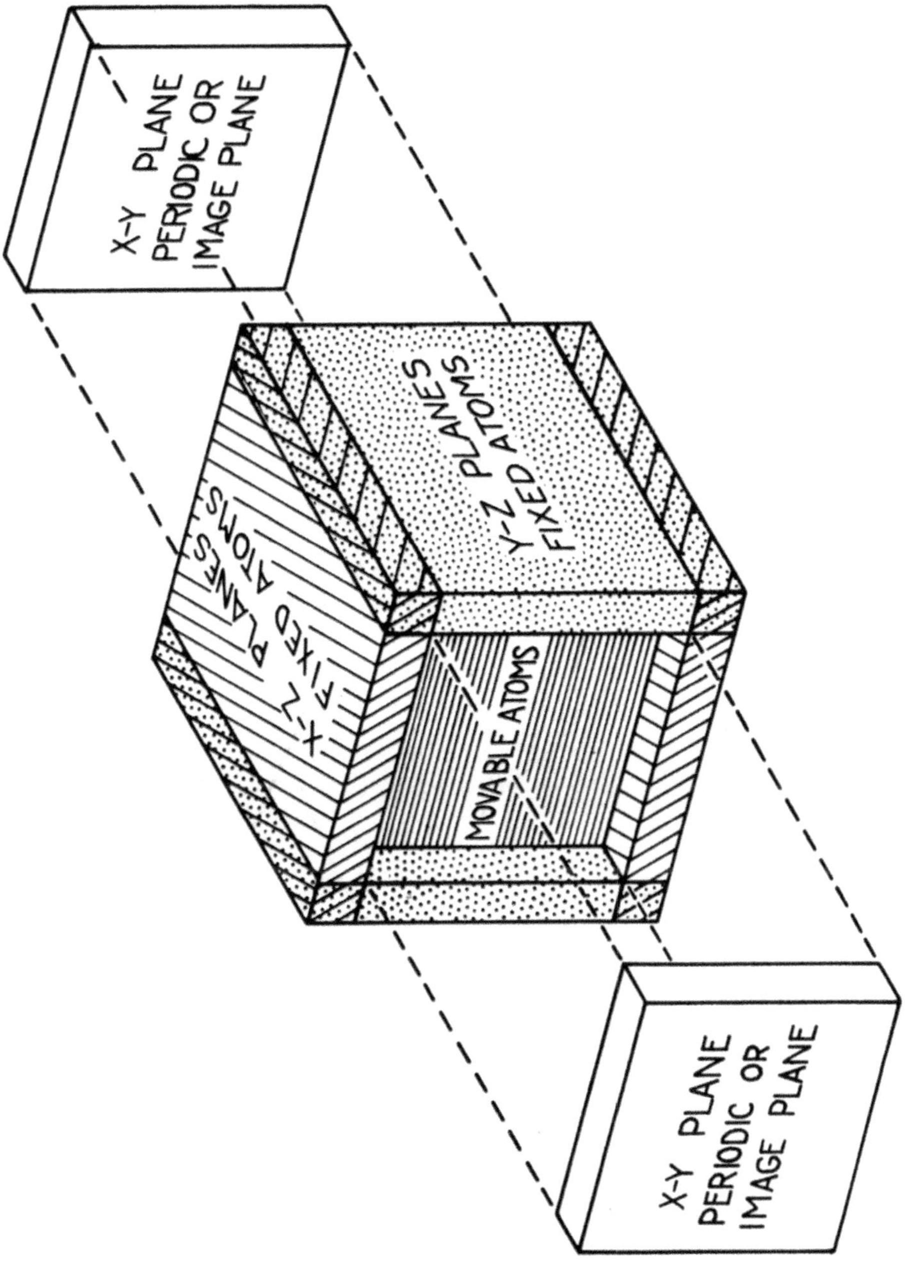

Fig. 1. Computational cell boundaries used in the GRAINS computer code.

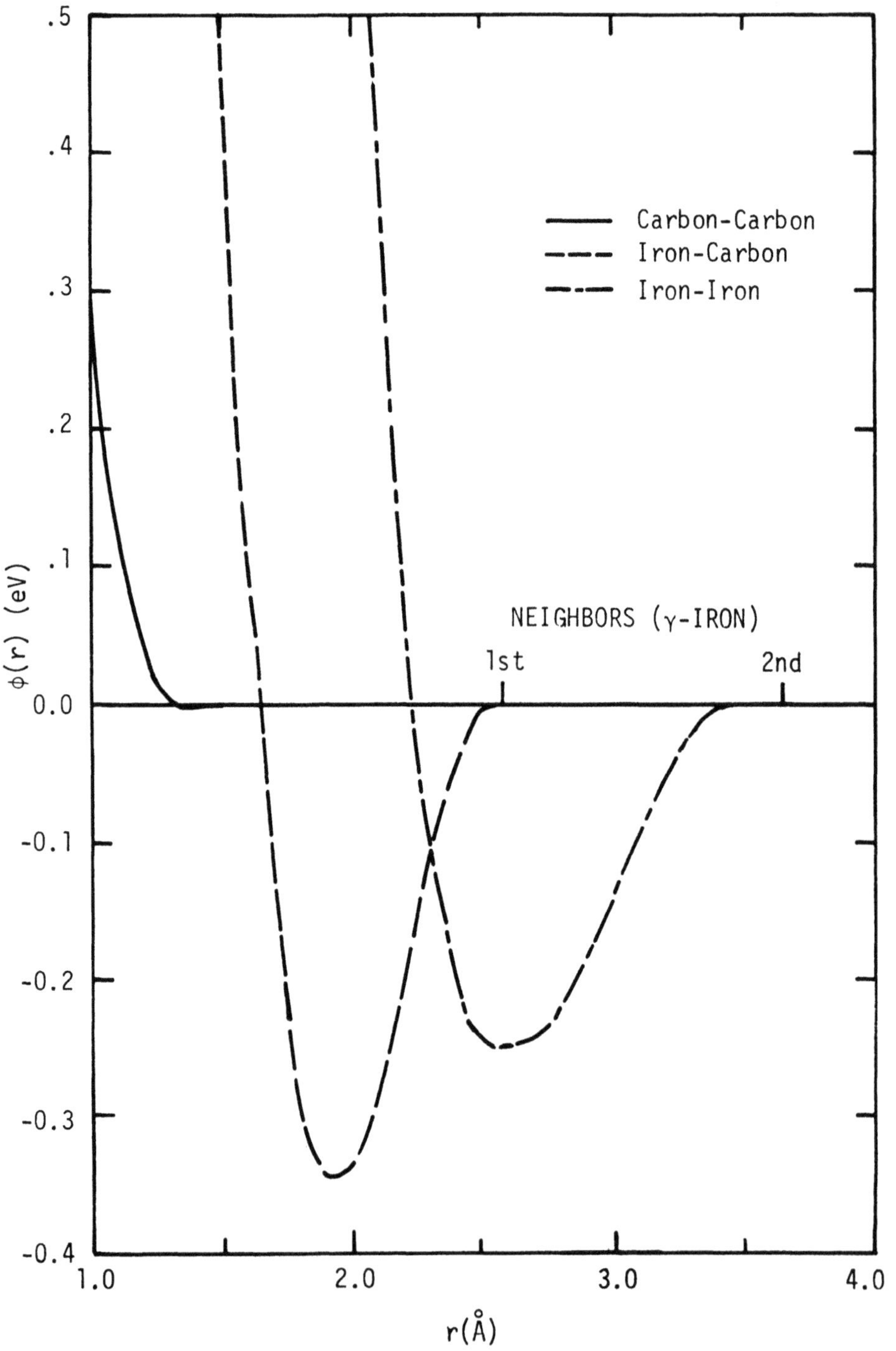

Fig. 2. Interatomic functions.

nature (i.e., repulsive or attractive forces can exist between first neighbors) since the cohesive forces which are long range in nature exert a pressure on the cell. These interatomic functions are the Johnson I iron-iron potential [11], the Johnson iron-carbon potential [12], and a carbon-carbon potential derived by Dahl [13]. These functions are shown in Figure 2. The iron-iron function does not extend to second neighbors in fcc iron and has the first neighbors nearly at equilibrium position. The iron-carbon function has both an attractive and repulsive region. The latter effectively allows no carbon-carbon bonding thereby treating their interactions as those of semi-hard spheres.

A plotting routine provides accurate graphical representation of the atomistic structure and is invaluable in analyzing data. Bad runs and mistakes can be recognized more readily and corrected. Structure or strain fields which would have been difficult or impossible to perceive from numerical data became immediately apparent when represented graphically. Important examples include structure of grain boundaries and edge dislocations, and the "healing" of grain boundaries with carbon atoms. Examples of graphical output are presented in figures throughout this report. The position of atoms in the figures are at the ends of the fiducial marks which extend into the symbols. Hexagons denote host atoms which had three degrees of freedom and squares the fixed, cell boundary atoms. Triangles represent carbon impurity atoms. The planes which appear in these figures are X-Y planes.

The most important parameters in the construction of a grain boundary in a computational cell are accurate simulation and conservation of:

- The lattice structure of both grains
- The appropriate alignment between grains
- The proper distance between grains
- The shape and size of the computational cell, so that irregularities in the cell boundary do not perturb the results.

The method of construction consists of first constructing the left grain (A,B,C,D in Fig. 3), and then superimposing a larger grain (P,Q,R,S) over it in such a way that the required size of the cell containing a grain boundary (A,D,E,F) is obtained. Atoms from the second grain are rejected if:

1) They lie outside the boundaries of the cell to be constructed.

2) They lie within the first grain.

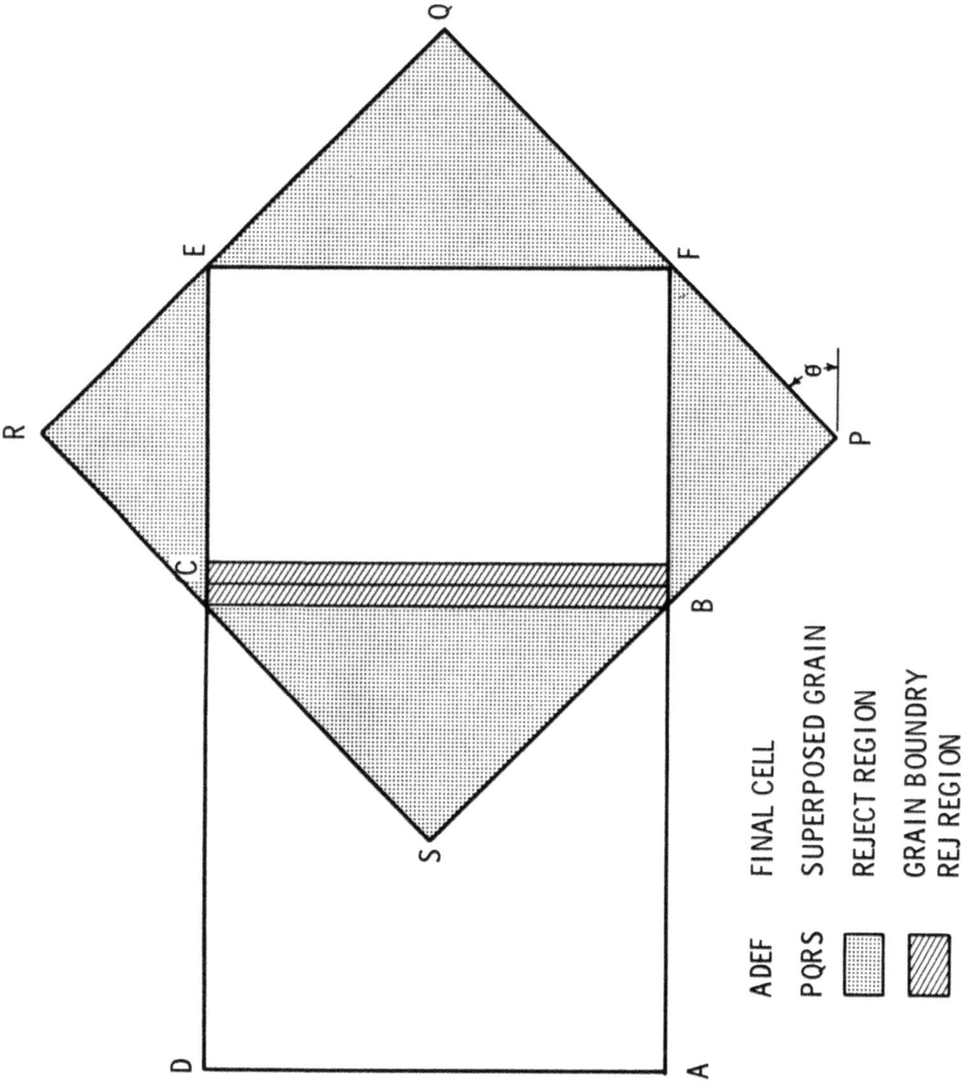

Fig. 3. Schematic of asymmetric grain boundary construction method.

3) They are too close to atoms in the first grain constructed.

(Refer to dark areas in Fig. 3.)

Rejection of atoms which violate principles (1) or (2) is quite simple since these criteria are precisely defined; however, criterion (3) appears somewhat arbitrary. A complete study was required before confidence in the results could be established. The interatomic spacing in the crystal (i.e., the first neighbor distance -- $\sqrt{2}a/2$ in fcc) was found to be the optimum criterion for (3). All atoms within range of the interatomic function of the grain boundary (shaded areas around the grain boundary CD) are tested for separation distance. If they are less than $\sqrt{2}a/2$ apart, the atom introduced through the addition of P,Q,R,S is rejected. Finally, the atoms are relaxed to equilibrium positions using the quasi-dynamic mode.

STUDIES OF VACANCY FORMATION AND MIGRATION IN AND NEAR GRAIN BOUNDARIES

The importance of vacancies in diffusion, quenching, annealing and radiation damage in metals motivated investigation of the influence of grain boundaries on the formation and migration of vacancies. Development of definitive models for polycrystalline metals has been impeded because of uncertainties concerning the influence of grain boundaries on the formation and migration of vacancies. Strength and activation of grain boundary sources and sinks of vacancies are difficult to estimate because of the complexity of grain boundary morphology and its influence on defects. The contributions of computer experimentation in this area will be valuable, since discrete lattice calculations can make these problems tractable.

The formation energy of a single vacancy, E_f^{IV}, is the change in energy of a crystal when an atom is removed from the interior and placed on a rough surface in such a manner so that the surface area does not increase, but the volume of the cell changes [11].

The migration energy of a single vacancy, E_m^{IV}, is equal to that energy which an atom in a first neighbor position must have to surmount the barrier and move into the vacant site. The vacancy moves in the opposite direction to the atom.

Diffusion calculations are the major application for calculated formation and migration energies of vacancies. The coefficient for self-diffusion in a vacancy mechanism is

$$D = D_0 \exp\,[-(E_f + E_m)/RT] \tag{1}$$

$$= D_0 \exp\,[-Q/RT]. \tag{2}$$

The sum of the formation and migration energies of vacancies (E_f and E_m, respectively) is the activation energy (Q) for diffusion in the vacancy mechanism, which is the predominant mechanism for diffusion [14]. Shewmon states that, although the concentration of grain boundary atoms is about one atom in 10^6, the diffusion coefficient is approximately doubled at 650°C by the presence of these few grain boundary atoms in silver [14]. The importance of accurately estimating the migration and formation energies of vacancies in the vicinity of, or in, grain boundaries is obvious.

Although the pre-exponential term and the activation energy terms D_0 and Q appearing in Equation (2) can be determined empirically from very careful experiments [15], accurate prediction of these terms for a polycrystalline metal or alloy is very difficult. Since parameters can be varied far more easily in the computer than materials can be made and tested physically, computer experimentation can play a vital role in the development of materials and in the prediction of their properties.

Method

The research on vacancies was principally conducted on a 6° boundary. This boundary angle was chosen since it was near the center of a range, in θ, in which the dislocation model should be valid, and in which computer experimentation with a practical size computational cell was feasible. Concurrent studies were made in perfect crystals to provide a reference and to compute values comparable to the results of experimental measurements and other computer experiments.

Vacancy formation energies were calculated by removing a chosen atom from a computational cell and relaxing the atoms to new equilibrium positions. The formation energy for the vacancy is then

$$E_f = E_c - E_{c+v} + E_R,$$

where E_c = energy of the cell without a vacancy

E_{c+v} = energy of the cell with a vacancy

E_R = recovery energy = 1.5 eV.

The recovery energy is that energy recovered by the cell when the removed atom is replaced on a surface of the cell in a manner such that the surface area is unchanged but the volume is increased. Because of the configuration necessary to conserve the surface area, the atom has six of its twelve bonds completed. Since each of the bonds has a strength of 0.25 eV, 1.5 eV is added to the cell. Vacancy formation energies computed for cells not subjected to external pressures were always positive and ranged from just above zero to 1.5 eV, the energy required to introduce a vacancy into a perfect crystal. Interaction energies ranged from zero to 1.5 eV. An atom possessing an interaction energy of zero would require the same energy for removal as an atom in a perfect crystal.

Vacancy migration energies were calculated by moving an atom toward a vacant site incrementally and calculating the configurational energy of the cell at each step. The moved atom was allowed two degrees of freedom so it could go to the lowest energy position for each incremental step. Thus, the lowest energy path between two positions, often a circuitous route, was followed. The energy barrier calculated by this technique is less but is more realistic than that which would be calculated if an atom were moved along a straight path from its site to the vacant site.

These studies were conducted in a computational cell composed of seven (100) plane layers. The dimensions in the [010] and [001] (Y and Z) directions were 19 and 21 half-lattice constants, respectively. The grain boundary was located at X = 11, just past the center of the cell [010]. All studies on defects were made in the grain to the left of the grain boundary. This grain was constructed to be the wider so defects could be located beyond the grain boundary region, and the results would not be influenced by the rigid boundary. The computer plot of the center (100) layer appears in Figure 4. Vacancies were introduced into the center plane, Z = 4. Note the structure in this plane. Beginning at the bottom of the plot at the grain boundary (X = 11), note that the boundary is indistinct. This is a region of "good fit". The energy of the atoms in this region is very close to that of atoms in the bulk, i.e., far into the lattice. A "hole" or "bad fit" region occurs at about $X \simeq 11.5$, $Y \simeq 9.5$, near atoms 391, 402, 1035, and 1038. Atom 402 has the highest energy, 1.05 eV, of any of the atoms in its grain. Above this region the structure repeats itself beginning with a good fit region. Finally, at about X = 11, Y = 18, the "bad fit" structure is again repeated. Adjacent planes have similar but not coincident structures. The misfit regions are displaced from one plane to the next. No distinguishable channel exists through the array.

Results

A necessary step in this investigation was to repeat experiments on systems which had been previously studied to substantiate

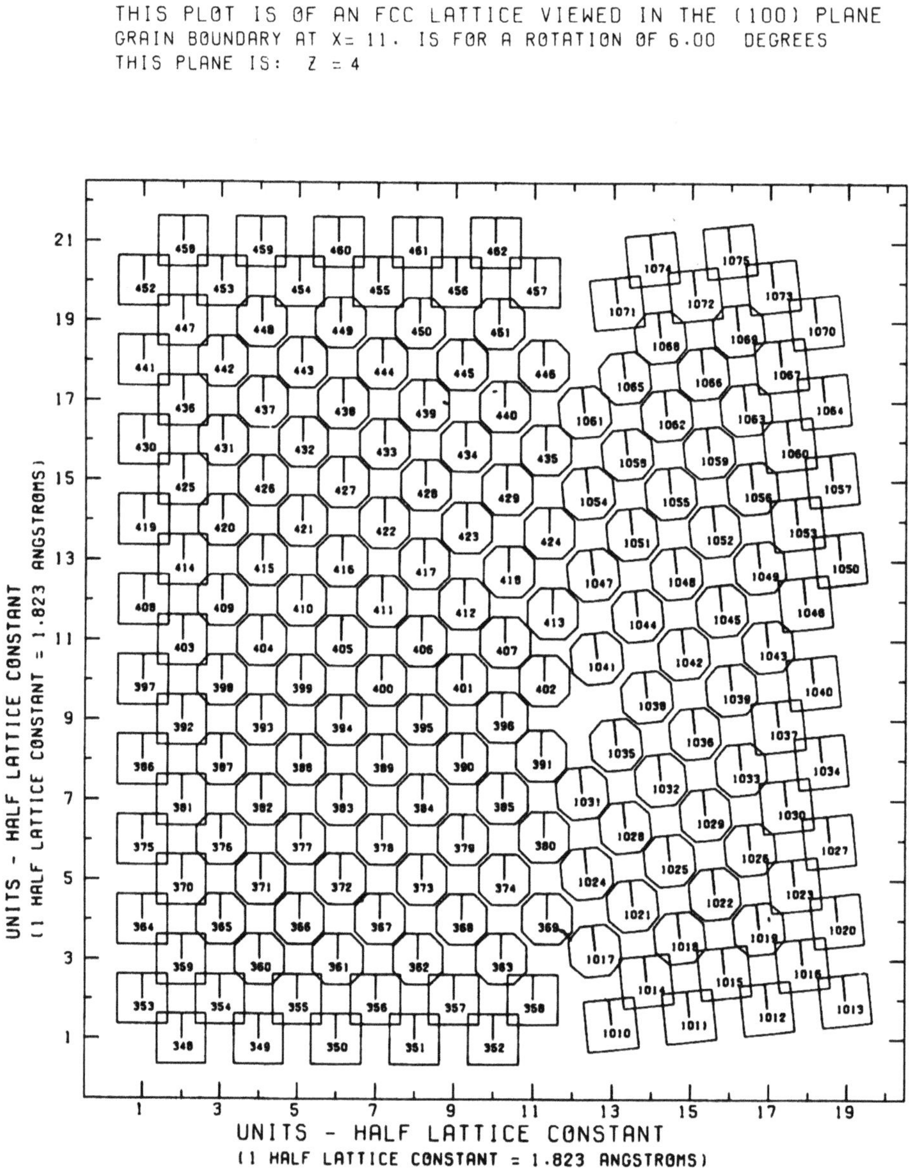

Fig. 4. A 6° tilt grain boundary in γ-iron [center (100) plane].

the validity of the methods and the computer code. Perfect crystals of bcc and fcc iron were chosen. Formation energies for single vacancies and atomic displacements, determined in experiments in which vacancies were introduced into perfect crystals, are presented in Table I.

Table I. Vacancy formation energies and atomic displacements in perfect crystals calculated using GRAINS.

<u>α-Iron (bcc) a = 2.86 Å</u>

Vacancy formation energy = 1.369 eV

Movements of atoms in vicinity of vacancy (units % a/2)

1st Neighbors	2nd	3rd	4th	5th
5.12	-5.18	0.37	-0.54	1.54

(Negative displacements indicate movement away from vacant lattice site)

<u>γ-Iron (fcc) a = 3.646 Å</u>

Atomic movement toward vacant lattice site

1st Neighbors	2nd Neighbors
0.981	0.007

Vacancy formation energy = 1.50 eV

The migration energy in a perfect crystal of γ-iron was found by Johnson [11] to be 1.32 eV. Thus, the activation energy for self-diffusion, obtained by summing the formation (1.5 eV) and migration energies of a single vacancy, is 2.82 eV. This result is in excellent agreement with the value of 2.8 eV obtained experimentally [16]. It is also close to the activation energy of 2.91 eV measured for tracer diffusion of ^{59}Fe in iron [17].

The results for vacancy formation energies and atomic displacements agree exactly with results obtained by Johnson and by Doran [18], who used the same potential but with a different method of computer experimentation, the variational technique. Exact symmetry was observed in the results for atoms lying at first and second neighbors to the vacancy.

Vacancy formation energies were determined near asymmetric tilt, low and high angle boundaries and near a coincidence site boundary. All formation energies for vacancies in these unstressed lattices were positive indicating that the systems are physically realistic and stable. The energy required to create vacancies differed greatly along the 6° grain boundary (Fig. 5). Since this energy required near the misfit regions was almost zero, it is concluded that atoms could be removed very easily at these positions. The formation energy at positions relatively deep into the grain was exactly equal to that determined in the perfect crystal experiments. This value, 1.5 eV, shall be referred to as the "bulk" value in the following discussions.

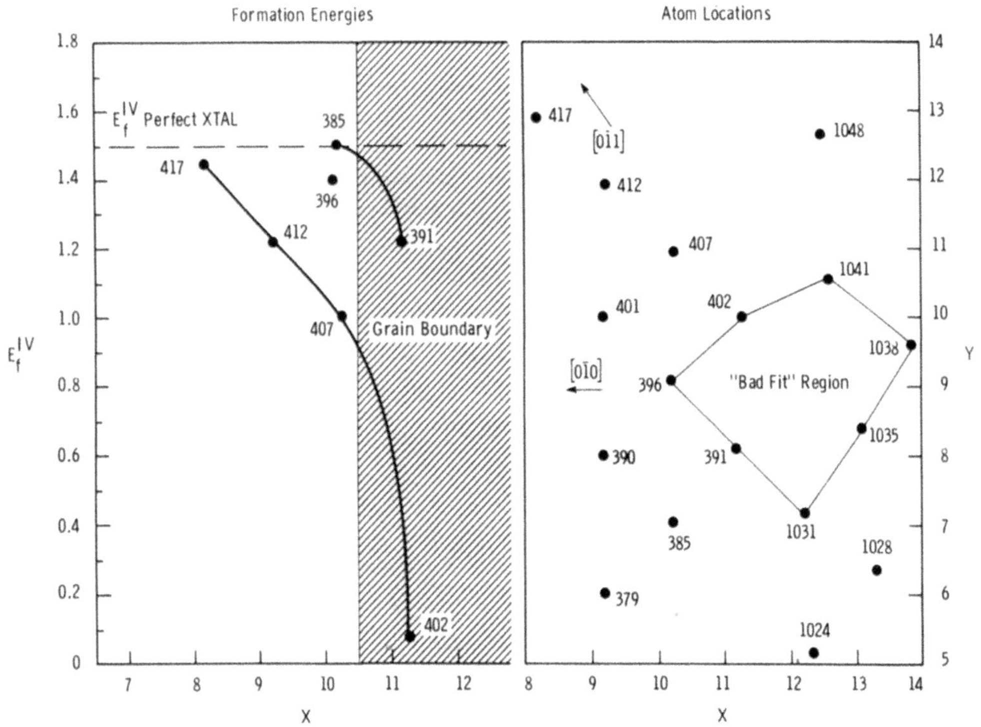

Fig. 5. Formation of energies of single vacancies in and near a grain boundary.

The vacancy formation energies at sites lying along close-packed [0$\bar{1}$0] lines from misfit regions of a 6° boundary are low and do not equal the bulk value for about eight interatomic distances into the grain. Conversely, formation energies at positions along directions [$\bar{1}$00] agree with the perfect crystal value within one atomic distance of the grain boundary. Low vacancy formation energies along the close-packed lines indicate that these are paths of "easy" diffusion which extend from the misfit regions of the boundary for some distance into the adjacent grains. Consequently, the influence of a grain boundary extends significantly farther into a grain than the distance normally considered to be the grain boundary width. However, as can be seen, this situation appears to occur along particular paths. The extent of the interaction zone, or "zone of influence" in certain directions, is less than the zone observed, in microscopy studies of radiation damage [19], to be denuded of loops and impurities; however, it is probably sufficient to establish concentration gradients and cause the observed denuding. Correlated motion of interstitials and defects with vacancies to the grain boundary misfit region would therefore contribute largely to the "cleanness" of the zone near grain boundaries.

Less extensive experimentation conducted for several high angle and a coincidence site boundary (36.9°) verified that conclusions drawn for the low-angle boundary applied to these boundaries also.

The energy required for vacancy migration into and away from the misfit region of a 6° boundary was determined by the series of experiments whose results are presented in Figure 6. In this series

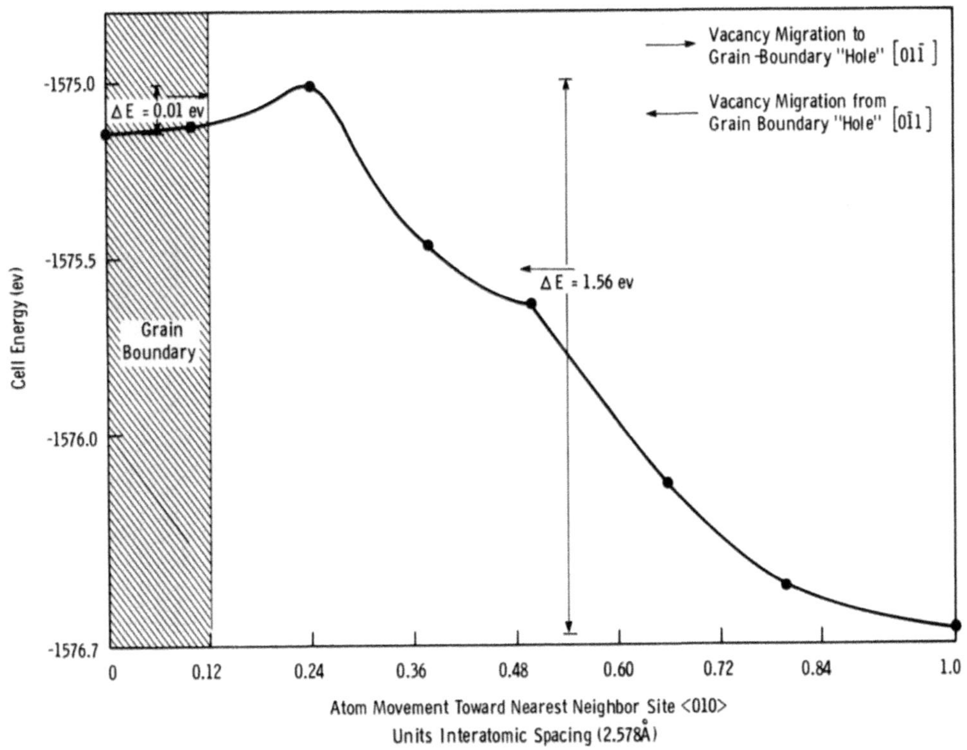

Fig. 6. Vacancy migration energy to a "bad fit" region of a 6° tilt boundary in γ-iron.

of experiments, atom number 407 was removed. Atom number 402 was moved from its position at the edge of the boundary hole along the $[0\bar{1}1]$ direction into the grain, thus simulating vacancy movement in the opposite direction. It is easily seen that the energy required for a vacancy to move into the grain boundary hole was found to be much less than that required for it to move away from it. Only about 0.01 eV was required for the vacancy to move to the hole while 1.5 eV was required for it to move into the grain. Thus, the grain boundary hole should be a sink for vacancies and a possible nucleation site for a large vacancy cluster. The disorder in the local structure is reflected in the asymmetry of the migration energy plot, in that the energy barrier lies only one-fourth the interatomic distance from the grain boundary. The path followed by the atom in migrating is seen in Figure 7. Local disorder caused some metastable positions and increased tortuosity in the path.

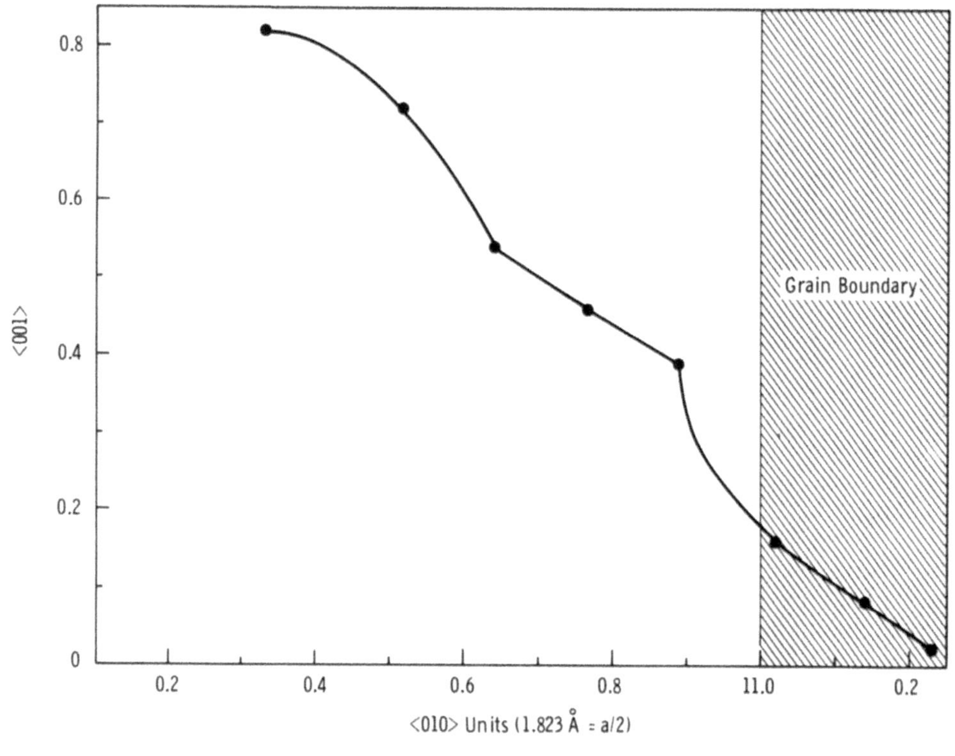

Fig. 7. Vacancy migration path near a "bad fit" region in a 6° boundary.

The investigation most closely related to this study is that of Beeler [20] on free surfaces in α-iron and their interactions with vacancies and interstitials. He demonstrated that there should be a vacancy deficient zone near a surface since vacancy migration to the surface is favored while the formation energies become approximately equal to the bulk value within a very few planes of the crystal. These conclusions and those of Beeler are in agreement in that the influence of the surface is very short ranged in a direction normal to the surface. However, along close-packed lines, the boundary or surface influence extends into the crystal up to six planes. The similarity of effects would indicate that the misfit regions of grain boundaries function as voids or internal surfaces.

INTERACTIONS OF CARBON IMPURITY ATOMS WITH A LOW-ANGLE GRAIN BOUNDARY

Small concentrations of interstitial impurity atoms can have a large effect on the properties of metals. Of these impurities, which include boron, nitrogen, hydrogen, carbon, and others, carbon is perhaps the most important, particularly for the iron-based alloys. This study was initiated to obtain a qualitative understanding of the effects that carbon atoms have on a grain boundary and to obtain data on the relative energy required to place and move carbon impurity atoms in the vicinity of a grain boundary.

The specific tasks of this study were:

- To determine the probable region of impurity atom segregation,
- To determine the effects upon grain boundary structure of impurity atoms, and
- To estimate the energy required for migration of an impurity atom in a grain boundary.

Method

A simple tilt grain boundary of 6° rotation was also chosen for these studies. The results apply to any high angle grain boundary since the misfit region is typical of those occurring in all but the very low angle boundaries.

The computational cell containing the grain boundary and impurity atom was larger than that used for the vacancy studies since impurity accommodation along the boundary was to be investigated. The cell used was composed of 1275 iron atoms in seven (100) planes. Impurity atoms were added in the center layer so that strain fields could be easily contained within this size cell.

Results

Carbon atoms were inserted individually at several locations in and near a misfit region in a 6° grain boundary to determine relative effects on the grain boundary structure, and the extent of the region of interaction between the grain boundary and impurity atoms. The center of this misfit region or "hole" is surrounded by atoms 391, 402, 1041, and 1035 (Fig. 4). Carbon atoms were added into the hole, at the edge of the hole, at 1 and 2 octahedral positions from the hole along a close-packed [0$\bar{1}$1] direction, and directly into the cell normal to the grain boundary.

When an impurity atom was inserted near the center of the hole, the structure of the grain boundary became very well ordered (compare Figs. 4 and 8) apparently because carbon atoms bonded with the neighboring iron atoms and drew them together. Iron-iron bonds were formed and others strengthened, so considerable relaxation took place in the entire configuration as shown by the absence of misfit regions in the plane containing the carbon atom and the adjacent plane (Fig. 9). The energy of the cell was reduced some 2.6 eV; of this, approximately 1.6 eV can be attributed to the relaxation of the lattice into a more stable configuration. The remainder resides with the impurity atom. The ordering, or "healing", of the grain boundary would certainly strengthen the grain boundary, facilitating transmission of stresses and dislocation movement from grain to grain.

Other small interstitial impurity atoms, such as boron or nitrogen, should cause similar effects. Small concentrations of boron located in grain boundary regions are known to increase low temperature ductility. The "healing" observed in these studies would cause such an effect.

Energy and structural changes resulting from the addition of carbon atoms in 1 and 2 octahedral positions away from the boundary in a close-packed [0$\bar{1}$1] direction (i.e., at coordinates 10, 10, and 4) were almost identical to those observed when a carbon atom was added to a perfect crystal. The strains caused by carbon atoms located in octahedral sites were so small that they did not influence the structure. Thus, carbon atoms located only a distance of one-half lattice unit from the boundary acted as though they were in solution in a perfect lattice and did not interact with the grain boundary.

A reference calculation was made in which a carbon atom was inserted into a perfect lattice, and the results compared with these of other investigations (Table II). A carbon atom, which was deliberately added at a nonequilibrium position, was forced to migrate. The atom finally assumed an octahedral site in the lattice.

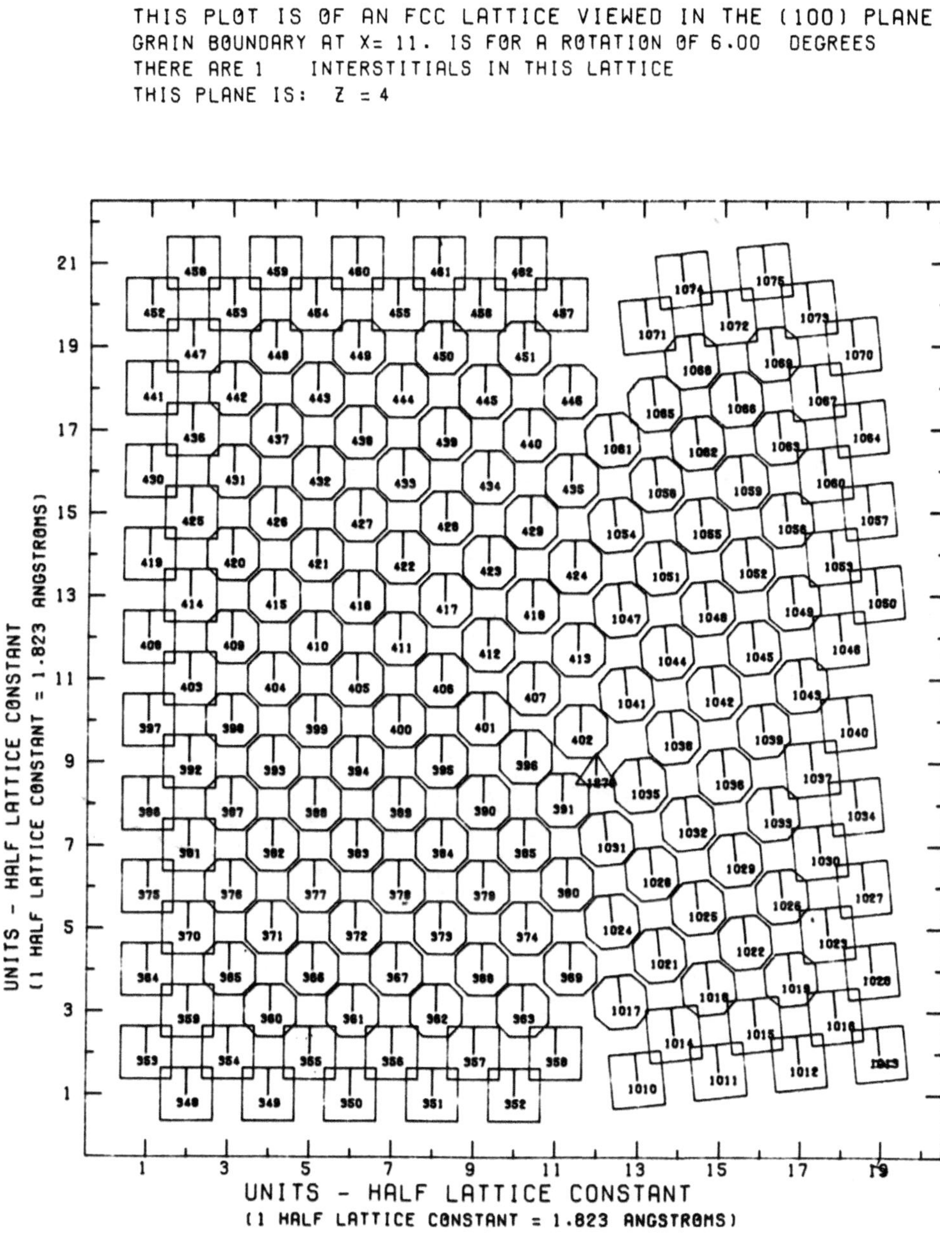

Fig. 8. A 6° tilt grain boundary with a carbon interstitial atom in a misfit region plane Z = 4.

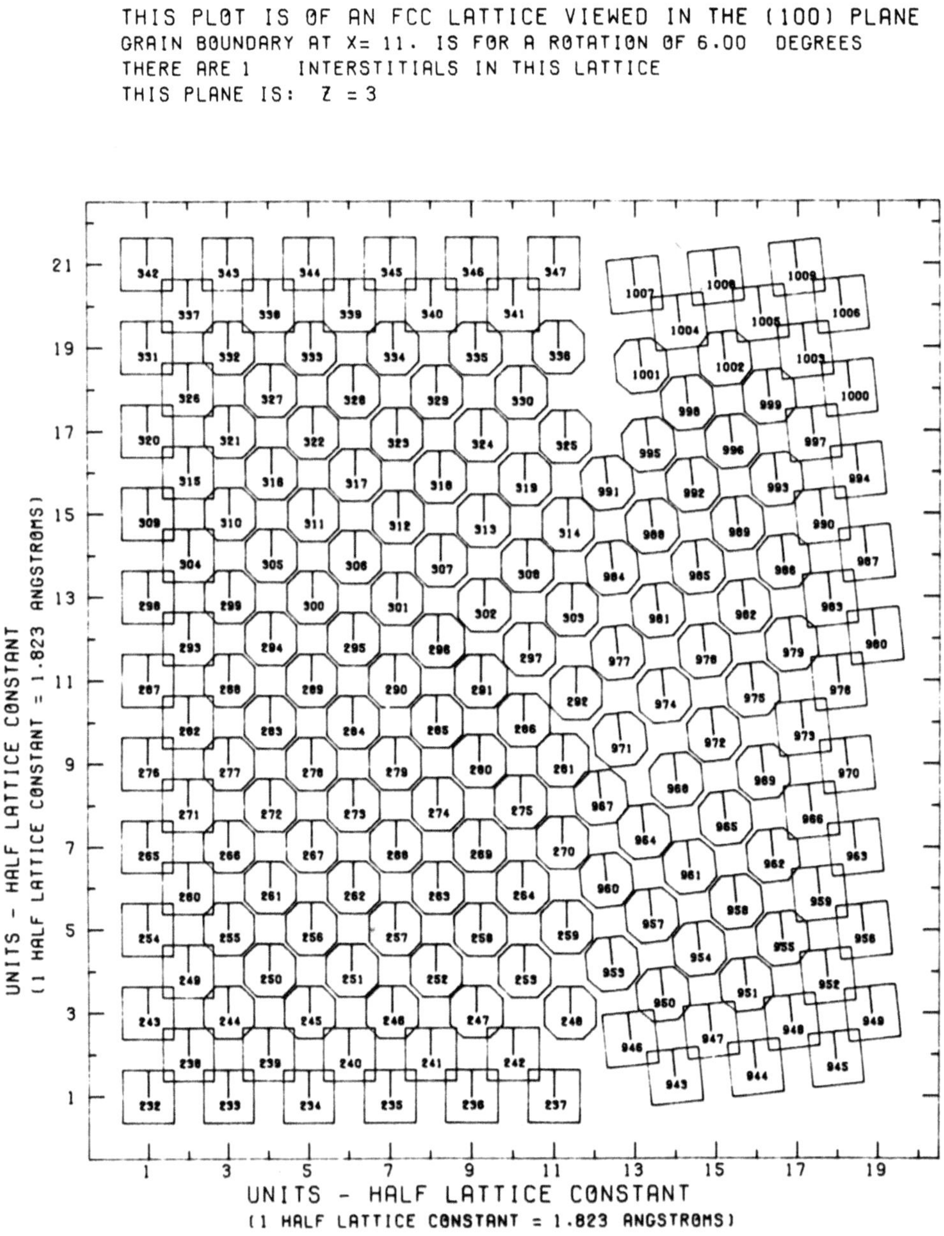

Fig. 9. The (100) plane adjacent to the plane containing a carbon interstitial.

Displacements of the first neighbor iron atoms were approximately 0.03 of a half-lattice unit. Iron atoms to which the carbon atoms were bonded were found to change their energy from -1.50 to -1.66 eV. Strains and energy changes were identical on all first neighbor iron atoms in the perfect crystal. Thus the configuration was symmetric. The bond length, the distance between the carbon atom and the adjoining iron atoms, was approximately 1.9 Å. Movement of iron atoms in second neighbor positions to the carbon interstitial was negligible. In this experiment, the carbon atom moved some 0.6 Å to the octahedral position, thus demonstrating that this is indeed the equilibrium position for the interstitial impurity in an fcc lattice.

Table II. Displacements and energy changes caused by addition of a carbon atom in a perfect crystal of γ-iron.

Atom	Distance from Carbon Impurity, Å	Displacement, Å*	Energy Change, eV
1st Neighbors	1.8651	0.04211	-0.1636
2nd Neighbors	3.1570	-0.00053	-0.00066
3rd Neighbors	4.0836	0.00725	+0.00103
4th Neighbors	5.4690	0.000024	+0.000001
5th Neighbors	6.0472	0.000947	+0.000017

NOTE: Displacements and energy changes were symmetrical within computer roundoff error, i.e., all first neighbor displacement and energy changes were the same in magnitude and direction. Carbon atom was added to a nonequilibrium position and migrated to the octahedral position.

*Positive displacements indicate movement away from impurity. Negative displacements indicate movement toward impurity.

Results of experiments in which single carbon atoms were added near the 6° grain boundary are shown in Figure 10. When the carbon atom was added at a position one plane into the left grain in a [0$\bar{1}$0] direction from the grain boundary hole, a very slight ordering occurred. The change in energy was approximately 0.1 eV greater than that observed for the perfect crystal. The carbon atom forced atom 402 slightly into the misfit region and caused a slight ordering of the structure. Because of the movement of atom 402, the carbon atom moved to a position slightly away from what could be considered a normal octahedral position.

The addition of a carbon atom at the edge of the grain boundary hole (coordinates 10.9, 9, and 4) caused an energy change of about 0.25 eV greater than that resulting from the addition of the carbon

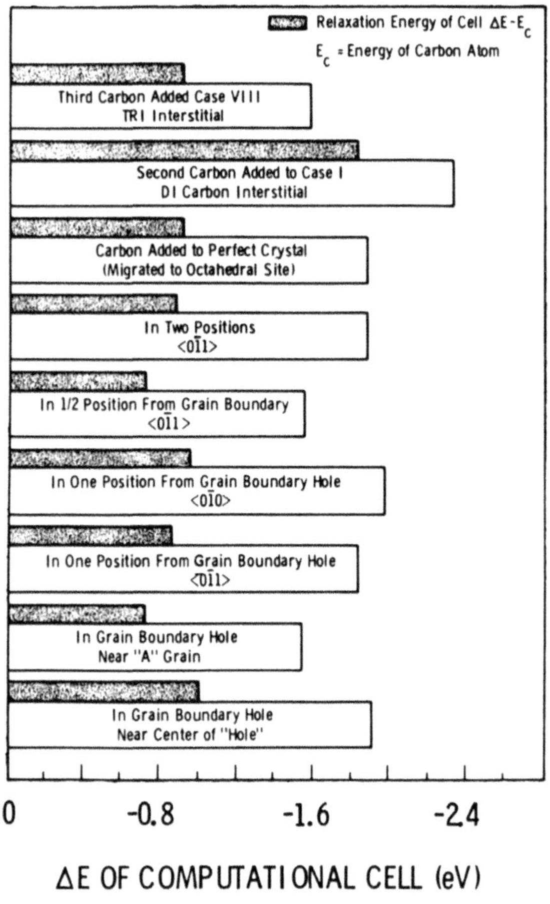

Fig. 10. Changes in energy caused by carbon interstitial atoms in various positions in and near a 6° tilt grain boundary.

atom to a perfect crystal. The carbon atom moved about 0.4 Å from the point of insertion into the misfit region. No ordering of the structure was apparent. Thus, the mid-point between two (100) planes was found to be a high energy position in an experiment in which a carbon atom was inserted in that position and migrated to an octahedral site.

Finally, a carbon atom was added to a position thought to be a saddle point (between the nearest octahedral site to the misfit region and the misfit region itself) in an attempt to obtain an estimate of the energy barrier for migration to the grain boundary. This proved to be a high energy position; however, it is not known if this is the saddle point for migration to the grain boundary since a migration study was not conducted. If this is the saddle point, a carbon atom would require 0.4 eV to migrate to the boundary, but 1.0 eV to migrate from the boundary. Migration into the misfit region is definitely favored.

The results of this study are in qualitative agreement with those by Beeler [20] regarding segregation of carbon near a free surface. In this study, the carbon atoms favored the center of the misfit region, where they could interact with atoms of both grains. There is no analogous position in a free surface. However, the configuration energy of the system was clearly much lower with the carbon atom located between the boundary plane and the first interior plane than with the carbon atom either on the edge of the misfit region or between the second and third planes. Thus, carbon would also segregate either into the misfit region or between the first and second plane, and produce a carbon-rich region similar to that near a free surface.

Di- and Tri-Interstitials in a Grain Boundary Hole. Experiments were conducted in which a second and then a third carbon impurity atom were added into the grain boundary hole. Each experiment was begun with the strain field of the experiment in which the carbon atom was added at coordinates 12, 9, and 4. Each subsequent carbon atom was added at exactly the same location and forced to seek an equilibrium position with complete freedom of motion. The results of these experiments are presented in Figure 11. A large change in energy, structural ordering, occurred upon the addition of the first carbon atom to the misfit region. The energy changes accompanying the addition of the second and third carbon atoms were approximately equal, and slightly less than that observed in the perfect crystal experiment. Thus, it seems certain that the tri-interstitial did not cause disordering or an extensive strain field extending into the grains. Apparently, the tri-interstitial, which assumed the shape of a platelet (Fig. 12), was easily accommodated in the misfit region of the grain boundary.

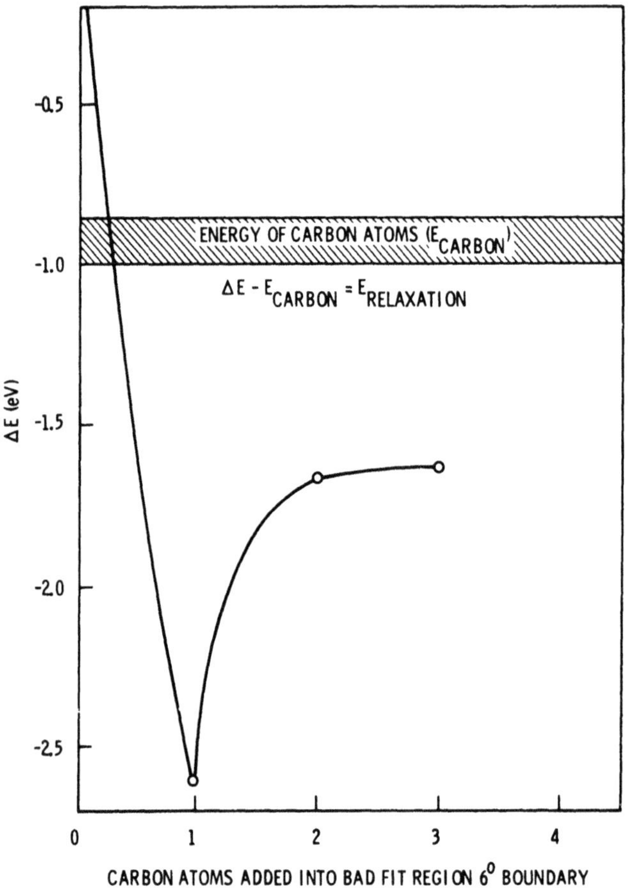

Fig. 11. Changes of the cell energy upon the addition of carbon atoms into a "bad fit" region 6° boundary [common (100) plane].

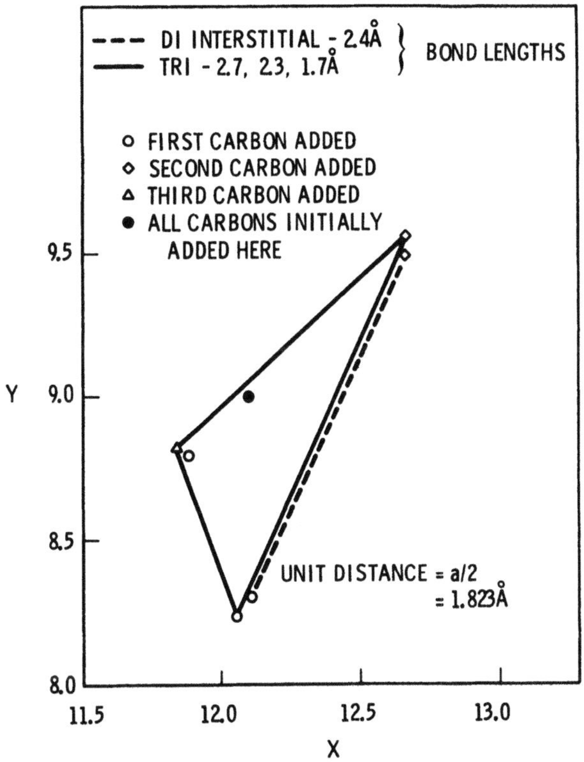

Fig. 12. Configuration of di- and tri-interstitial clusters in grain boundary "hole".

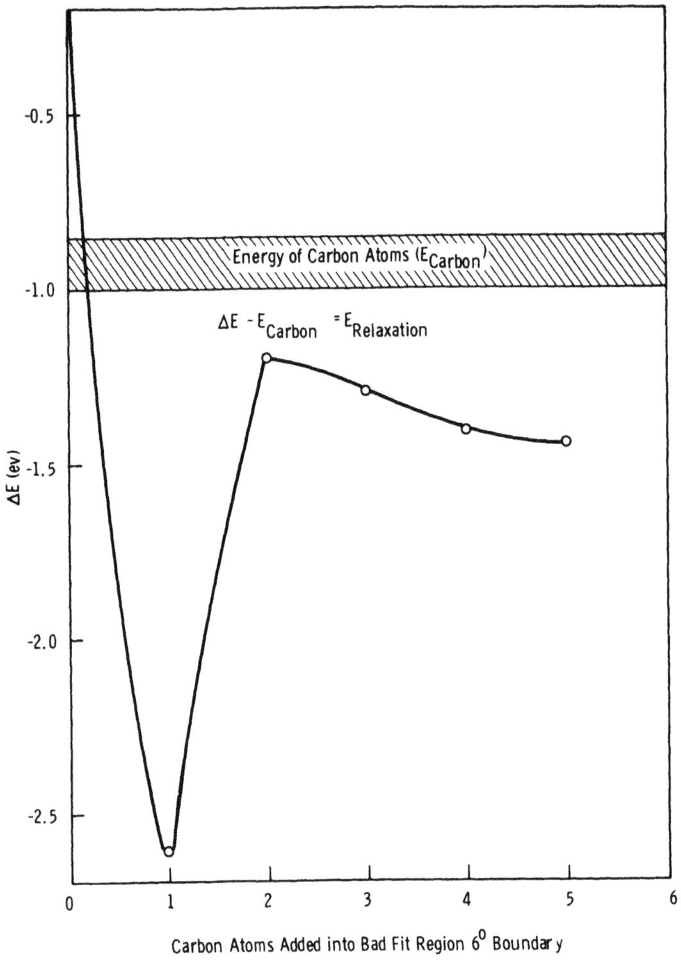

Fig. 13. Reduction in cell energy with addition of carbon atoms into the grain boundary plane (010).

Accommodation of Impurities in a Grain Boundary. The capacity of a grain boundary to accommodate interstitial atoms is of interest since carbide precipitates occur preferentially at grain boundaries. Carbon atoms were added in this series of experiments at locations in the grain boundary plane. The first experiment was the addition of a carbon atom to the misfit region in the center (100) plane, and the second atom in the hole occurring in the (100) plane above the central layer. The third atom was added at the corresponding place in the (100) layer beneath the central plane. The fourth and fifth atoms were added at mid-points along the directional vector between the hole in the center layer (Z = 4), and the holes in the (100) layers above and below it. The choice of the positions, particularly the X and Y coordinates of the mid-point locations, was arbitrary, and in all cases the carbon atom migrated away from the point of insertion to a more favorable location.

The energy changes resulting from the addition of carbon atoms into the grain boundary, as seen in Figure 13, show a reduction in all cases. The addition of the third, fourth, and fifth carbons to the array caused a progressively increasing reduction in the energy of the cell. Thus, the five impurity atoms inserted within two interplanar (100) distances were easily accommodated.

Apparently, rather sizable carbide precipitates can be accommodated within a grain boundary region without inducing strains into the adjacent grains. These precipitates, however, would serve as pins which would impede grain boundary motion and thus impair plastic deformation of the metal.

SUMMARY

The most important conclusions and demonstrations are:

• Determination of values for migration and formation energies of vacancies in the vicinity of a grain boundary so that diffusion can be more accurately calculated.

• The extent of a grain boundary-vacancy interaction zone. The easy diffusional paths extend much farther from misfit regions of the grain boundary into the adjacent grain than would be anticipated from the grain boundary width.

• Grain boundary "healing" caused by an impurity atom. The most significant observation in the studies of carbon atom interactions with grain boundaries is the ordering of a grain boundary by a single impurity atom. A single carbon atom added to the misfit region of the grain boundary caused nearly complete local ordering of the structure.

• Accommodation of impurity clusters in a grain boundary region. Relatively large carbon clusters can be accommodated in misfit regions of a grain boundary without stressing the adjacent grains, and thus their formation is energetically favorable. These clusters would unquestionably function as pins to impede grain boundary motion but could facilitate dislocation motion from one grain to the next, and thus would affect the strength of a metal.

• Determination of the zone of grain boundary-impurity atom interaction. The interaction region where the formation and migration of interstitial impurities is influenced by the presence of a grain boundary is quite small and certainly less than the interaction region observed between vacancies and grain boundaries. If a carbon atom is located more than one-half lattice unit from the grain boundary, it is--in essence--in solution in a perfect crystal.

ACKNOWLEDGEMENTS

This paper is based on work performed by Battelle-Northwest under U. S. Atomic Energy Commission Contract AT(45-1)-1830 [Dahl and Bourquin]; by the Hanford Engineering Development Laboratory, Richland, Washington, operated by WADCO Corporation, a subsidiary of Westinghouse Electric Corporation, under U. S. Atomic Energy Commission Contract AT(45-1)-2170 [Dahl]; and by the North Carolina State University under U. S. Atomic Energy Commission Contract AT(40-1)-3912 and Air Force Materials Laboratory Contract F33615-68-C-1012 [Beeler].

REFERENCES

1. M. Weins, B. Chalmers, H. Gleiter, and M. Ashby, Scripta Met., 3, 601-604, (1969).
2. W. Bollmann and A. J. Perry, Phil. Mag., 20, 33, (1969).
3. J. Levy, Phys. Stat. Sol., 31, 193, (1969).
4. M. Weins, H. Gleiter, and B. Chalmers, Scripta Met., 4, 235-238, (1970).
5. R. E. Dahl, Jr., Scripta Met., 4, 977-980, (1970).
6. T. Schober and R. W. Balluffi, Observations of Dislocation Arrays in Low Angle Tilt Boundaries in Gold, NYO-3504-44, Materials Science Center, Cornell Univ., Sept. 22, 1970.
7. T. Schober and R. W. Balluffi, Dislocations in Symmetric High Angle [001] Tilt Boundaries in Gold, NYO-3504-49, Materials Science Center, Cornell Univ., Oct. 7, 1970.
8. J. R. Beeler, Jr., The Role of Computer Experiments in Materials Research, Lecture Notes, North Carolina State Univ., Raleigh, North Carolina, (1968).
9. R. E. Dahl, Jr., J. R. Beeler, Jr., and R. D. Bourquin,

Computer Phys. Comm., 2, 301-321, (1971).

10. J. B. Gibson, A. N. Goland, M. Milgram, and G. H. Vineyard, Phys. Rev., 120, 1229-1253, (1960).
11. R. A. Johnson, Phys. Rev., 145, 423-433, (1966).
12. R. A. Johnson, Acta Met., 13, 1259-1262, (1965).
13. R. E. Dahl, Jr., A Computer Simulation Study of Tilt Grain Boundaries in Gamma-Iron and Their Interaction with Point Defects, Ph.D. Thesis, North Carolina State Univ., (1970).
14. P. G. Shewmon, Diffusion in Solids, McGraw Hill Book Co., New York, (1963), pp. 164-175.
15. F. L. Vogel, W. G. Pfann, H. E. Corey, and E. E. Thomas, Phys. Rev., 90, 489-490, (1953).
16. F. W. Buffington, K. Hirano, and M. Cohen, Acta Met., 9, 434, (1961).
17. T. H. Heumann and R. Imm, J. Phys. Chem. Solids, 29, 1613-1621, (1968).
18. D. G. Doran, Unpublished work, Battelle-Northwest, Richland, Washington, (1969).
19. R. Carlander, S. P. Harkness, and F. L. Yaggee, Nucl. Appl., 7, 67-75, (1969).
20. J. R. Beeler, Jr., "Interactions of Vacancies and Interstitials with Free Surfaces and Grain Boundaries," International Conference on Vacancies and Interstitials in Metals, Kernforschungsanlage, Jülich, Germany, (1968) p. 598.

ENERGETICS OF GRAIN BOUNDARIES

GRAIN BOUNDARY PHASE TRANSFORMATIONS

Edward W. Hart

General Electric Research and Development Center

Schenectady, New York 12301

INTRODUCTION

Metals and alloys are commonly employed and processed in the form of polycrystalline aggregates. The mechanical strength of such aggregates depends greatly on the strength of the interfaces or grain boundaries that join the individual crystal grains. In general a grain boundary can be expected to be weaker than the grain matrix since the degree of atomic order in the boundary is lower than that in the crystal. Furthermore, the chemical composition with respect to solute components and the diffusive properties of a grain boundary differ from the crystalline matrix.

The current theoretical understanding of grain boundaries is limited almost entirely to description of low temperature structure. In these descriptions models are developed that minimize the energy associated with the grain boundary. The principal useful ideas in such treatments have been the dislocation description of low angle tilt and twist boundaries[1,2,3] and the idea of coincidence boundaries[4,5] for larger orientation differences. Beyond the low temperature régime, however, the problem of detailed description becomes considerably more difficult. It is important then that the few thermodynamic tools that are available for describing more general states of a grain boundary be utilized to the fullest.

It has only recently been realized[6,7] that grain boundaries are, in principle, capable of existing in more than one distinct phase structure, and that a grain boundary may undergo phase transformations in much the same way as can bulk phases. Such phase transformations can result in discontinuities in the structure, strength, and chemical and kinetic properties of boundaries. It is the purpose of this paper to show how such phase transformations can be described thermodynamically in terms of the measurable properties of the boundaries.

For clarity of presentation we shall restrict our discussion to the boundaries that separate chemically identical grains. The general methods that we shall describe are, of course, applicable also to interfaces in multi-phase alloys, however, some of the simple arguments and conclusions that we shall adduce for the grain boundaries of single phase metals and alloys cannot be extended to the multi-phase interfaces without some modification.

GRAIN BOUNDARY THERMODYNAMICS

The first fully consistent description of the equilibrium thermodynamic properties of two phase interfaces was developed by J. W. Gibbs[8]. Gibbs showed how interfaces could be described as distinct phases of a two-dimensional character. The descriptive variables of such phases satisfy equations of state just as do those of bulk phases.

Now, a grain boundary is a special type of two phase interface for which the two contiguous phases are generally identical chemically and structurally. The phases are distinguished from each other only by their relative crystallographic orientation and relative spatial displacement. The grain boundary is further distinguished by its orientation relative to the orientation of the crystals. These distinguishing properties are generally subject to external constraint. Within such constraint, grain boundaries can be described by Gibbs' methods as two-dimensional phases.

The virtue of Gibbs' method is that all the variables employed in the treatment are, at least in principle, measurable. The method rests on two main features: (1) an expression for the internal energy of a heterogeneous system containing an interface, and (2) a systematic method for comparing that energy with that of suitably selected homogeneous phases.

Description of a Boundary Phase

The system that we shall consider first is two contiguous grains of a pure substance and the grain boundary joining them. The descriptive variables will be the total volume V, the total number of atoms N, the total energy and entropy, E and S respectively, the grain boundary area A, all as extensive variables, and the intensive variables T, μ, P and σ that are respectively the absolute temperature, the chemical potential of the component species, the pressure, and the surface tension of the grain boundary. It is, of course, the last item, σ, in which we are specially interested. Gibbs showed, by an analysis that we shall not duplicate, that these variables are related by a fundamental relationship

$$E = TS + \mu N - PV + \sigma A. \tag{1}$$

This is the starting point for our analysis.

The second element of the analysis consists of comparing these quantities with the corresponding ones of a homogeneous sample of the bulk phase. In the classical treatment of interfaces carried out by Gibbs, the comparison was made with a reference quantity of bulk phase of the same total volume as that of the system of interest. We shall follow an alternative procedure here and make a comparison instead with a bulk phase sample containing the same number of atoms as the heterogeneous sample. The two samples are shown schematically in Fig. 1. The utility of such alternative comparison methods seems to have been recognized first by Hansen[9] and has been given a general treatment by Goodrich[10]. In our present problem the significance of considering a fixed number of atoms is that we proceed as though we considered the change in all extensive variables when an interface is introduced into a homogeneous phase by some reversible mechanical process. We need not consider the process itself but only the end states.

If for a homogeneous phase at the same values of T, μ, and P, the energy, entropy, and volume per atom is ε, s, and v respectively, we have the usual fundamental relation

$$\varepsilon = Ts + \mu - Pv. \tag{2}$$

The total energy of a homogeneous reference phase of N atoms is then

$$N\varepsilon = TNs + \mu N - PNv. \tag{3}$$

We now define the superficial excess quantities associated with the presence of the grain boundary as

$$E_\sigma \equiv E - N\varepsilon, \tag{4}$$

$$S_\sigma \equiv S - Ns, \tag{5}$$

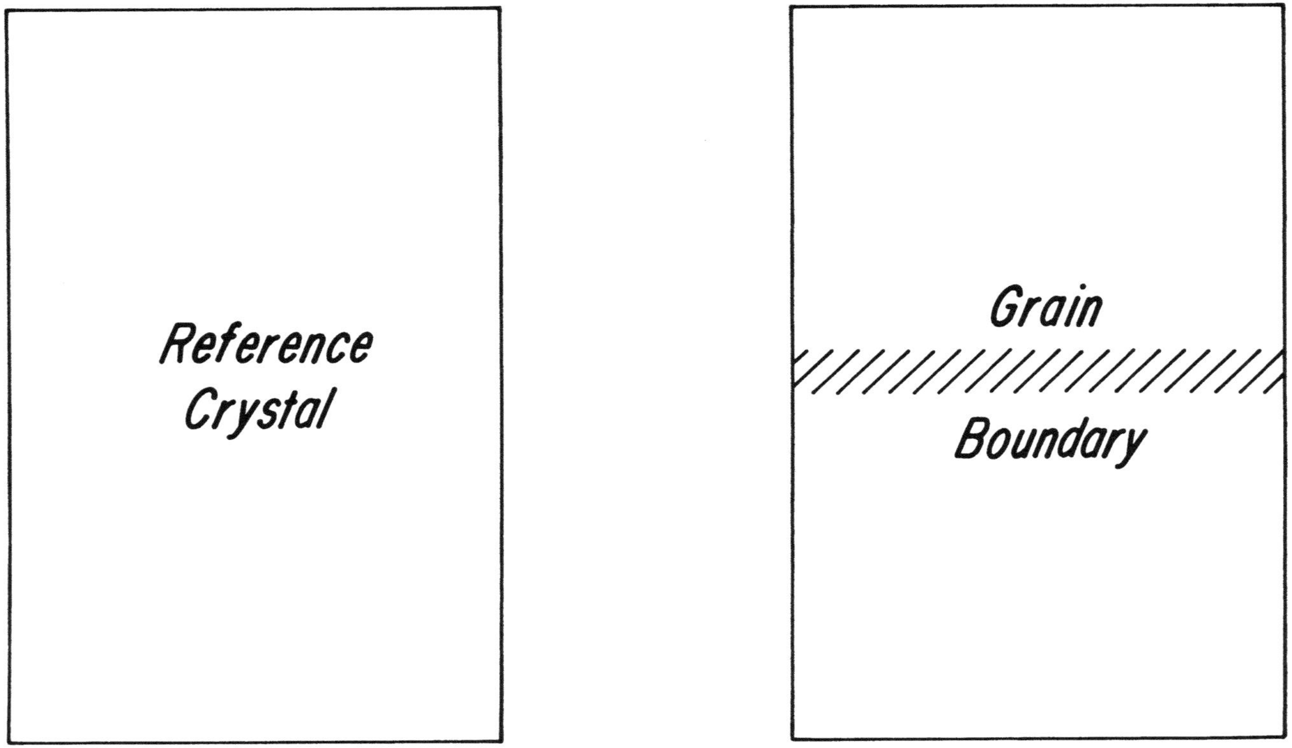

Fig. 1. Heterogeneous bi-crystal with grain boundary of area A and homogeneous reference mono-crystal. Both specimens have the same number N of atoms.

$$V_\sigma \equiv V - Nv. \tag{6}$$

The superficial excess quantities are well defined because the corresponding quantities for the heterogeneous and homogeneous samples are well defined. Now comparing Eqs.(1) and (3) we obtain the fundamental relation for the surface

$$E_\sigma = TS_\sigma - PV_\sigma + \sigma A\,, \tag{7}$$

or, if we let $E_\sigma/A \equiv \varepsilon_\sigma$, $S_\sigma/A \equiv s_\sigma$, and $V_\sigma/A \equiv v_\sigma$, we have

$$\varepsilon_\sigma = Ts_\sigma - Pv_\sigma + \sigma. \tag{8}$$

We shall next show that, just as in the bulk phase T, μ and P are not all mutually independent, so in the "surface phase" T, P, and σ are mutually dependent.

A reversible infinitesimal variation of the variables of the heterogeneous sample follows the general "work equation".

$$dE = TdS + \mu dN - PdV + \sigma dA. \qquad (10)^*$$

This equation together with the full differential of Eq.(1) implies the Gibbs-Duhem type equation

$$0 = SdT + Nd\mu - VdP + Ad\sigma \tag{11}$$

Furthermore, the well-known Gibbs-Duhem equation for the homogeneous phase is

$$0 = sdT + d\mu - vdP. \tag{12}$$

If Eq.(12) is multiplied by N and subtracted from Eq.(11) we obtain the desired relation for the "surface phase",

$$0 = S_\sigma dT - V_\sigma dP + Ad\sigma, \tag{13}$$

or

$$0 = s_\sigma dT - v_\sigma dP + d\sigma. \tag{14}$$

This last equation together with Eq.(8) constitute the justification for our term, surface phase.

---

* - No Equation for (9)

Since Eq. (14) can be considered as an expression for the differential $d\sigma$ in the form

$$d\sigma = v_\sigma dP - s_\sigma dT, \tag{15}$$

we have an intrinsic representation for v_σ and s_σ as

$$v_\sigma = (\partial\sigma/\partial P)_{T,N}, \tag{16}$$

and

$$s_\sigma = -(\partial\sigma/\partial T)_{P,N}. \tag{17}$$

We shall return to a further consideration of the results of this section in our discussion below on the use of detailed microscopic models.

The Boundary Phase Transformation

If the grain boundary undergoes a phase transformation at some temperature below the melting temperature of the matrix we can describe the two phase equilibrium at the transformation point in a manner analogous to the usual description of phase transformations of bulk matter.

At the transformation point there must be a full thermodynamic equilibrium between the high temperature phase and the low temperature phase of the grain boundary. This requires that the surface tension σ be the same for both phases. There will, in general, be discontinuities in s_σ and v_σ between the two phases. Designate those discontinuities by Δs_σ and Δv_σ such that

$$\Delta s_\sigma \equiv s_\sigma^H - s_\sigma^L, \tag{18}$$

$$\Delta v_\sigma \equiv v_\sigma^H - v_\sigma^L, \tag{19}$$

where the superscripts H and L refer to the high and low temperature phases respectively, and the differences are evaluated at a point of two phase equilibrium.

Then, since Eq. (14) holds separately for each phase along the P-T locus of transformation points, that locus is characterized by the equation

$$0 = \Delta s_\sigma dT - \Delta v_\sigma dP \tag{20}$$

which is the Clausius-Clapeyron relation for the transformation locus. Thus at any point of the locus the transformation temperature will be displaced by pressure variations according to

$$(dT/dP)_{\text{transf.}} = \Delta v_\sigma/\Delta s_\sigma \ . \qquad (21)$$

Furthermore, from Eqs. (16) and (17) we can immediately deduce the discontinuities of the slopes of the σ-T and σ-P curves. Thus

$$\Delta(\partial\sigma/\partial P)_{T,N} = \Delta v_\sigma \ , \qquad (22)$$

and

$$\Delta(\partial\sigma/\partial T)_{P,N} = - \Delta s_\sigma \ . \qquad (23)$$

We can generally expect that s_σ, v_σ, Δs_σ, and Δv_σ will all be positive. Schematic curves for a P-T transformation locus and for σ as function of T for constant P and of P for constant T are shown in Figs. 2 and 3.

The Effect of a Solute

So far we have described a grain boundary and its phase transformation for a pure substance. We shall now show that the equations of the previous sections hold substantially unchanged when a solute is present.

We shall consider a sample containing two grains and a grain boundary and consisting of N atoms of the solvent crystal and N^* atoms of a solute. We shall then compare this with a reference homogeneous sample containing the same N and N^* of solvent and solute atoms. Now, since the ratio of solute atoms to solvent atoms in the immediate vicinity of the boundary need not be the same as N^*/N, the atom ratio in the surrounding matrix will not be the same as that in the homogeneous sample. Therefore, the chemical potentials of solvent and solute, μ and μ^*, will in general differ in the heterogeneous sample and the reference one. Let us use the subscript h to designate the homogeneous state. Then for homogeneous and heterogeneous states respectively

$$E_h = TS_h + \mu_h N + \mu_h^* N^* - PV_h \qquad (24)$$

and

$$E = TS + \mu N + \mu^* N^* - PV + \sigma A. \qquad (25)$$

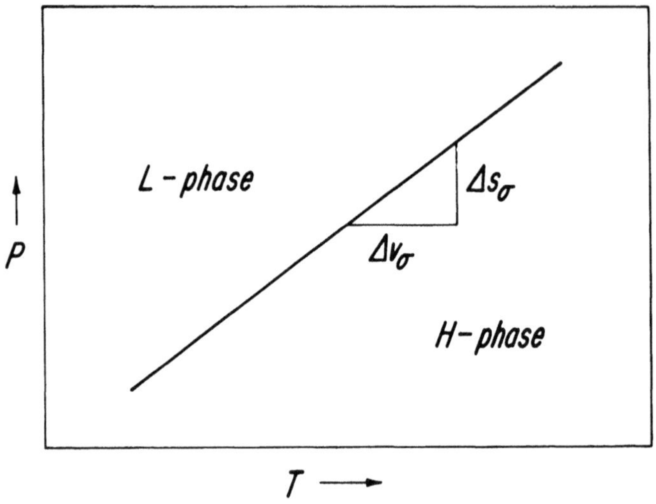

Fig. 2. The P-T locus of a grain boundary phase transformation.

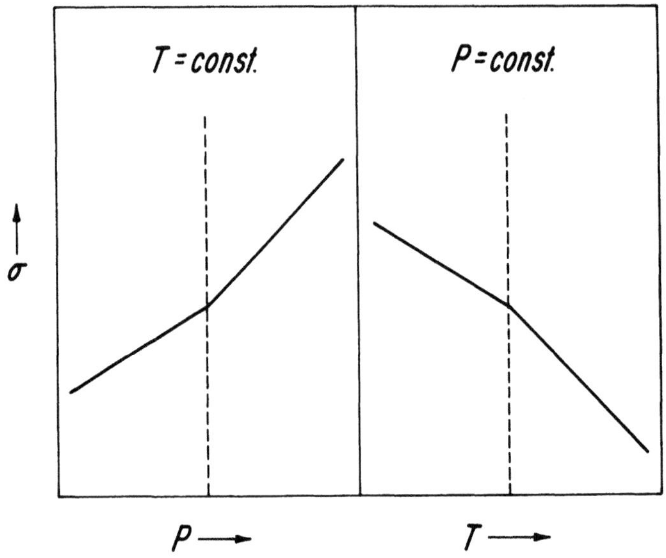

Fig. 3. Schematic curves showing discontinuities of slope for σ as function of P and T at a transformation. The vertical dashed lines represent the transformation pressure and temperature.

Now, in a manner analogous to the earlier treatment, we define

$$E_\sigma \equiv E - E_h , \quad (26)$$

$$S_\sigma \equiv S - S_h , \quad (27)$$

$$V_\sigma \equiv V - V_h \quad (28)$$

Subtracting Eq. (24) from Eq. (25) we obtain

$$E_\sigma = TS_\sigma - PV_\sigma + \sigma A + (\mu-\mu_h)\, N + (\mu^*-\mu_h^*)N^* . \quad (29)$$

This equation is the same as Eq. (7) except for the terms in N and N^*. Now if, as is proper for our mode of treatment, we take N and N^* to be very large compared to the atomic content of the boundary, the differences $\mu-\mu_h$ and $\mu^*-\mu_h^*$ will become small and of the order of 1/N. Then they may be treated as infinitesimals and we have, from the Gibbs-Duhem of the homogeneous phase, that at constant P and T, which is the condition of our comparison,

$$0 = N\, \delta\mu + N^*\, \delta\mu^* , \quad (30)$$

and so the extra terms in Eq. (29) vanish identically and we recover Eq. (7) exactly. Then all the relations for a pure substance hold unchanged for the solute case so long as all variations are for fixed solute content as well as fixed solvent atom content. Of course the actual values of surface excess quantities depend on the quantity and nature of the solute, but so long as external chemical interactions are excluded the solute does not appear as an explicit state variable.

Boundary Adsorption

How then do we describe the boundary solute adsorption that we have already acknowledged can occur? Such description can be accomplished only by a suitable choice of comparison reference state.

Let us choose then a heterogeneous sample as in the preceding section, but for the reference homogeneous sample select the same N atoms of solvent but a number N_h^* of solute atoms such that μ^* is exactly the same in both samples. Then, because of the Gibbs-Duhem relation for the homogeneous phase, μ will also be the same in both samples. Using notation similar to before, we now have for the reference state

$$E_h = TS_h + \mu N + \mu^* N_h^* - PV_h , \quad (31)$$

and for the sample with boundary,

$$E = TS + \mu N + \mu^* N^* - PV + \sigma A. \qquad (32)$$

Now the differencing procedure yields

$$E_\sigma = TS_\sigma + \mu^* N_\sigma^* - PV_\sigma + \sigma A, \qquad (33)$$

where

$$N_\sigma^* \equiv N^* - N_h^* , \qquad (34)$$

and on a unit area basis

$$\varepsilon_\sigma = Ts_\sigma + \mu^* n_\sigma^* - Pv_\sigma + \sigma . \qquad (35)$$

Now, although we have not used a different notation for them, the quantities ε_σ , s_σ , and v_σ of this section are not numerically the same as those obtained by the earlier procedure. This is illustrated more clearly by the surface phase Gibbs-Duhem relation

$$0 = s_\sigma dT + n_\sigma^* d\mu^* - v_\sigma dP + d\sigma, \qquad (36)$$

and even more especially by the implied relations analogous to Eqs. (16) and (17),

$$s_\sigma = - (\partial\sigma/\partial T)_{P,\ \mu^*,\ N} , \qquad (37)$$

$$n_\sigma^* = - (\partial\sigma/\partial\mu^*)_{T,\ P,\ N} , \qquad (38)$$

$$v_\sigma = (\partial\sigma/\partial P)_{T,\ \mu^*, N} . \qquad (39)$$

The important difference, of course, is the appearance of μ^* as an independent variable instead of N^* as in the preceding section.

The surface excess n_σ^* as defined here is a convenient one and is experimentally accessible. It corresponds to investigating the solute phase always as a ratio to the solvent phase on an atomic basis.

The new relation, Eq. (38), is of course a Gibbs adsorption isotherm.

At a phase transformation, n_σ^* can exhibit a discontinuity Δn_σ^*. We cannot, however, decide easily whether Δn_σ^* will be positive or negative. In fact the same uncertainty holds with respect to n_σ^* itself. This is simply because either case may hold depending upon the nature of the solute and the solvent. A detailed investigation of this question is beyond the scope of this paper. Nevertheless, a reasonable expectation is that for a dilute solute with positive volume of mixing in the solvent both n_σ^* and Δn_σ^* are positive when v_σ and Δv_σ are positive.

The phase transformation locus is now a surface in P-T-μ^*-space, and satisfies the relationship

$$0 = \Delta s_\sigma dT + \Delta n_\sigma^* d\mu^* - \Delta v_\sigma dP \, , \tag{40}$$

which is the generalized Clausius-Clapeyron equation.

Some Precautions

Before proceeding with the main theme of the paper, we shall make a few peripheral comments about the general thermodynamic theory given above.

It must be clear by now that there is a large number of choices that can be made in defining surface excess quantities associated with the extensive variables. These surface excess quantities are always well defined and can be given experimental meaning. Nevertheless, none of them are unique, nor are the corresponding "free energy" excesses that can be formed from them unique. On the other hand, the intensive variables are invariant with respect to changing methods of bookkeeping. This applies especially for the variable σ that is properly called the surface (or interfacial) tension.

Most experiments that are designed to measure the mutual mechanical equilibrium among several grain boundaries are concerned with the value of the surface tension σ. It is incorrect and misleading to describe such results as measurements of interfacial "energy" or "free energy", as is commonly done in the metallurgical literature, unless the appropriate free energy has been clearly defined and has been shown to be compatible with the conditions of the experiment.

The critical comments of this section are especially important when an attempt is made to compute the surface tension and the interfacial excesses from microscopic models.

The Use of Microscopic Models

Although it is not the purpose here to discuss the role of microscopic models in any detail, we shall consider briefly some of the problems connected with microscopic calculations.

As noted in the introduction, the microscopic models are addressed principally to the description of the boundary structure and properties at T = 0. A survey of the principal models that have been employed has been published recently in a short review by Fletcher[11]. It is reasonable to expect that such calculations can be made sufficiently precise to yield fair values for the surface volume excess in the low temperature range.

The problem of accurately predicting a phase transformation at some elevated temperature is considerably more difficult. The simplest approach to that problem involves the identification of some unit defect in the low temperature ordered boundary structure that might ultimately lead to large scale disorder. Such unit defects might be vacancies at boundary atom sites or displacements of boundary atoms to alternative unoccupied sites. Each such unit defect or "disorder element" involves an energy for formation of the defect. One can also expect to find an "interaction energy" for the defects because of the greater ease of creating a second defect next to an existing one instead of at an isolated site. Considerations of this sort are very similar to the simple models for considering melting and order-disorder phenomena in more regular crystal lattices. Now, although all such methods are highly approximate they may give some indication of whether a boundary transformation temperature lies lower than the melting temperature of the crystal matrix. It was on the basis of such consideration that Hart[6] concluded that a high angle boundary might undergo a disordering transformation in the gross vicinity of one-half the crystal melting temperature. His argument was based on the general feature of lattice statistics that critical temperatures for lattice transformations scale proportionally to the energy of a unit of disorder. Then, since the bond energies and the coordination numbers for atoms in a boundary are smaller than in the matrix, the corresponding critical temperatures should also be lower for the boundary.

We emphasize that great caution must be observed in relating conclusions from microscopic models to the measurements of real grain boundaries. The only reliable way of characterizing the experimental results is through the thermodynamic formalism that we have discussed above. Model calculations should be addressed to the task of computing the well defined thermodynamic parameters. The model calculations must also be demonstrably consistent with the requirements of the thermodynamic equilibrium.

CHARACTERISTICS OF A BOUNDARY TRANSFORMATION

The characteristic phenomena that can be associated with a grain boundary phase transformation can be separated into two general classes: (1) discontinuities of equilibrium properties, and (2) discontinuities of non-equilibrium or kinetic properties. While the phenomena of the first class are more fundamental, it is probably the second class that is more easily accessible to experiment. After all, the solid-liquid phase transformation of a bulk phase is generally identified experimentally by the drastic difference in mechanical properties of the two phases.

For both classes of phenomena the crucial feature of the transformation is discontinuity. Many properties that are difficult to measure in a range of slow variation can frequently be distinguished readily when they change discontinuously. Of course, if the observations are carried out on a polycrystalline specimen containing many boundaries, the transformation temperature may not be the same for boundaries of all orientations. Nevertheless, even then, it is possible that for many boundaries the transformation temperatures occur within a narrow temperature range, and so an anomalously rapid change of the measured property can be observed instead of an actual discontinuity.

It is important, furthermore, to emphasize that the property changes associated with a boundary transformation should be reversible.

Equilibrium Properties

The equilibrium properties that are most obvious candidates for observation are the ones we have discussed in the thermodynamic treatment above. Those property changes Δs_σ , Δv_σ , and Δn_σ^* can be measured either directly in terms of the defining difference equations or indirectly by observing σ as a function of T, P, and μ^* and measuring the discontinuity of the derivatives of σ with respect to the control variables.

A direct measurement of Δs_σ can be done only by measuring the quantity $T\Delta s_\sigma$ which is the latent heat l_σ of the transformation. Since, even in a fine grain specimen, the grain matrix represents a large associated thermal capacity, the observable temperature effects are especially small. It is unlikely, therefore, that a measurement of l_σ could be successful.

It seems reasonable that Δv_σ could be detected in sufficiently fine grain specimens by sensitive dilatometry. If this can be done, the dependence of the transformation temperature on P can be readily investigated as well. Such a measurement is also well adapted to verifying reversibility.

The presently available methods for measuring n_σ^* (and therefore Δn_σ^*) are essentially destructive. The boundary must be ruptured and then analyzed for the relative amounts of solute and solvent by some technique such as Auger spectroscopy[12,13] or mass spectrometry during controlled dissolution of the exposed surface. This measurement has the advantage over the Δv_σ measurement that it can be exercised on individual boundaries. On the other hand, it is also subject to the problems connected with the diffusion kinetics of the solute species in the solvent matrix.

It is probable that improved methods for measurement of σ itself as a function of T, P, and μ^* will provide a valuable tool for identifying boundary phase transformations and for measuring Δs_σ, Δv_σ, and Δn_σ^*. Some of the problems connected with such measurements are inadvertantly illustrated in Gleiter's[14] measurements on a lead tri-crystal. While it is likely that Gleiter identified the phase transformation temperature, it is almost certain that the measurements below the transformation temperature were carried out on boundaries that had not fully equilibrated mechanically. This was probably due to the slow migration kinetics of the low temperature boundary phase. This judgment is based on the circumstance that a discontinuity was claimed for σ. As we have shown above, such a conclusion is inconsistent with a state of thermodynamic equilibrium at the transformation temperature. Gleiter's structural explanation of the transformation ignores this inconsistency and so is thermodynamically unsound. Nevertheless, the partial success of the method is an encouraging evidence for potential future success.

Non-Equilibrium Properties

Discontinuous changes of kinetic properties of grain boundaries at boundary phase transformations can in general be expected to be more dramatic than the changes of equilibrium properties. Because of this, investigations of transformations in single boundaries are probably best carried out by the observation of kinetic discontinuities.

Some of the non-equilibrium properties of grain boundaries that can be expected to be affected by the change in boundary structure at a phase transformation are: grain boundary migration mobility and atomic diffusivity, intergranular cohesive strength, and grain boundary sliding resistance. Because of their importance we shall discuss each of these a bit further.

Possibly the clearest evidence for a grain boundary phase

transformation is the measurement reported by Aust[15] of a discontinuity in the activation energy for grain boundary mobility of a large angle grain boundary in zone refined lead. Aust[15] also reports similar results by Simpson for Pb-0.01 a/o Au. The transformation temperature is in both cases about 0.77 times the melting temperature of lead. This is substantially the same temperature as that of the discontinuities observed by Gleiter that we discussed above.

Measurements of grain boundary diffusivities should also show discontinuities at boundary phase transformations. There is at present, so far as we are aware, no available data on this point. The reason is probably that grain boundary diffusivities are difficult to distinguish from the bulk diffusion process at high temperatures.

It is, of course, well known that grain boundary sliding is a prominent feature of high temperature metal deformation, and that a characteristic mode of high temperature mechanical failure is intergranular fracture. We should like to speculate at this point that these phenomena, which roughly distinguish the high temperature mechanical behavior of polycrystalline metals from the low temperature behavior, are due to a real change of grain boundary structure that occurs as a boundary phase transformation in a rather narrow range of temperatures for most of the boundaries. The direct effect of grain boundary sliding on the flow stress of a metal depends in a somewhat complex way[6] on strain rate and so it is not clear from the available data whether such a sharp distinction has yet been exhibited experimentally.

The most direct way to test this hypothesis would be to load a single grain boundary in shear and to monitor its shearing rate as the temperature was raised. A discontinuous rise in that rate at some temperature would be at least partial confirmation.

An alternative approach would be an investigation of grain boundary anelasticity or of internal friction as a function of temperature. Such tests have been made in the past but have generally employed only one frequency for measurement. Unequivocal conclusions would require somewhat more detailed experimentation and analysis.

The results of all kinds that are presently available are not inconsistent with the hypothesis, but they are not sufficiently detailed to test it crucially.

CONCLUSIONS

The main theme of this paper has been that, if our general understanding of the character and effects of grain boundaries is to be advanced, it is necessary to take proper account of the possibility that high and low temperature behavior may be distinguished in large part by the existence of distinct high and low temperature grain boundary states. These states are distinct equilibrium phase structures that can be in mutual equilibrium at some intermediate transformation temperature. We have shown how such a transformation can be described thermodynamically in complete analogy to bulk phase transformations.

A consideration of some of the available experimental reports suggests that such a transformation has been observed in lead.

We have discussed some of the further experimental implications of grain boundary phase transformations and have shown that there is a wide range of equilibrium and non-equilibrium phenomena that can depend upon such transformations.

REFERENCES

1. J. M. Burgers, Proc. Phys. Soc., 52, 23, (1940).
2. W. L. Bragg, Proc. Phys. Soc., 52, 54, (1940).
3. W. T. Read and W. Shockley, Phys. Rev., 78, 275, (1950).
4. M. L. Kronberg and F. H. Wilson, Trans. AIME, 185, 501, (1949).
5. D. G. Brandon, Acta Met., 14, 1479, (1966).
6. E. W. Hart, in Ultrafine-Grain Metals , J. J. Burke and V. Weiss, Eds., Syracuse University Press, Syracuse, New York, (1970), p.255.
7. E. W. Hart, Scripta Met., 2, 179-182, (1968).
8. J. W. Gibbs, Collected Works, Vol. 1, Yale University Press, New Haven (1948), p.219 et seq.
9. R. S. Hansen, J. Phys. Chem., 66, 410, (1962).
10. F. C. Goodrich, Trans. Faraday Soc., 64, 3403, (1968).
11. N. H. Fletcher, in Advances in Materials Research, Vol. 5, H. Herman, Ed., Wiley-Interscience, New York (1971), p.281.
12. H. L. Marcus and P. W. Palmberg, Trans. TMS-AIME, 245, 1664, (1969).
13. D. F. Stein, A. Joshi, and R. P. LaForce, Trans. ASM, 62, 776, (1969).
14. H. Gleiter, Z. Metallkde, 61, 282, (1970).
15. K. T. Aust, Can. Met. Quart., 8, 173, (1969).

BEHAVIOR OF GRAIN BOUNDARIES NEAR THE MELTING POINT

C.L. Vold and M.E. Glicksman

Naval Research Laboratory

Washington, D.C. 20390

ABSTRACT

The absolute grain boundary energies of $\{01\bar{1}\}$ tilt boundaries in bismuth at a temperature very near the melting point were measured over the range of misorientations 0.5° to 14.5°. A study of the structure of these boundaries was extended as far as 27°. The results at small tilt angles ($\theta \leq 6^{\circ}$) can be described accurately in terms of a heterophase dislocation model. This model also correctly predicts the grain boundary energies of [001] tilt boundaries in copper near the melting point, again for tilt angles less than about 6°. The heterophase dislocation model is unique, inasmuch as it permits the calculation of absolute grain boundary energies in terms of usually available thermodynamic and elastic quantities and a simple macroscopic parameter related to the boundary structure. In addition, the theory provides a basis for interpreting the structural transition observed in bismuth tilt boundaries at intermediate misorientations ($\theta = 15^{\circ}$). Finally, the failure of current theories to predict the correct energy-misorientation dependence over a wide range of misorientations is ascribed to linear and nonlinear interactions among the misfit dislocations—interactions which increase rapidly in importance for misorientations above about 5°. For a quantitative description of the energetic behavior of higher angle grain boundaries than are treated at present, a theory which accounts for such interactions appears to be required.

INTRODUCTION

The properties of a polycrystalline metal at temperatures

near the melting point are, in large measure, controlled by the behavior of the grain boundaries. The ability to predict the behavior and response of a grain boundary to applied stresses or to specified chemical environments depends strongly on one's knowledge of the excess free energy per unit area of the boundary.

The problem of determining the free energy of an arbitrary grain boundary is extremely difficult, since five independent spatial parameters must be specified just to define the orientation of, and the crystallographic mismatch across, such a boundary. As the symmetry of a boundary's orientation and crystallography increases, the number of independent parameters decreases. For the simplest case, that of the symmetric tilt or twist boundary, only one parameter (the misorientation angle θ) need be specified for an unambiguous description of the boundary's macroscopic configuration. However, even for these simple cases, a detailed microscopic model of a grain boundary is required to interpret the observed energetic behavior.

The purpose of this paper is to present a theoretical basis on which the energy of a symmetric tilt boundary can be calculated in terms of known thermodynamic quantities. This formulation utilizes some of the major aspects of earlier dislocation models which are based upon linear elastic theory, but also employs the full conditions of thermodynamic equilibrium inherent in the concept of heterophase* dislocations. In common with other linear theories, the new model does not accurately predict grain boundary energies over a very large range of misorientation; it does, nonetheless, lead to accurate estimates of the absolute excess free energy density of low-angle grain boundaries, and also permits prediction of boundary behavior at larger misorientations, using information derived only from the behavior at small misorientations, where our knowledge of grain boundary structure is on firmer experimental and theoretical footings.

THEORY

For the present purposes, the crystal geometry and thermodynamic state of the grain boundaries considered here are sufficiently defined by a linear, periodic array of parallel edge dislocations, each of which has a liquid-like second phase at its

*The term heterophase is used here to connote the facts that an additional phase is present at the core of the dislocation in minute quantities, and that this phase exists in constrained (metastable) equilibrium with respect to the matrix crystal.

core. The thermodynamics of heterophase dislocations have already been examined in some detail (1), and we shall draw on the results of that treatment as required. The present analysis of grain boundary energetics will be somewhat restricted by the condition that the temperature is close to the melting temperature, in which case (1) the equilibrium radius r_o of the heterophase dislocation core is given by

$$r_o = \frac{bE_o}{2\pi\gamma_c}, \tag{1}$$

where E_o is the elastic energy factor at the temperature of interest for the boundary misfit dislocations with Burgers vector $\vec{b}$, and γ_c is the specific interfacial free energy of their cores, which for a liquid heterophase core (near the melting point T_e) should be close but not necessarily equal in value to that of the macroscopic solid-liquid interfacial free energy γ_{SL}.

The free energy of a tilt boundary constructed from an infinite, linear, periodic array of these dislocations was calculated (1) from the properties of an isolated heterophase dislocation. The reversible work required to form a unit area of grain boundary, including the local energy associated with the cores, is

$$\gamma_B = \frac{bE_o}{h}\left[\alpha_o \operatorname{ctnh}(\alpha_o) - \frac{\alpha_o^2}{2}\operatorname{csch}^2(\alpha_o) - \ln 2\sinh(\alpha_o)\right] + 2\alpha_o\gamma_c + \frac{\alpha_o^2 h P_o}{\pi}, \tag{2}$$

where h is the misfit dislocation spacing, $\alpha_o = \pi r_o/h$ is a dimensionless geometrical parameter related to the misorientation across the boundary, and P_o is the equilibrium hydrostatic pressure inside the dislocation cores. This equation is valid only under the conditions that the dislocations are widely spaced, since it fails to account for any interactions among them. Under this rather severe restriction, that is $\alpha_o \ll 1$, equation 2 reduces to the asymptotic form

$$\gamma_B = E_o\theta\left[3/2 - \ln(E_o/\gamma_c) + \frac{bE_o\Delta S_f(T - T_e)}{4\pi\gamma_c^2\Omega_\ell} - \ln\theta\right], \tag{3}$$

where the third term in the brackets of equation 3 arises from using the Gibbs-Duhem relationships to evaluate P_o as an explicit function of temperature. In equation 3, the tilt misorientation angle θ is defined by $\theta = b/h$, and $\Delta S_f/\Omega_\ell$ is the entropy of fusion per unit volume of the core phase. On imposing the additional restriction that the temperature T is very near the melting point, i.e., $T \to T_e$, the expression for the boundary energy reduces further to

$$\gamma_B = E_o\theta[3/2 - \ln(E_o/\gamma_c) - \ln\theta], \qquad (4)$$

which is a form very similar to the well-known expression for grain boundary energy derived by Read and Shockley (2,3), where A, a parameter depending on the (inelastic) core energy of the dislocation, is replaced in this theory by the terms $3/2-\ln(E_o/\gamma_c)$. To calculate γ_B explicitly as a function of the tilt angle θ, one needs good estimates for E_o and γ_c. Unfortunately, the quantity γ_c, which is a microscopic energy parameter, is usually unknown, so that equation 4 cannot be utilized directly for calculations of grain boundary energies. We now seek a more applicable form of equation 4, i.e., one requiring only macroscopic parameters for the calculation of grain boundary energies.

Let us consider first the consequences of increasing the tilt misorientation θ of a heterophase dislocation boundary beyond several degrees. As the spacing between the misfit dislocations decreases, a critical angle of crystallographic mismatch will be reached eventually where the heterophase dislocation cores in the boundary will merge, forming a continuous slab of "core phase" at the grain boundary. The dislocation spacing at this critical tilt angle $\theta^\star$ will then be approximately $h=2r_o$, where again, r_o given by equation 1 is the equilibrium radius of the liquid core. The energy of the grain boundary must, of course, be single valued whether the critical misorientation $\theta^\star$ is approached from above or below. However, the method of calculating the energy $\gamma^\star_B$ at the transition misorientation does depend upon the structure of the boundary, i.e., whether the misorientation is just above or below the critical value $\theta^\star$. The interface balance equation for the misorientation angle <u>just below</u> $\theta^\star$ is given by the usual macroscopic equilibrium relation (see for example Figure 1), namely,

$$\gamma^\star_B = 2\gamma_{SL}\cos(\Psi^\star/2), \qquad (5)$$

where $\Psi^\star$ is the minimum equilibrium dihedral angle subtended by the liuqid phase for boundaries below the critical misorientation $\theta^\star$. Now, as θ is increased to <u>just above</u> the critical angle $\theta^\star$, the specific boundary energy is given by

$$\gamma^\star_B = 2\gamma_c, \qquad (6)$$

where we note that for misorientations greater than $\theta^\star$ the individual misfit dislocations in the grain boundary merge, and the boundary misfit becomes accommodated by a continuous slab or film of liquid-like "core phase", with an energy $2\gamma_c$ per unit area of grain boundary. Equations 5 and 6 can be combined to form a relationship between the <u>microscopic</u> quantity γ_c and the

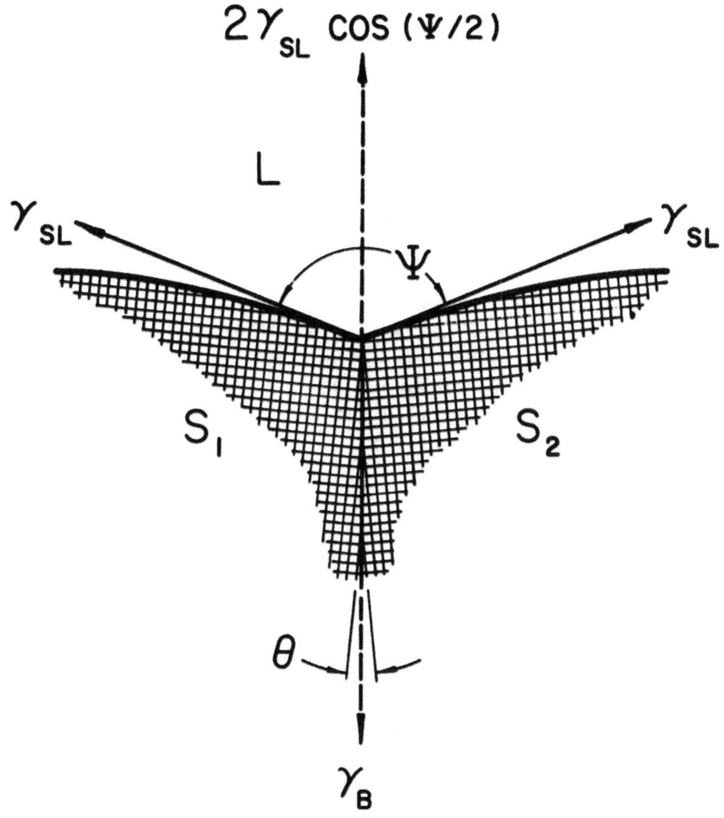

Fig. 1 - Interaction model used to interpret the "balance" of grain boundary surface tension, γ_B, against solid-liquid surface tension, γ_{SL}. The equilibrium dihedral angle Ψ is the limiting angle subtended by the liquid phase L at the root of the grain boundary groove. θ is the lattice tilt misorientation across the boundary separating crystals S_1 and S_2.

<u>macroscopic</u> quantity γ_{SL}, viz.,

$$\gamma_c = \gamma_{SL}\cos(\Psi^*/2), \tag{7}$$

and the critical equilibrium dihedral angle associated with the critical misorientation θ^* is, therefore,

$$\Psi^* = 2\cos^{-1}(\gamma_c/\gamma_{SL}). \tag{8}$$

Replacing γ_c in equation 4 with equation 7 yields the desired expression for the grain boundary energy

$$\gamma_B = E_o\theta\{3/2-\ln[(E_o/\gamma_{SL})\sec(\Psi^*/2)]-\ln\theta\}. \tag{9}$$

In equation 9, all terms are generally known or measurable by independent means, except for the term $\sec(\Psi^*/2)$. However, the critical dihedral angle Ψ^* must, of course, be greater than zero, and, in general, will have a value less than about 90°. The corresponding term $\sec(\Psi^*/2)$ should, therefore, assume values somewhere between the limits 1.0 to about 1.4. When evaluating γ_B with equation 9, a maximum error in γ_B of only about 20% would occur for even the worst possible choice for Ψ^*. The situation is usually much better than this, because Ψ^* is a macroscopic parameter of the grain boundary, which may be measured directly when θ approaches θ^* from below; thus, equation 9 can be fully evaluated with experimentally measurable quantities. For example, the term E_o in equation 9 is given by the general expression $E_o = bK(S_{ij})/4\pi$, where $K(S_{ij})$, a specific function of the elastic compliance constants at the melting temperature, depends on the symmetry of the crystal and the crystallographic orientation of the edge dislocations in the tilt boundary. Eshelby (4) has considered the general problem of the elastic behavior of edge dislocations in an anisotropic body. A specific $K(S_{ij})$ for any material can be extracted from Eshelby's general expressions, providing that the misfit dislocations and their associated elastic fields do not display too low a degree of symmetry in the crystals surrounding the grain boundary.

EXPERIMENTAL

In an earlier study (5,6) the authors developed techniques for directly viewing the equilibrium interaction of grain boundaries with the solid-liquid interface. The purpose of that study was to measure the quantity E_o/γ_{SL}, where E_o is the anisotropic energy factor for the boundary dislocations (described just above in the theory section) and γ_{SL} is the ordinary macroscopic solid-liquid interfacial free energy. The determination of E_o/γ_{SL} was based upon measurements of the equilibrium dihedral angle formed at a solid-liquid interface intersected by low-angle tilt boundaries. A schematic of such a boundary and the perturbed (grooved) interface is shown in Figure 1, depicting the equilibrium disposition of the balanced surface tensions. The interface balance equation defining this equilibrium is given by the relation

$$\gamma_B(\theta) = 2\gamma_{SL}\cos[\Psi(\theta)/2] , \tag{10}$$

derived from Herring's general formulation of the equilibrium

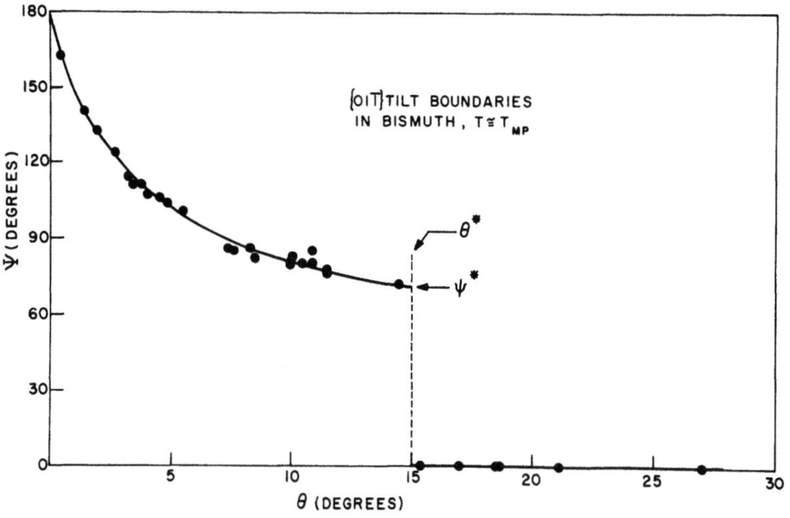

Fig. 2 - Dihedral angle Ψ versus tilt misorientation θ in bismuth. The conspicuous discontinuity occurring in the dihedral angle at the coordinates (θ*, Ψ*) can be interpreted as a first-order structural transition in the boundary.

among three intersecting interfaces (7). For present purposes, measurements of the equilibrium dihedral angle formed by {01$\bar{1}$}* tilt boundaries in pure bismuth have been obtained as a function of θ over the extended range of misorientations 0.5° to 27° (1); the new measurements also were obtained at temperatures only a small fraction of a degree below the melting point, 545°K, and are presented in Figure 2. The discontinuity appearing at $\theta = 15 \pm 0.5°$ corresponds to the theoretically predicted transition in the boundary structure, which results from the merging of the linear array of misfit dislocation cores. As described early in the theory section, merging of the dislocation cores is expected as the dislocation spacing approaches the value $h = 2r_o$, which corresponds to the critical misorientation $\theta^* = 15°$. Up to the occurrence of this structural discontinuity, the experimental grain boundary energy can be calculated from the observed dihedral angle with equation 10, with $2\gamma_{SL}$ replaced by $-E_o/m$. Here, m is an experimental parameter defined as the slope at small θ of the

*All crystallographic indices and directions are referred to the primitive rhombohedral unit-cell vectors.

plot $(1/\theta)\cos(\Psi/2)$ vs. $\ln\theta$. See reference (5).

For the specific case of bismuth, we are considering only pure tilt boundaries of the type $\{01\bar{1}\}$, with misfit dislocations parallel to [111], having Burgers vectors in the directions $\langle 01\bar{1}\rangle$. E_o, which is given by $bK(S_{ij})/4\pi$, was calculated from the known magnitude of the Burgers vector ($|b| = 4.546$ Å) and the functional form of $K(S_{ij})$ derived from Eshelby's general theory. Specifically, it was shown (6) for the misfit dislocations under discussion that $K(S_{ij}) = [2(S_{11} - S^2_{13}/S_{33} - S^2_{14}/S_{44})]^{-1}$, where the S_{ij}'s are the isothermal elastic compliance constants evaluated at the melting point. These constants have recently been measured accurately at our laboratory (8) and yield a value for E_o of 830 ± 9 ergs/cm^2 (6). The precision determination of E_o, together with our experimental measurement of $m = -5.05 \pm 0.2$, permits unequivocal application of equation 10 to the conversion of the Ψ vs. θ data to absolute excess grain boundary free energy vs. misorientation.

The specific excess free energy of $\{01\bar{1}\}$ grain boundaries in bismuth as a function of the tilt misorientation is presented in Figure 3. If we use the fact that $E_o/\gamma_{SL} = -2m$ in equation 9, the theoretical dependence of γ_B predicted from that equation

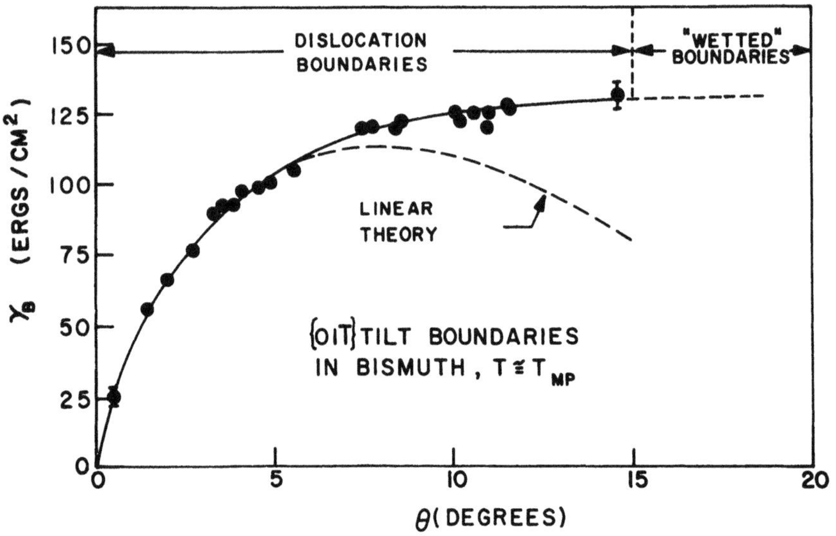

Fig. 3 - Absolute grain boundary energy in bismuth versus misorientation θ. The linear heterophase dislocation theory describes the data in the low-angle region $\theta \lesssim 5^o$. Error bars indicated for the two end points are estimates of the error.

accurately describes the observed data over the misorientation range $0^\circ < \theta < 5^\circ$, and then deviates markedly from the observed energy dependence at larger θ. Gjostein and Rhines (9) have found similar deviations for <u>all</u> of the linear theories that they compared with their measurements of tilt and twist boundary free energies in copper at elevated temperatures. It appears, as a general conclusion, that a grain boundary theory which takes into account the interaction among the misfit dislocations is required to improve the present situation.

APPLICATION OF HETEROPHASE DISLOCATION THEORY TO TILT BOUNDARIES IN COPPER

In 1959, Gjostein and Rhines (9) determined the excess free energies of [001] tilt and twist boundaries in copper at 1065°C, by measuring the solid-vapor dihedral angles during the thermal grooving process. They showed that an equation of the Read-Shockley form describes the <u>relative</u> dependence of small-angle grain boundaries up to misorientations of 5° to 6°, but that for larger misorientations Read-Shockley theory predicts an incorrect variation with misorientation. We now show that the energy dependence of the tilt boundaries can be predicted accurately (within the misorientation limitations normally imposed by any linear non-interaction theory) with the heterophase dislocation formulation developed in the theory section.

Before these formulae can be evaluated, certain thermodynamic quantities and crystallographic properties for copper must be known in advance, and, for this particular application, they are summarized below:

a. $\gamma_{SL} = 177$ ergs/cm² (Ref. 10)

b. $\vec{b} = (\sqrt{2}/2)b\,[110]$ (Ref. 9)

c. $E_o = 1550$ ergs/cm² (Ref. 9)

The theoretical excess boundary free energy was calculated as a function of misorientation angle, using equation 9 and assigning rather disparate values (0° and 70°) to Ψ^*, the critical solid-liquid dihedral angle *; these results are presented in

*It should be emphasized that the quantity Ψ^* is <u>not</u> a solid-vapor dihedral angle such as measured by Gjostein and Rhines, but is the critical dihedral angle that would be subtended by <u>liquid</u> copper in equilibrium with a grain boundary having a maximum density of misfit dislocations—i.e., one with a misorientation just short of the point where Ψ actually falls to zero.

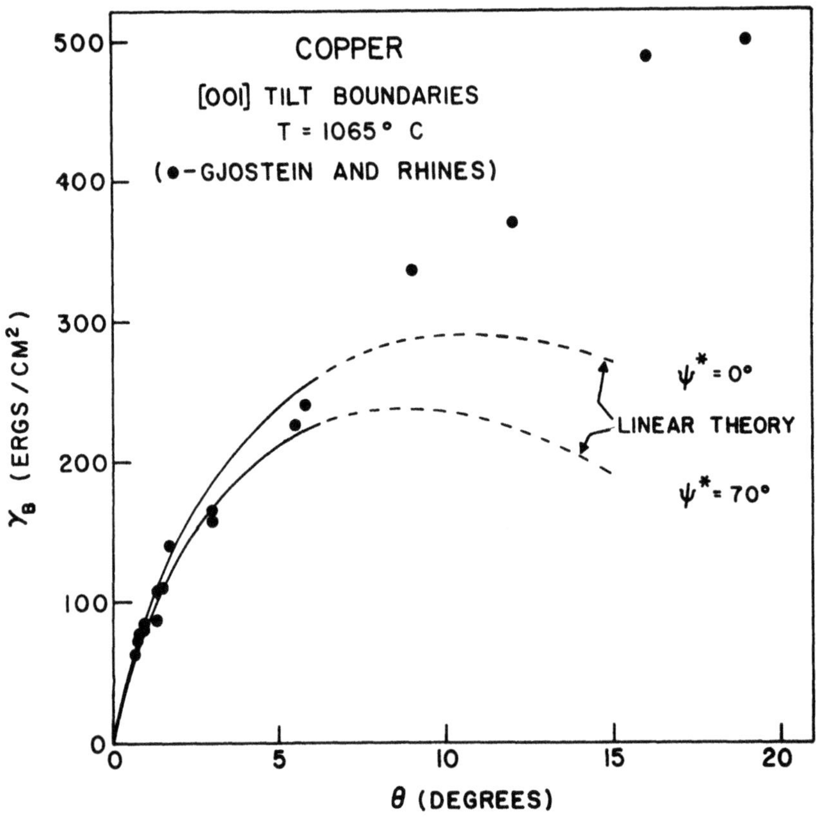

Fig. 4 - Absolute grain boundary energy in copper versus misorientation θ. The linear heterophase dislocation theory describes the data in the low-angle region $\theta \lesssim 6^\circ$, and depends only slightly on the choice of Ψ^*, shown here for two extreme cases of boundary behavior

Figure 4. It is seen that this equation also describes quantitatively in a satisfactory manner the observed low-angle data of Gjostein and Rhines, their data having been placed on an absolute scale by employing the surface free energy value γ_S for copper of 1670 ergs/cm^2 (9). The plot also serves to emphasize the relative insensitivity of γ_B to the choice of Ψ^*, particularly at small misorientation angles. In common with the results found for bismuth, the linear heterophase theory also deviates sharply from the observed energy data for copper at approximately $\theta = 6^\circ$. The measured grain boundary free energy for copper continues to rise with increasing misorientation and reaches a broad energy maximum (not shown here) peaking at $\theta = 45^o$. This energy maximum observed by Gjostein and Rhines reflects only the required crystallographic symmetry of these boundaries about $\theta = 45^o$, and bears no relationship to the energy peak predicted from the linear theory—a point which shall be considered further in the following section.

DISCUSSION AND SUMMARY

1) The theory of heterophase dislocations provides a thermodynamic basis for rationalizing states of constrained equilibria, which correspond to local phase transitions within grain boundaries. The theory also provides a useful model for predicting the free energy variation of a grain boundary with misorientation, providing that the choice of the local phase transition is correct. In the cases of bismuth and copper considered above, the grain boundaries were at temperatures quite near their respective melting points, so that the choice of a liquid-like core phase—one having properties akin to those of bulk liquid, such as entropy density, atomic volume, response to stress, chemical potential, etc.—was rather clear.

2) A major result of this theory is the formulation of the absolute energetic behavior of grain boundaries solely in terms of macroscopic energy and interaction parameters such as surface energy and elastic moduli.

3) The present theory also describes the structural behavior of higher angle boundaries in terms of thermodynamic data supplemented by information obtained from measurements of low-angle boundaries. For example, the model was used to predict the observed onset of a structural transition in grain boundaries in bismuth. It was reasoned that this transition arises from the merging of the heterophase misfit dislocation cores as the misorientation angle reaches a critical value at which a slab of "core phase" forms.

4) The theoretical constant A that appears in the Read-Shockley form for grain boundary energy attains satisfactory agreement with experiment when the conventional dislocation description is modified to that of the liquid-core heterophase dislocation model. The improvement found is attributed to the fact that the heterophase model is based on considerations of full thermodynamic equilibrium.

5) As noted earlier in reference to Figures 3 and 4, the new theory correctly predicts the absolute energy of grain boundaries with small misorientations ($\theta \lesssim 6^o$) but for larger misorientations yields too small an energy. To provide a rigorous experimental

test for the Read-Shockley theory, as first stressed by Gjostein and Rhines (9), it is necessary to compare the experimental values of <u>both</u> E_o and A with those derived from dislocation theory. This was done for the case of copper, and it was shown that the absolute energy of the grain boundaries was again in accord with theory only when the misorientation angle θ was below 6°. In the case of bismuth, the theoretical value for $A = 3/2 - \ln(E_o/\gamma_c)$ when $\gamma_c = \gamma_{SL}$ agrees to within 20% of its value determined experimentally. The disparity between the theoretical and experimental values of A was interpreted as an indication that the microscopic parameter E_o/γ_c is slightly different from the macroscopic parameter E_o/γ_{SL}, implying that the core phase of the misfit dislocations in bismuth boundaries has thermodynamic properties slightly different from those of ordinary bulk liquid. Indeed, the difference between γ_c and γ_{SL} uncovered by these experiments is of major importance in interpreting the sudden structural transition occurring in bismuth tilt boundaries when $\theta \approx 15^o$. (See for example equation 8, which shows that when $\Psi^*>0$, then $\gamma_c/\gamma_{SL}<1$.)

6) Earlier attempts (2, 11-13) to fit Read-Shockley theory over a wide range of θ produced values for A which are generally far too large when compared with any reasonable microscopic (dislocation) theory. The present work in bismuth and that of Gjostein and Rhines on copper lend strong support to the suggestion of Brooks (13) and Shaw, et al. (14) that A is actually a function of θ. As a consequence of this finding, the positive A values calculated for other metals are merely artifacts of the erroneous application of a low-angle theory to high-angle grain boundary data. It is concluded that current linear non-interaction theories of grain boundary energy predict sharp maxima in the γ_B vs. θ plot that bear no correspondence with observation, and arise as a direct result of the failure of these theories to account for interactions occurring at larger misorientations. A new grain boundary theory which takes into account linear (or non-linear) core-core interactions among the misfit dislocations is required for a satisfactory description of the energetic behavior of grain boundaries at larger misorientations.

REFERENCES

1. M.E. Glicksman and C.L. Vold, "Heterophase Dislocations—An Approach Towards Interpreting High Temperature Grain Boundary Behavior", Proceedings International Conference on Structure and Properties of Grain Boundaries and Interfaces, Surface Science, in Press.

2. W.T. Read and W. Shockley, Phys. Rev., <u>78</u>, p. 275 (1950).

3. W. Shockley and W.T. Read, Phys. Rev., 75, p. 692 (1949).

4. J.D. Eshelby, Phil. Mag., 40, p. 903 (1949).

5. M.E. Glicksman and C.L. Vold, Acta Metallurgica, 17, p. 1 (1969).

6. M.E. Glicksman and C.L. Vold, Scripta Met., 5, p. 493 (1971).

7. C. Herring, chapter in The Physics of Powder Metallurgy, edited by W.E. Kingston, McGraw-Hill (1951).

8. E.W. Kammer, L.C. Cardinal, C.L. Vold, and M.E. Glicksman, submitted for publication in J. Phys. Chem. Solids.

9. N.A. Gjostein and F.N. Rhines, Acta Metallurgica, 7, p. 319 (1959).

10. D. Turnbull, J. Appl. Phys., 21, p. 1022 (1950).

11. K.T. Aust, AIME Trans., 206, p. 1026 (1956).

12. W.T. Read, Dislocations in Crystals, McGraw-Hill (1953).

13. H. Brooks, chapter in Metal Interfaces, American Society for Metals (1951).

14. R.B. Shaw, T.L. Johnston, R.J. Stokes, J. Washburn, and E.R. Parker, Mineral Research Laboratory Report, Ser. 27, Issue 14, University of California (May 1956).

THE INTERACTION OF MIGRATING LIQUID INCLUSIONS WITH GRAIN BOUNDARIES IN SOLIDS

T. R. Anthony and H. E. Cline

General Electric Research and Development Center

Schenectady, New York

INTRODUCTION

A liquid droplet on a planar grain boundary in a solid is in a position of minimum energy since the grain boundary area must increase if the droplet moves away. Thus a droplet migrating through a polycrystalline solid may either penetrate or be trapped by a grain boundary, depending on whether or not the applied driving force on the droplet is sufficient to propel the droplet out of the energy valley associated with a grain boundary.

To examine this droplet-boundary interaction directly, a transparent KCl bicrystal containing liquid water droplets was used as a model system. Droplet driving forces were generated by thermal gradients and accelerational fields.

MIGRATION THROUGH BOUNDARIES IN ACCELERATIONAL FIELDS

An examination of bubbles in a glass of Vichy water reveals an upper limit to the size of the bubbles held to the bottom of the glass. As bubbles grow larger than a critical size, they separate from the bottom of the glass and float upwards. The relative force holding various size bubbles to the bottom of the glass can be found by rapping on the glass and observing that only the larger of the bubbles float upwards.

The interesting behavior of carbon dioxide bubbles in sparkling mineral water can be explained in terms of a surface tension force which holds the bubbles to the bottom of the glass and a bouyancy force which pulls the bubbles upwards. Because the surface tension force increases only linearly while the buoyancy force increases as the cube of the bubble radius, only bubbles below a critical size can be held to the bottom of the glass. The interaction of liquid droplets with grain boundaries in KCl is a direct analogy of this sparkling mineral water demonstration. The buoyant carbon dioxide bubbles on the bottom of a glass are replaced by buoyant liquid water droplets on a grain boundary surface in KCl. From the accelerational field necessary to detach the saturated water droplets from the grain boundary surface, the grain boundary tension of the KCl may be determined.

Experiment

Liquid water droplets were introduced into a KCl bicrystal by thermomigration.[1,2] An accelerational field of sufficient magnitude was applied in order to migrate a selected droplet through the 15° twist boundary of the KCl bicrystal. On piercing the grain boundary plane, the crystallographic orientation of the forward dissolving face of the droplet changed from that of the lower grain to that of the upper grain. (Since the saturated water in the droplet is less dense than KCl, a buoyant force on the droplet propels it against the accelerational field--i.e. upwards in the photographs used in this paper.)[3] In contrast, salt atoms depositing on the rear face of the droplet continue to adopt the orientation of the lower grain of the bicrystal. Consequently, as the droplet moves through and away from the twist boundary plane, a square column of crystal with the orientation of the lower grain is deposited behind the droplet in the upper grain of the bicrystal[4,5] (Fig. 1). From Fig. 1, it is evident that the square columnar grain boundary between the liquid droplet and the twist boundary plane is a simple 15° tilt boundary about the <100> direction. The simple rectangular geometry and pure tilt character of this connecting boundary remain intact throughout the experiment because of the immobility of grain boundaries in KCl at room temperature.

The columnar grain boundary exerts a force on the droplet pulling it back towards the twist boundary plane. The velocity of the droplet in this situation was measured in accelerational

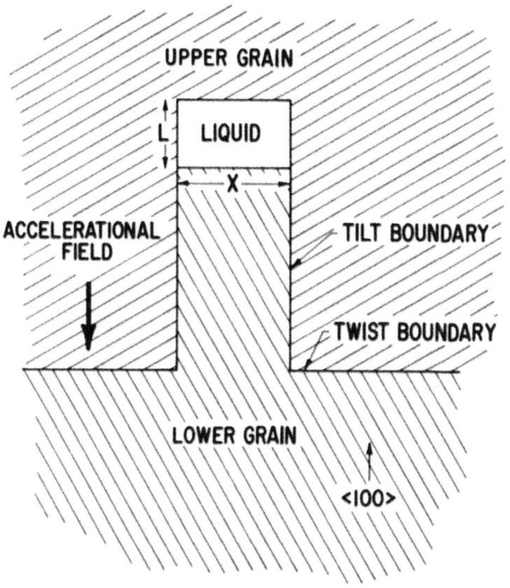

Fig. 1 A schematic diagram of the penetration of a brine droplet through a twist boundary in a KCl bicrystal in a large acceleration field. The grain boundary linking the droplet and the twist boundary plane is a square columnar tilt boundary which exerts a force on the droplet pulling it back towards the twist boundary plane. The accelerational field exerts a buoyant force on the droplet pulling it away from the twist boundary plane.

fields ranging from zero to 6.5×10^7 cm /sec^2(66,000 times earth's gravity). In the absence of an accelerational field, the droplet is drawn back to the twist boundary plane with a negative velocity in the convention of Fig. 2. With increasing accelerational fields, the droplet migration rate increases as shown in Fig. 2 until the droplet begins to migrate away from the twist boundary plane at an accelerational field of about 8×10^6 cm /sec^2. With a further increase of the accelerational field, the droplet velocity increases at less than a linear rate as can be seen in Fig. 2.

Discussion

The velocity V of a liquid droplet with dimensions X by X by L may be determined by multiplying the sum of the various forces F_D acting on a droplet times the mobility of the droplet M_D.

$$V = M_D F_D \qquad (1)$$

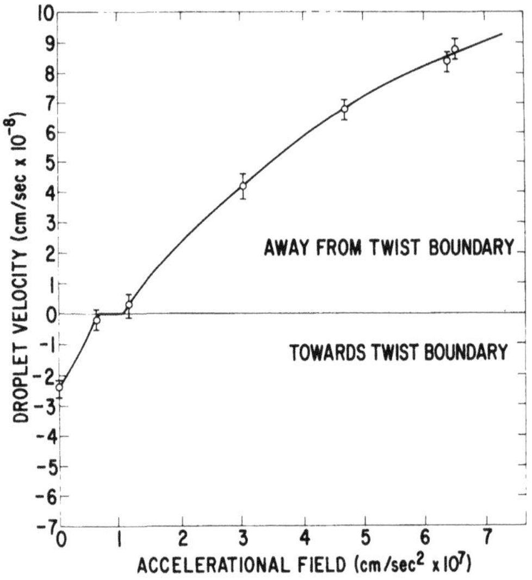

Fig. 2 The velocity of a droplet in the situation depicted in Fig. 1 versus accelerational field. A positive velocity is away from while a negative velocity is towards the twist boundary plane.

The net force on the droplet F_D is given by the sum of the buoyancy F_g produced by the accelerational field, the backward pull of the columnar grain boundary F_{GB} and a pseudofrictional force F_K caused by the kinetics of transfer of salt atoms across the solid-liquid interfaces of the droplet.[4]

The buoyancy force F_g acting on the droplet is given by $g\Delta p\ V_D$ where g is the accelerational field, Δp is the difference in density between solid KCl and the saturated water in the droplet and V_D is the droplet volume.[3]

The backward pull of the columnar grain boundary F_{GB} is given by the tilt boundary tension γ_{Ti} times the length of the perimeter of the droplet, 4X, along which the boundary tension pulls.[4,5]

The pseudofrictional force F_K is given by the sum K of discontinuities in salt chemical potential at the dissolving and depositing interfaces of the droplet (Fig. 3) divided by the length of the droplet L over which these chemical potential discontinuities are effectively distributed times a numerical factor V_D/V_S where V_S is the molar volume of solid KCl.[4,5]

The mobility of the droplet M_D is given by

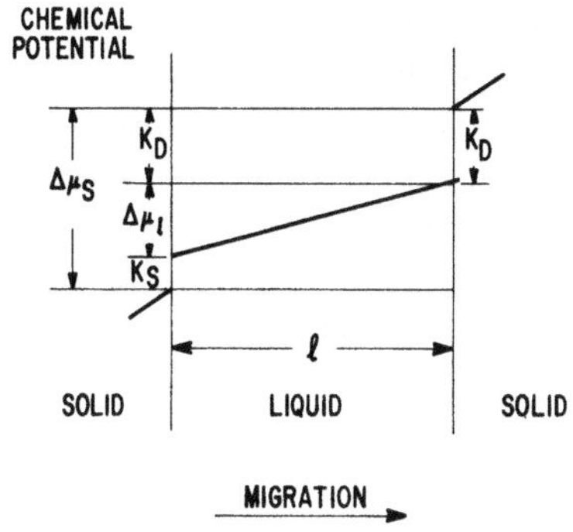

Fig. 3 The chemical potential μ across the liquid droplet, illustrating the kinetic potentials of dissolution K_d and solidification K_s at the solid-liquid interface. These chemical potential discontinuities are required to cause KCl to dissolve and deposit at a rate appropriate to the droplet velocity. The consequential lowering of the chemical potential gradient in the liquid decreases the flux of KCl across the droplet and results in the pseudo-frictional force F_K.

$\frac{C_L}{C_S} \frac{D}{RT} \frac{V_S}{V_D}$ where C_L and C_S are the concentrations of KCl in the liquid and solid respectively, D is the interdiffusion coefficient of KCl in the liquid and RT is the gas constant times the absolute temperature.[1-3,6] Inserting the forces and mobility into equation (1), we have:

$$V = \frac{C_L}{C_S} \frac{D}{RT} \left[g\Delta p \, V_S - \frac{4Y_{Ti} V_S}{XL} \pm \frac{K}{L} \right] \tag{2}$$

where the plus sign before K is used when V is negative and the minus sign when V is positive due to the resistive character of this term.

Because the force F_K resists both positive and negative displacements of the droplet, a range of accelerational fields exist over which the droplet will not move. As indicated in

Fig. 2, this range for the particular droplet studied in the present investigation extends from about 0.6 to 1.1 x 10^7 cm /sec^2. By taking the average of these values, the critical g* at which the buoyancy force on the droplet exactly balances the pull of the columnar tilt boundary is estimated to be 0.8 x 10^7cm /sec^2. The energy of the columnar <u>tilt</u> boundary linking the droplet and the planar twist boundary can then be determined by setting K and V equal to zero in equation 2 to obtain

$$\gamma_{Ti} = \frac{g^*\Delta\rho XL}{4} \tag{3}$$

From the constants listed in Table I, the energy of the 15° pure tilt boundary in KCl is found to be 75 ± 20 ergs/cm^2.[4]

LIQUID INCLUSION SHAPES AND SHAPE STABILITY ON GRAIN BOUNDARIES IN ACCELERATIONAL FIELDS

In a given accelerational field, the shape of a liquid droplet on a grain boundary will conform to a minimum system energy if the solid-liquid system is in equilibrium. The energy of the droplet-solid system shown in Fig. 4 in an accelerational field is given by,[4]

$$E = (A_o - X^2)\gamma_{TW} + (2X^2 + 4XL)\gamma_{SL} - g\Delta\rho V_D \frac{L}{2} \tag{4}$$

where A_o is the area of the twist boundary plane, X^2 is the area of the twist boundary plane occupied by the solid-liquid interface of the droplet, γ_{TW} is the energy per unit area of the twist boundary, $2X^2+4XL$ is the area of the solid-liquid interface of the droplet, γ_{SL} is the solid-liquid surface energy, $-g\Delta\rho V_D \frac{L}{2}$ is the potential energy of the droplet in an accelerational field and $V_D = X^2L$ is the constant volume of the droplet. By eliminating L from Eq. (4), the minimum energy of the system is found by setting

$g^*(V=0) = 0.8 \times 10^7$ cm /sec^2	$\gamma(15°$ tilt$) = 75\pm20$ ergs/cm^2
$\Delta\rho = 0.652$ gm/cm^3	$\gamma(15°$ twist$) = 120$ ergs±15 ergs/cm^2
$V_S = 31.45$ cm^3/mole	$\gamma(15°$ twist$)/$ (solid-liquid)$=3/2$
$X = 9.6 \times 10^{-3}$cm	
$L = 6.0 \times 10^{-3}$cm	

<u>Table I</u> List of constants used to calculate the grain boundary energy of the 15° columnar tilt boundary in KCl.

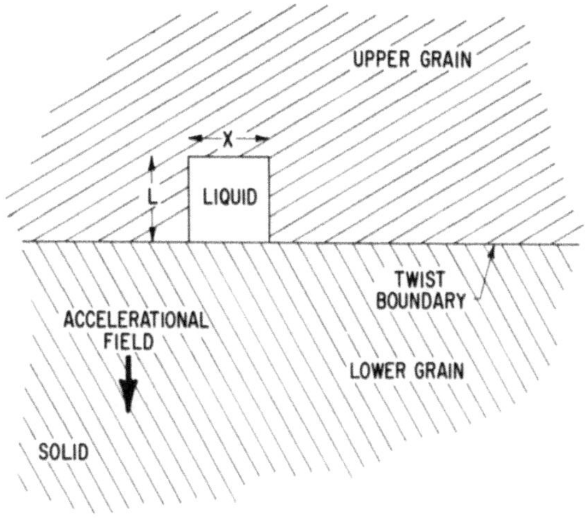

Fig. 4 A saturated H_2O droplet on a grain boundary in KCl being distorted by an applied accelerational field.

the first derivative of E equal to zero

$$\frac{\partial E}{\partial X} = 0 \tag{5}$$

or

$$2X^4(2\gamma_{SL}-\gamma_{TW})-4V_D\gamma_{SL}X + gV_D^2\Delta p = 0$$

For a zero accelerational field, the equilibrium droplet aspect ratio X/L is found to be:

$$\frac{X}{L} = \frac{1}{1 - \frac{\gamma_{TW}}{2\gamma_{SL}}} \tag{6}$$

Equation (6) indicates that the equilibrium aspect ratio should be independent of droplet size whereas Table II shows an increasing X/L aspect ratio with decreasing droplet size. A previous investigation has shown that because of interface kinetics only the smaller droplets in KCl reached their equilibrium shapes.[7] Briefly, the smaller droplets with their larger surface-to-volume ratio were more strongly driven to their equilibrium shapes and were thus able to overcome the resistance of interface kinetics. Similarly, only the shapes of the smaller droplets on a grain boundary should conform to the equilibrium aspect ratio. Thus, the aspect ratio of the smallest droplet in Table II indicates that the ratio γ_{TW}/γ_{SL} is approximately 3/2.

A number of droplets escaped from the energy valley of the twist boundary by a mechanism illustrated in Fig. 5. As the accelerational field was increased, the droplets commenced to neck down at their point of attachment to the grain boundary and adopted a modified tear drop shape (Fig. 5a-5c). When the liquid neck linking the main body of the droplet to the twist boundary plane had shrunk below a critical size, the main body of the droplet moved rapidly away from the twist boundary plane and stretched the thin liquid neck to relatively long lengths in some cases (Fig. 5d). At this point, the liquid neck usually sealed itself off at the twist boundary plane as salt atoms deposited on the solid walls of the liquid neck at its junction with the twist boundary (Fig. 5e). With the neck no longer held to the twist boundary, the neck contracted into the main body of the droplet, driven by solid-liquid surface tension forces (Fig. 5f, 5g). The mode of escape illustrated in Fig. 5 generally occurred with the larger liquid droplets. The smaller droplets with their much larger surface-to-volume ratio demonstrated more resistance to such shape deformations.

As is described above, droplet deformation provides a second way for the droplet to escape from the twist boundary plane in an

Aspect Ratio X/L	4.0	3.8	4.0	3.4	3.1	3.2	2.1
Droplet Volume $\times 10^{-8} cm^3$ V_D	1.5	2.4	7.9	10.6	19	34	115

$\gamma_{TW}/\gamma_{SL} = 3/2$

Table II The aspect ratios X/L of saturated H_2O droplets on a 15° twist boundary in KCl in a zero accelerational field.

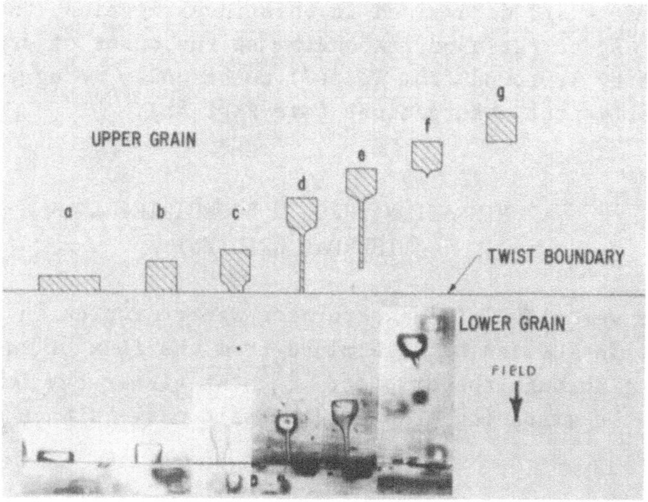

Fig. 5 Buoyant droplets breaking away from the twist boundary in an accelerational field by the tear-drop deformation mode described in the text. The microphotographs are of a number of different bubbles at various stages of the breakaway process.

accelerational field. Up to this point, only the necessary but not sufficient condition, $\frac{\partial E}{\partial X} = 0$, was used for a minimum energy system. If the shape of a droplet on a grain boundary is to be stable, then the additional condition for a minimum energy system, $\frac{\partial^2 E}{\partial X^2} > 0$, must also be satisfied.[4]

$\frac{\partial^2 E}{\partial X^2} > 0$: Requirement for droplet shape stability on the twist boundary

or

$$\frac{X}{L} > \frac{1}{2\left(2 - \frac{\gamma_{TW}}{\gamma_{SL}}\right)} \tag{7}$$

From equation (5), it can be seen that as the accelerational field g increases, X and the aspect ratio X/L of a droplet of constant volume both must decrease. Hence with increasing accelerational fields, the aspect ratio X/L decreases below the critical limit given in equation (7). At this point, the droplet shape becomes unstable. Subsequently, the droplet forms a neck and detaches from the grain boundary as shown in Figs. 5a-5g. By using the value of γ_{TW}/γ_{SL} = 3/2 determined in this investigation, the critical droplet aspect ratio corresponding to the onset of droplet shape instability is found from Eq. (7) to be unity[4] in agreement with the experimental observations (see Fig. 5b).

MIGRATION THROUGH BOUNDARIES IN A THERMAL GRADIENT

The velocity V_T of a saturated water droplet in a thermal gradient in KCl can be determined from the flux of salt atoms diffusing through the droplet.[1,2] At either the front or rear face of the migrating droplet, the salt mass-balance equation requires

$$J = -V_T C_S \tag{8}$$

where J is flux of salt passing from the hot to the cold face of the droplet and C_S is the concentration of salt in the solid. The salt flux J across the droplet results from the composition gradient in the droplet. This composition gradient is generated by the change of the solubility of KCl in water with temperature and the Soret effect, the interaction between mass and heat flow.[1,2]

$$J = -D\left(\frac{\partial C_E}{\partial T} + \sigma C_L\right) \nabla T_L \tag{9}$$

Here D is the diffusion coefficient of KCl in the saturated water, C_E is the equilibrium concentration and C_L is the actual concentration of KCl in saturated water and σ is the Soret coefficient. From Eq. (8) and (9), the velocity V_T of the droplet in a thermal gradient is then

$$V_T = \frac{D}{C_S}\left(\frac{\partial C_E}{\partial T} + \sigma C_L\right) \nabla T_L \tag{10}$$

Since $V_T = M_D F_T$ from equation (1), the effective thermodynamic force F_T on a droplet in a thermal gradient is[5]

$$F_T = \left(\frac{1}{C_L}\frac{\partial C_E}{\partial T} + \sigma\right)\frac{V_D}{V_S} RT \nabla T_L \tag{11}$$

In Table III, the droplet mobility and various forces on a droplet are summarized.

Symbol	Name
$M_D = \frac{C_1}{C_S} \frac{D}{RT} \frac{V_s}{V_D}$	Droplet Mobility
$F_g = g \Delta\rho V_D$	Force in an Accelerational field.
$F_T = \left(\frac{1}{C_1} \frac{\partial C_E}{\partial T} + \sigma\right) \frac{V_D}{V_S} \nabla T_L RT$	Thermal Force
$F_K = \pm \frac{K}{L} \frac{V_D}{V_S}$	Frictional Force of Interface Kinetics
$F_{GB} = -4\gamma X$	Grain Boundary Tension Force

C_1 = concentration of salt in brine droplet
C_E = equilibrium concentration of salt in brine in contact with salt
C_S = concentration of salt in solid salt
D = diffusivity of salt in brine
R = gas constant
T = absolute temperature
K = kinetic potential
γ = grain boundary tension
L = dimension of droplet parallel to thermal gradient
X = dimension of droplet perpendicular to thermal gradient
V_S = molar volume of solid salt
V_D = X^2L = volume of droplet
∇T_L = temperature gradient in the brine droplet
σ = Soret coefficient of salt in brine
g = accelerational field
Δp = difference in density between the solid and liquid

Table III The droplet mobility, gravity force, thermal force, frictional force, and grain boundary tension force expressed in terms of common parameters.

Experiment

A selected saturated water droplet with the shape of a square platelet was migrated toward a 15° twist boundary along an applied thermal gradient perpendicular to the twist boundary plane. The thermal gradient was large enough to force the droplet through the twist boundary plane. On piercing the twist boundary, the crystallographic orientation of the hot dissolving face of the droplet changed from that of the colder grain to that of the hotter grain of the bicrystal. In contrast, the atoms depositing on the colder rear face of the droplet continued to adopt the orientation of the colder grain of the bicrystal. Thus as the droplet pulled away from the twist-boundary plane, a square column of crystal with the orientation of the colder grain was deposited behind the droplet as it migrated through the hotter grain of the bicrystal up the thermal gradient (Fig. 6).[5]

From Fig. 6, it can be seen that the type of grain boundary between the piercing square column of the cold grain and the hotter grain is a simple 15° tilt boundary about the <100> direction. Because grain boundaries in KCl are immobile at room temperature, the trailing grain boundary retains its simple rectangular geometry and pure tilt character throughout the experiment.

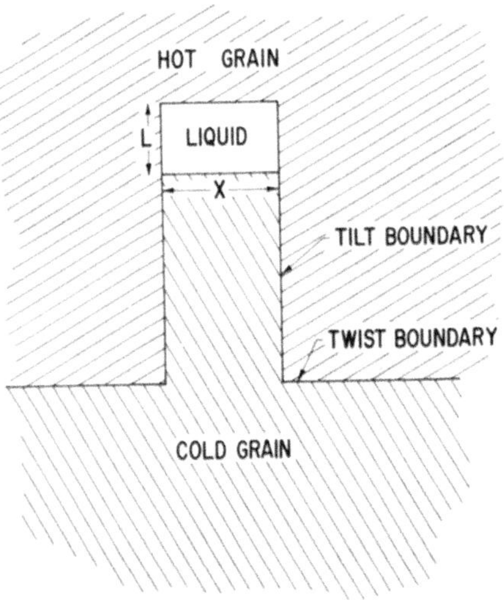

Fig. 6 A schematic diagram of a twist boundary in a KCl bicrystal after penetration by a liquid droplet migrating up a thermal gradient perpendicular to the twist boundary. The secondary grain boundary generated behind the migrating droplet is a square, columnar, pure-tilt boundary.

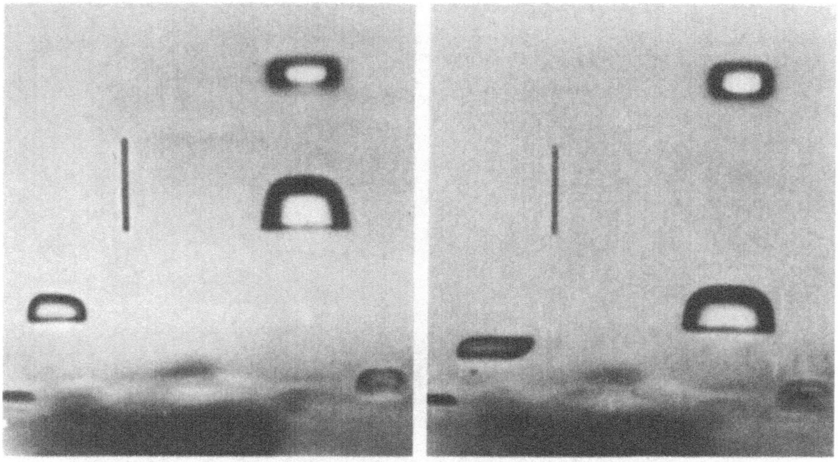

(a) (b)

Fig. 7 (a) A brine droplet which had penetrated the twist boundary plane and migrated into the hot (upper) grain shortly after the thermal gradient had been turned off. The grain boundaries are not contrasted in these photographs. The twist boundary plane is decorated by trapped droplets.

(b) The same brine droplet as Fig. 7(a) after 36 hours with no thermal gradient. The grain boundary tension of the columnar tilt boundary between the droplet and twist boundary plane has pulled the brine droplet back towards the twist boundary plane.

The tilt boundary behind the droplet exerts a force on the droplet pulling it back towards the twist boundary plane. Figure 7(a) shows a droplet which had been driven through a 15° twist boundary plane in a large thermal gradient. In Fig. 7(a), the thermal gradient has already been turned off for several hours. The initiation of droplet motion back towards the twist boundary plane can be surmised from the droplet shape since the dissolving interface of a liquid droplet in a solid is flat and faceted while the depositing interface tends to become rounded.[1-5,7] In Fig. 7(b), taken 36 hours after Fig. 7(a), the droplet has been pulled back approximately three droplet diameters towards the boundary by the columnar tilt boundary between the droplet and the twist boundary plane.

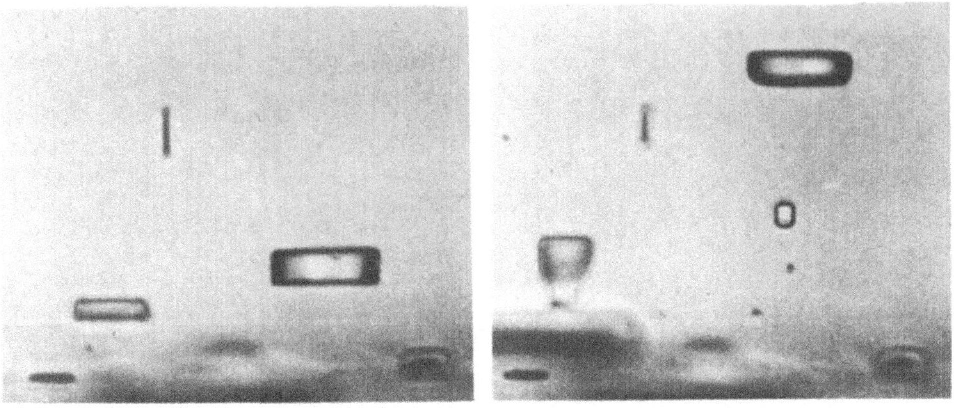

(a) (b)

Fig. 8 (a) The same brine droplet as Fig. 7 in a thermal gradient of 8°C/cm.

(b) The brine droplet of Fig. 8(a) after 21.5 hours in a thermal gradient of 8°C/cm. The thermal force on the droplet has overcome the restraining tension of the columnar grain boundary between the droplet and the twist boundary plane and has migrated away from the twist boundary.

Figures 8(a) and (b) show the same droplet in a thermal gradient of 8°C/cm. Figure 8(b), taken 21.5 hours after Fig. 8(a), clearly shows that the thermal force on the droplet has overcome the backward-pulling tilt-boundary forces and that the droplet as a consequence has migrated away from the twist boundary plane. By varying the thermal gradient until the droplet remained stationary (Fig. 9), the thermal gradient which just balanced the grain-boundary tension forces was found to be 3°C/cm (Fig. 10). Figure 10 also shows the droplet velocity vs the applied thermal gradient. The plotted velocity is positive away from and negative towards the twist boundary plane.

Discussion

The velocity V of a brine droplet of dimensions X by X by L (see Fig. 6) may be found by multiplying the total force F_D on the

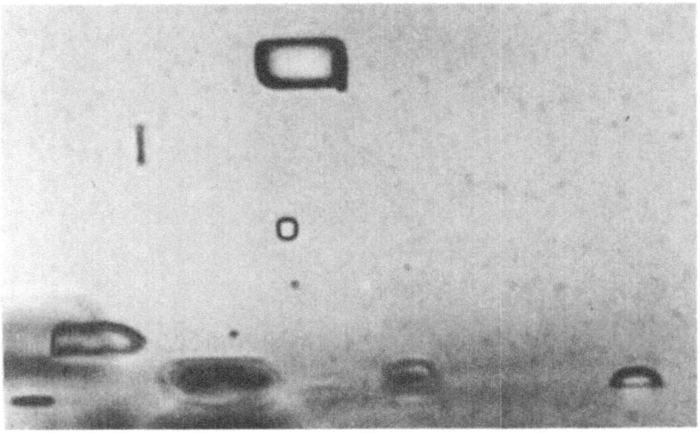

Fig. 9 The same brine droplet shown in Figs. 7 and 8 in a thermal gradient of 3°C/cm. The thermal force on the droplet just balances the backward pull of the columnar tilt boundary and the droplet is stationary.

droplet by the droplet mobility M_D (Equation 1). The total force F_D on the droplet is given by the sum of the thermal force F_T produced by the thermal gradient, the frictional force of interface kinetics F_K, and the grain boundary tension force F_{GB}. By inserting these forces listed in Table III into equation (1), the droplet velocity V is found to be[5]

$$V = \frac{C_1}{C_S}\frac{D}{RT}\left[\left(\frac{1}{C_1}\frac{\partial C_E}{\partial T} + \sigma\right)\nabla T_L RT - \frac{4\gamma W_s}{XL} \pm \frac{K}{L}\right] \qquad (12)$$

where the plus sign before the K is used when V is negative and the minus sign when V is positive; ∇T_L is the temperature gradient in the brine droplet and is a function of the aspect ratio X/L of the droplet as well as the ratio of the thermal conductivities of the solid salt and the brine.[1,2]

From Eq. (12), it can be seen that in a given thermal gradient the ability of a boundary to trap a migrating droplet increases with increasing boundary surface tension and decreasing droplet size. When the droplet is stationary (V=0), the frictional

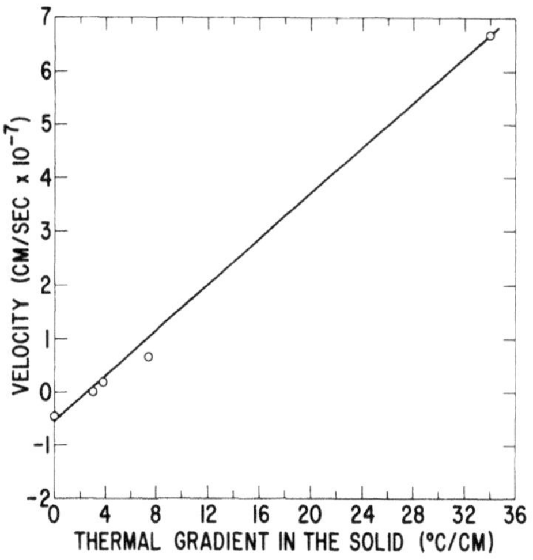

Fig. 10 The variation of the velocity of the brine droplet shown in Figs. 7 through 9 with different thermal gradients.

force of interface kinetics goes to zero and Eq. (12) may be solved for the surface energy of the columnar grain boundary between the droplet and the planar twist boundary.[5]

$$\gamma = \frac{RT(XL)\nabla T_L}{4V_s}\left(\frac{1}{C_1}\,\frac{\partial C_E}{\partial T} + \sigma\right) \qquad (13)$$

C_1 = 4.5 moles/liter	$L = 1.5 \times 10^{-3}$ cm
$\frac{\partial C_E}{\partial T}$ = 0.039 moles/liter-°C	V_s = 31.45 cm^3/mole
$\sigma = +1.35 \times 10^{-3}$/°C	$\nabla T_L = 1.75\nabla T_S = 5.1$ °C/cm
$X = 2.2 \times 10^{-3}$ cm	$RT = (8.31 \times 10^7)(300)$ ergs/mole

Table IV Values of the parameters used in Eq. (13) to calculate the energy of a 15° tilt boundary in KCl

Substituting values of the parameters listed in Table IV, we find that the energy of a 15° pure tilt boundary in KCl is 32 ergs/cm^2. Experiments in metal systems have shown that a 15° tilt boundary has an energy of 0.30 times the free surface energy of the crystal.[8-10] Using the value of 110 ergs/cm^2 found for the free surface energy of (100) planes of KCl,[11,12] one would predict that the grain boundary energy of a 15° tilt boundary about the <100> direction in KCl would be 33 ergs/cm^2. This estimate is in good agreement with the value of 32 ergs/cm^2 determined in the thermal gradient experiment. However, in comparing the tilt boundary tension obtained in the accelerational field experiment (75 ergs/cm^2) with that obtained in the thermal gradient experiment (32 ergs/cm^2), agreement is not so good. It is believed that this disagreement stems from the interface kinetic factor which in reality varies sharply from one droplet to another[1,2,7] and which has a nonzero value even when the droplet is stationary.[1-3]

Acknowledgement

This work was supported by the U.S. Army Research Office, Durham.

References

1. T. R. Anthony and H. E. Cline, J. Appl. Phys. 42, 3380 (1971).
2. H. E. Cline and T. R. Anthony, in press J. Appl. Phys. (1972).
3. T. R. Anthony and H. E. Cline, Phil. Mag. 22, 893 (1970).
4. T. R. Anthony and H. E. Cline, Phil. Mag. 24, 695 (1971).
5. H. E. Cline and T. R. Anthony, Acta Met. 19, 491 (1971).
6. F. A. Nichols, J. Nucl. Mat. 30, 143 (1969).
7. H. E. Cline and T. R. Anthony, Acta Met. 19, 175 (1971).
8. N. A. Gjostein and F. N. Rhines, Acta Met., 7, 319 (1959).
9. A. P. Greenough and R. King, J. Inst. Metals, 79, 415 (1951).
10. W. T. Read, Jr., Dislocations in Crystals, McGraw-Hill Book Co. Inc., New York (1953), p. 194.
11. A. R. C. Westwood and T. T. Hitch, J. Appl. Phys., 34, 3085 (1963).
12. M. Born and O. Stern, Sitzber Preuss Akad. Wiss. 48, 901 (1919).

AN ELECTRON MICROSCOPE STUDY OF CONFIGURATIONAL EQUILIBRIUM AT TWIN-GRAIN BOUNDARY INTERSECTIONS IN FCC METALS*

R.J. Horylev and L.E. Murr

Department of Materials Science

University of Southern California, Los Angeles

ABSTRACT

A transmission electron microscopy study of 304 stainless steel films has been undertaken to systematically study the interrelationships of the degrees of freedom characterizing a grain boundary. From this study a configurational theory has been developed which is useful in explaining the existence of interfacial torques at twin-grain boundary intersections. The grain boundary misorientation (Θ) is defined as the relative rotation of the $<110>$ directions in the adjacent grains of identical (110) orientation. The two remaining degrees of freedom are represented by the tilt or inclination (θ) and the asymmetry (Φ) of the grain boundary plane. Torques arise because of a difference in grain boundary energy with a change in misorientation or tilt. 90° twin configurations (twin plane along a $<112>$ direction) are essentially high-torque situations, as a result of the change in misorientation ($\Delta\Theta$) between the twinned grain and its neighboring grain. 35° twins (twin plane along a $<110>$ direction) are low-torque configurations, but can exhibit high torque anomalies when there is a sufficient variation in tilt across the intersection, $\Delta\theta$. Misorientation, Θ, appears to be the dominant torque producing parameter for high-torque configurations, and dominates the variations in grain boundary free energy. Also, a functional relationship between $\Delta\theta$ and $\Delta\Phi$ is observed for both high-torque occurrences. Spreads in the histograms for twin boundary-grain boundary energy ratios are due to torque terms or variations in grain boundary energy with changes in grain boundary parameters.

*This paper is based upon a dissertation submitted by R.J. Horylev in partial fulfillment for the Ph.D. degree.

INTRODUCTION

The concept of surface free energies or surface tensions, and their relationships to the geometrical arrangement of interfaces at an equilibrated interfacial junction was first introduced by C.S. Smith.[1] Applying the principle of virtual work to an equilibrated interfacial boundary junction, Herring[2] demonstrated the necessary addition of torque terms to the equilibrium equations. These torque terms accounted for variations in interfacial energy with geometry. Herring's equation was then applied in a very approximate form by Fullman[3] to determine twin boundary-grain boundary energy ratios based on two-dimensional (optical microscope) observations of the twin boundary-grain boundary intersection; assuming torque to be negligible as in the original work of Smith.[1] The results exhibited a large distribution in twin boundary-grain boundary energy ratios. Inman and Khan[4] suggested that transmission electron microscopy be used to facilitate a three-dimensional analysis of twin boundary-grain boundary intersection systems. True dihedral angles between intersecting boundary planes were determined, with a resultant reduction in the scatter of the data.[5] Additionally, Murr[6] examined coherent twin-grain boundary intersections in thin films of annealed 304 stainless steel by transmission and diffraction electron microscopy utilizing three-dimensional methods to reduce experimental errors.

However, with a true three-dimensional approach and a minimizing of experimental errors, there still existed a residual spread in the energy ratio values. Part of the scatter can be attributed to a variation in grain boundary energy due to a change in any of the five degrees of freedom associated with the planar grain boundary. Furthermore, any torque terms, defined by Herring's equations, and which were assumed to be negligible in previous twin-grain boundary energy analyses, would also contribute to the residual statistical spread of the energy-ratio histograms.

Few, if any, attempts have been made to seriously study torques associated with grain boundaries characterizing solid-solid interfaces, or even to identify the existence of torques at interfacial intersections. There is a need to investigate energies and torques for practical materials of engineering importance with the various degrees of freedom considered. In addition, it would seem expedient to physically demonstrate the existence of the torque concept associated with solid-solid equilibration, particularly at interfacial junctions as implicitly stipulated in the equilibrium equation of Herring.[2]

The characterization of a grain boundary in a thin film section by transmission electron microscopy will be described in this work.

In addition, the geometrical configurations resulting from interfacial torques at twin-grain boundary intersections will be described conceptually, and observed physically by thin film electron microscopy. The effect of interfacial torque on the geometry of twin-grain boundary intersections will be observed in thin metal films, and the interrelationships of the degrees of freedom characterizing a grain boundary in a random three-dimensional thin section by electron microscopy will be studied.

DEVELOPMENT OF THE ANALYTICAL TECHNIQUES

Interfacial Equilibrium - General Case

It is always mathematically possible to identify a force per unit length, resolved parallel to a solid interface, with an associated free energy per unit area of interface. It is this equivalent interfacial force which is considered in tension at a pseudoequilibrated junction. From a measurement of the dihedral angles and using a triangle-of-forces analysis,[1] the relative energy ratios can be determined.[7,8]

Most energy analyses assume that the surface tension γ_s, has a constant value. In reality the boundary free energy may depend upon the orientation on either side of the boundary.[9] In other words, the degrees of freedom characterizing the boundary may influence the energy associated with the boundary. The simple balance of forces method allows for no variation in interfacial energy and the force triangle must be replaced by a more general condition originally derived by Herring,[2] namely,

$$\sum_{i=1}^{m} [\gamma_i \underset{\sim}{t_i} + (\underset{\sim}{t_i} \times \underset{\sim}{z})\, \partial \gamma_i / \partial \Omega_i] = 0. \tag{1}$$

This equation states that the sum of the forces plus the moments will be zero at the line of intersection in order to preserve the equilibrium of the junction. If the derivative terms vanish, Eq. (1) reduces to the normal triangle-of-forces equation. The terms in $\partial\gamma_i/\partial\Omega_i$ imply that each interface strives to contract with a tension γ_i and to rotate to an orientation of lower γ_i with an effort measured by a torque $\partial\gamma_i/\partial\Omega_i$ per unit area. The torque per unit area of interface thus describes the variation of grain boundary free energy with respect to some orientation parameter, and in the perfectly general case there are five degrees of freedom or orientation parameters describing the grain boundary.[10] If equilibrium of interfaces between two grains having identical surface orientation is considered, then the associated torque will depend primarily on the relative crystallographic mismatch (misorientation), Θ, and the inclination of the boundary plane, θ; and secondarily on the boundary asymmetry, Φ. If it can be assumed that Θ and θ dominate the stabilization of a grain boundary, then even for

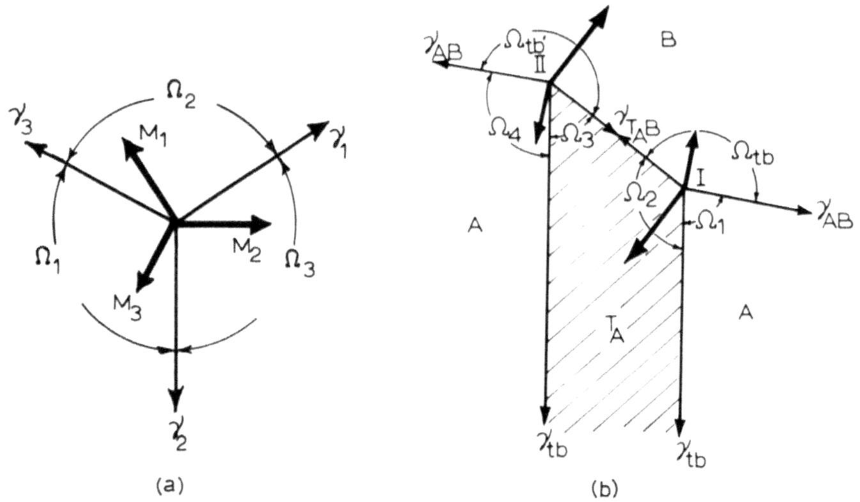

Fig. 1. (a) Intersection equilibrium of three general interfaces. (b) Equilibrated twin-grain boundary intersection.

random grain associations, the grain boundary torque will be approximately described by $\partial\gamma_{gb}/\partial(\theta, \Theta)$.

Figure 1(a) represents a general intersection equilibrium where the interfacial forces γ_1, γ_2, and γ_3, are considered in tension at a pseudoequilibrated junction. As a result, the interfacial free energies are necessarily positive. The associated torques which arise at the interfacial junction, M_1, M_2, and M_3, are also necessarily positive on invoking a uniform convention describing their origin. These associated torques result using a right-handed convention in performing the cross-product in Eq. (1). For this general case, shown in Fig. 1(a), the interfacial free energies need not be equal; whereupon the associated torques will be unequal. In this case only the sum of forces and moments together will be equal to zero. The Ω_i in Eq. (1) bear a direct relationship to the inclination of the boundaries,[11] θ, and can be considered to be indirectly related to the associated misorientation, Θ. In addition, specification of the crystallography and the associated dihedral angles is sufficient to characterize the asymmetry of the boundary trace, Φ.

Twin Boundary-Grain Boundary Interfacial Equilibrium, Crystallography and Geometry

The intersection of a twin band with a continuous grain boundary can be considered as a special case of two interconnected triple junctions as shown generally in Fig. 1(b). Using the conditions of energy balance implicit in Herring's equation, Eq. (1), and the previous arguments of Murr, et al.,[12] the twin boundary-grain boundary equilibrium conditions are described by

$\gamma_{tb}/\gamma_{gb} = (C_{AB} + C_{T_AB})/2$ (2) and $\Sigma M/\gamma_{gb} = |(C_{AB} - C_{T_AB})/2|$ (3);

where

$$C_{AB} = \frac{\cos\Omega_2 \cos\Omega_4 - \cos\Omega_1 \cos\Omega_3}{\cos\Omega_3 - \cos\Omega_2}, \quad C_{T_AB} = \frac{\cos\Omega_2 \cos\Omega_4 - \cos\Omega_1 \cos\Omega_3}{\cos\Omega_1 - \cos\Omega_4}.$$

Equation (3) above is considered to represent a relative interfacial torque ratio; and the absolute magnitude expresses the fact that the torques ΣM, (moments), like the interfacial free energies, γ_{tb} and γ_{gb}, (tensions) are positive. It is therefore unnecessary to uniquely characterize the twin-grain intersections. Only the difference ΔC, need be considered. By evaluating the energy ratios from Eq. (2), any variation in grain boundary free energy is averaged for the specific equilibrium system; and conversely Eq. (3) will supply an approximate measure of the average torque which arises at a junction as a result of the grain boundary free energy variation.

There are numerous arbitrary notations for describing a grain boundary in a general sense, and these assume the boundary to be a plane. In a real solid section, a grain boundary is generally plane only over short segments, and the parameters which describe its location with respect to the crystal portions it separates will vary. For the case of twin boundary-grain boundary intersections, tangents are taken along the grain boundary at the intersection pole so that even if the boundary is not planar, it will still be treated as such at the intersection; the actual parameters at the junctions of interest will be considered.

In the present study, the concern is for a description of twin-grain boundary intersection systems within a random solid slice. The grain boundary-twin boundary system is fully described as represented by Fig. 2. Figure 2(a) depicts the geometric parameters associated with a general intersection. Figure 2(b) shows the crystallographic and geometric parameters which characterize the twin-grain boundary intersection, as seen in a bright-field transmission electron micrograph. In Fig. 2(a) the grain boundary inclination or tilt is denoted by θ_1 and θ_4 for the boundary segments between the matrix grain A and grain B; and θ_3 for the portion between the twinned grain, T_A, and B grain. θ_2 or θ_T represents the inclination of the twinning plane. The trace direction of the twinning plane is simply determined relative to any surface direction in the matrix grain. Ω_1, Ω_2 and Ω_{tb} represent the true dihedral angles formed at the intersection designated as I, while Ω_3, Ω_4 and $\Omega_{tb'}$ are the dihedral angles at intersection II. The misorientation across the grain boundary, Θ, is defined as the angle between common crystallographic directions in each grain. In Fig. 2(b), $\Theta = \Theta_{AB} = \Theta_{T_AB}$ is the

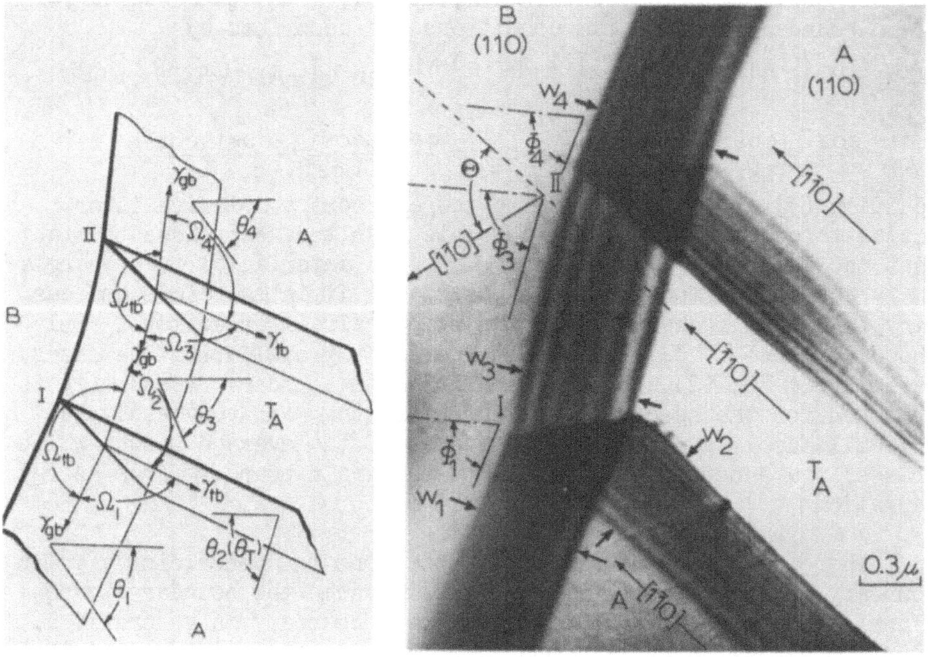

(a) (b)

Fig. 2. (a) Schematic section view of a twin-grain boundary intersection depicting the geometric parameters characterizing the interface. (b) Bright-field electron micrograph showing the crystallographic and geometric parameters which characterize the twin-grain boundary intersection.

angle between the [1$\bar{1}$0] direction in grains A and T_A and the [1$\bar{1}$0] direction in grain B. The angle between the perpendicular bisector of the misorientation angle, Θ , and the boundary trace direction is denoted the asymmetry angle, Φ. The asymmetry angle is a measure of the direction of the boundary in the surface of the system. This directional variable, Φ, is also determined for each grain boundary segment giving Φ_1, Φ_3 and Φ_4. The remaining two degrees of freedom are accounted for by the crystallographic orientations of each crystal portion $(HKL)_A$ and $(HKL)_B$; or more specifically, the deviation of the zone axes $[HKL]_A$ and $[HKL]_B$. W_1, W_3 and W_4 in Fig. 2(b) are the measured projected widths of the grain boundary, and W_2 is the projected width for the twin boundary.

$(HKL)_A$, $(HKL)_B$ and Θ completely define the orientation difference between the two grains, while θ and Φ specify the orientation of the grain boundary plane. This investigation will deal with twin-grain boundary systems that possess only three degrees of freedom; grains A and B will have the same surface crystallography.

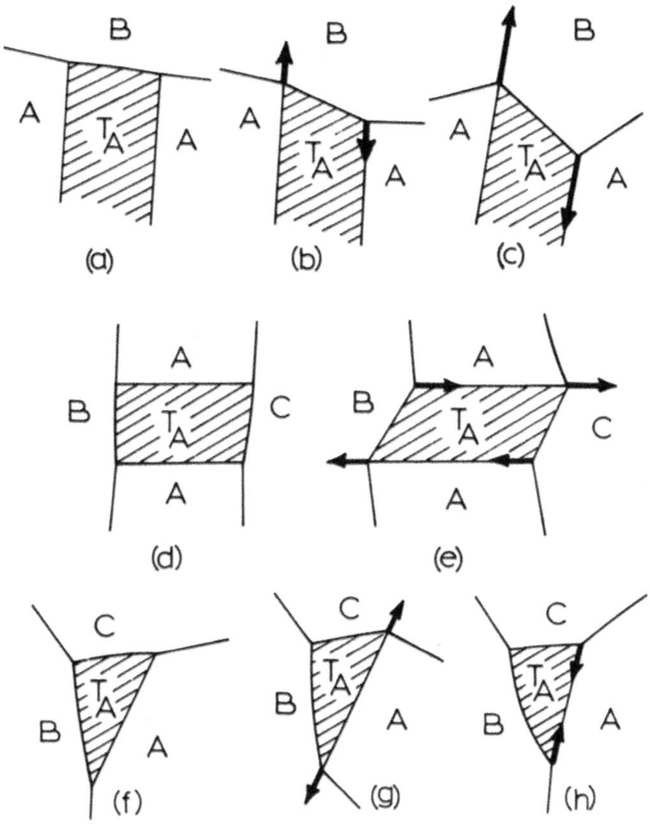

Fig. 3. Equilibrium configurations at twin-grain boundary intersections.

Twin-Grain Boundary Equilibrium Configurations: Theory

For the conventions which have been previously shown in Fig. 1, and by considering Eq. (1), the possible equilibrium configurations which can result at twin-grain boundary intersections can be constructed. These are shown in Fig. 3. Note, Fig. 3 only depicts the two-dimensional surface views. In general, this figure can be modified by considering variations in the grain boundary inclinations. From Figs. 3(a)-(c) can be seen the common types of twin-grain boundary intersections where the twin band intersects a continuous grain boundary. Figures 3(d) and (e) depict double intersections, where a twin band extends between two grains B and C; and Figures 3(f), (g) and (h) show the common idealized configurations of grain-corner twins. The net effective (resultant) torques shown in Fig. 3 occur by resolving the torques of Fig. 1(b) along the twin boundaries. The torque directions result by convention, Eq. (1) and Fig. 1(a); and their magnitudes arise primarily because of a difference in γ_{AB} and γ_{T_AB}. The conditions for creating the difference between γ_{AB} and γ_{T_AB} arise when the parameters upon which the grain boundary energy depends change at the junction. It should now be recalled that the magnitude of the torque depends on the variation of grain boundary energy with respect to a variation in an energy-determining parameter. Energy differences for systems with the same surface crystallography can result because of a variation in relative crystallographic misorientation, Θ , between grains A and B, and T_A and B, or they may arise because of a variation in the grain boundary inclination, θ , or asymmetry, Φ , at the junctions.

In the present investigation the misorientation of <110> directions as depicted in Fig. 4 will be considered. In Fig. 4(a), a rotation of 180° about the twin direction coincident with <110> causes no change in the misorientation, Θ , between grain A and B, and T_A and B. This particular configuration, where $\Theta_{AB} = \Theta_{T_AB}$ will be designated a low-torque configuration. In Fig. 4(b), however, a 180° rotation about the twin direction produces a mirror reflection of the <110> direction. The misorientation of <110> between A and B, and T_A and B is now generally different; that is, $\Theta_{AB} \neq \Theta_{T_AB}$. For $(\gamma_{AB}, \gamma_{T_AB}) = f(\Theta)$, it is now observed that a rather high likelihood exists for the occurrence of large, unbalanced torques (of equal magnitude and opposite sense at each junction) for crystallographies conducive to the situation depicted in Fig. 4(b). This particular situation, where $\Theta_{AB} \neq \Theta_{T_AB}$, will be designated a high-torque configuration. It should be borne in mind, however, that even though a <110> mirror reflection occurs, the corresponding energies may still be approximately equal. This will be particularly true if the <110> direction in grain B approximately coincides with, or is normal to, the twin plane trace direction. This latter feature would produce a situation where $\Theta_{AB} \cong \Theta_{T_AB}$.

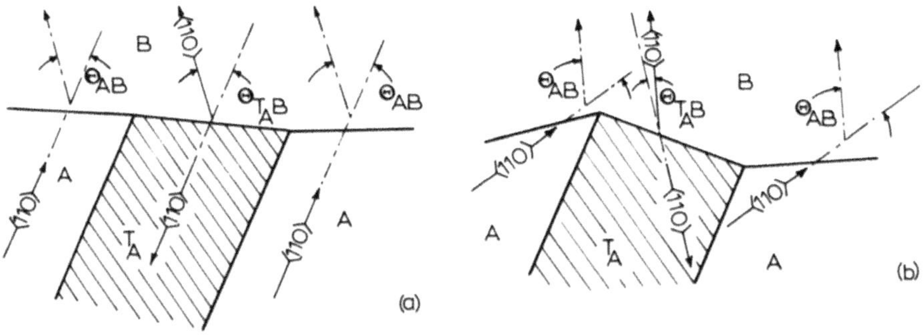

Fig. 4. Crystallographic misorientation at twin-grain boundary intersections. (a) Continuous, (b) discontinuous misorientation.

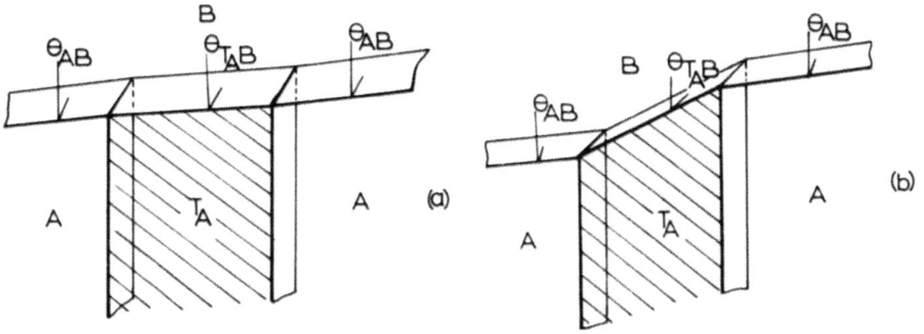

Fig. 5. Schematics for low-torque configurations. (a) Continuous tilt, (b) discontinuous tilt.

There are cases where low-torque configurations display significant torques. Whereas, high-torque configurations in general exhibit large unbalanced torques, low torques do not always result from low-torque configurations. Figures 5 (a) and (b) represent two low-torque configurations; both having a continuous misorientation. The twin-grain boundary intersection represented in Fig. 5(a) has a fairly constant grain boundary inclination, that is $\theta_{AB} \cong \theta_{T_AB}$. In Fig. 5(b), however, the inclination or tilt of the grain boundary is different between grains A and B and T_A and B; that is $\theta_{AB} \neq \theta_{T_AB}$. This variation in inclination across the intersection may be sufficient to produce a high-torque anomaly when considering the conventions of Fig. 4. The accompanying torque is mainly caused by the variation in grain boundary energy with respect to the change in grain boundary inclination. These low-torque situations which have large torque values will be designated anomalous high-torque configurations.

Finally, these same torques arising from a variation in grain boundary inclination can also contribute to the overall resultant torque for high-torque configurations.

EXPERIMENTAL PROCEDURE

Thin electron transparent specimens were prepared from type 304 stainless steel identical in composition to that of previous investigations;[7,8,12] and electrolytically thinned using the technique described by Murr.[13] The stainless steel thin films were examined by conventional three-dimensional transmission image evaluation combined with selected-area electron diffraction microscopy. This three-dimensional method of analysis was carried out using a Hitachi Perkin-Elmer H.U. 125 electron microscope fitted with a goniometer tilt stage, employing double condenser beam illumination, and operated at 125 kV accelerating potential.

Overall transmission images of the twin-grain boundary junctions were obtained at magnifications ranging normally from 9,000 to 25,000 times. Images were only considered if the intersection system was clear for both junctions; the trace directions for the grain boundary segments and the twinning planes had to show sufficient contrast to be unambiguously determined at the intersections. In addition, the transmission image had to be distinct enough to distinguish the sense of the grain boundary inclinations relative to the twin boundary inclination. Selected-area electron diffraction patterns were then taken in an area of the parent grain of the twin band (Grain A of Fig. 2), near one of the triple junctions; and in the adjacent grain (Grain B of Fig. 2) also near the same triple junction. If the relevant selected-area diffraction patterns contained Kikuchi lines, the patterns were accepted only if two or more pairs of low index Kikuchi lines were observed to fall within the vicinity of the corresponding Bragg reflection spot. In actual observations in the electron microscope, the transmission image was first viewed for its contrast clarity at the junctions. Secondly, selected-area electron diffraction patterns were observed to insure that both grains had (110) surface crystallographies. Finally, the specimen was oriented in the tilting stage to a position near exact orientation. An arbitrary criterion of acceptance was stipulated that the selected-area diffraction patterns in the A grain did not deviate by more than 10° in the $[1\bar{1}0]$ direction when compared with the image, and considering the fully calibrated lens rotations between the bright-field image and the selected-area diffraction pattern.

Boundary plane tilts were determined as indicated previously.[11-14] In addition, θ_3 was calculated at both junctions giving θ_{3I} and θ_{3II}. Misorientation, Θ , was defined as the angle between $[1\bar{1}0]$ directions, as measured from the two selected-area diffraction patterns. The asymmetry angle, Φ, was obtained by measuring the angle between the boundary direction as seen in the bright-field image and the superimposed perpendicular bisector of the misorientation angle. In a similar manner as for the boundary inclinations, the asymmetry was determined at both junctions yielding Φ_I, Φ_{3I}, Φ_{3II} and Φ_4. The crystallographic direction of the twin plane, and dislocation plane, for systems with the twin plane perpendicular to the surface, were then identified. This operation allowed the particular (111) twin plane and dislocation plane to be identified, and the angles of tilt with respect to the surface plane to be calculated. Having obtained $\theta_T(\theta_2)$ and the grain boundary tilt angles, the surface trace intersection angles ω_1, ω_2, ω_3 and ω_4 , the projected angles in Fig. 2(b) corresponding to the true dihedral angles represented in Fig. 2(a), were measured directly on photographic enlargements using a finely ruled protractor.

A computer program was used to evaluate the true dihedral angles from measured values of the θ's and ω's; and to automatically compute the interfacial energy ratios, γ_{tb}/γ_{gb}, Eq. (2), and provide C_{AB} and C_{T_AB}, enabling the relative interfacial torque ratios, $\Sigma M/\gamma_{gb}$, to be easily calculated, Eq. (3). In addition, the computer assigned to each relative interfacial energy ratio the label, low- or high-torque. The meaningful energy ratios were those labeled low-torque values, and which were arbitrarily limited to situations where C_{AB} and C_{T_AB} differed by less than 10%. Twin-grain boundary intersection systems having 1% or greater error in the computed dihedral angles at either junction were disregarded as in previous work.[6,8,12]

RESULTS AND DISCUSSION

Computation of the Relative Interfacial Energy Ratios, Torque Ratios and Grain Boundary Parameters

Following the procedures described in the previous section, observations and photomicrographic records were made of coherent twin boundary-grain boundary intersections in 304 stainless steel thin films. The average grain size of the specimens observed ranged from 15 to 20μ in diameter. Approximately 150 acceptable electrothinned foils were prepared which provided 300 nodes;[13] each with an area of about 0.3 mm^2, sufficiently thin to be observed in the electron microscope. One hundred and seventy-five twin systems fulfilled the visual criteria set forth in Section III and were permanently recorded. Of these, a total of 39 twin-grain boundary intersection systems met the geometrical and crystallographic criteria outlined in Section III with overall average errors of 0.2% and 4.7° for the geometric and rotational criteria. Fourteen were low-torque configurations, 15 were high-torque configurations, and 10 were anomalous high-torque configurations.

The low-torque and anomalous high-torque situations were observed to have composition planes meeting the (110) surface at an inclination angle of 35° and a [1$\bar{1}$0] trace direction, while the high-torque configuration twins had an inclination angle of approximately 90° with respect to the (110) surface plane; and either a [1$\bar{1}$2] or [$\bar{1}$12] trace direction. As a consequence of these unique geometrical criteria, the above systems will sometimes also be referred to as 35° low-torque twins, 35° anomalous high-torque twins and 90° twins, respectively. All surviving twin-grain boundary intersection systems were the common cases where a twin band intersects a continuous boundary. However, two double-junction twins, as well as a system with both the matrix and adjacent grain surfaces (100), and a high-torque configuration which was off exact orientation were recorded as test cases to be used as a comparison with the results for the situations described above.

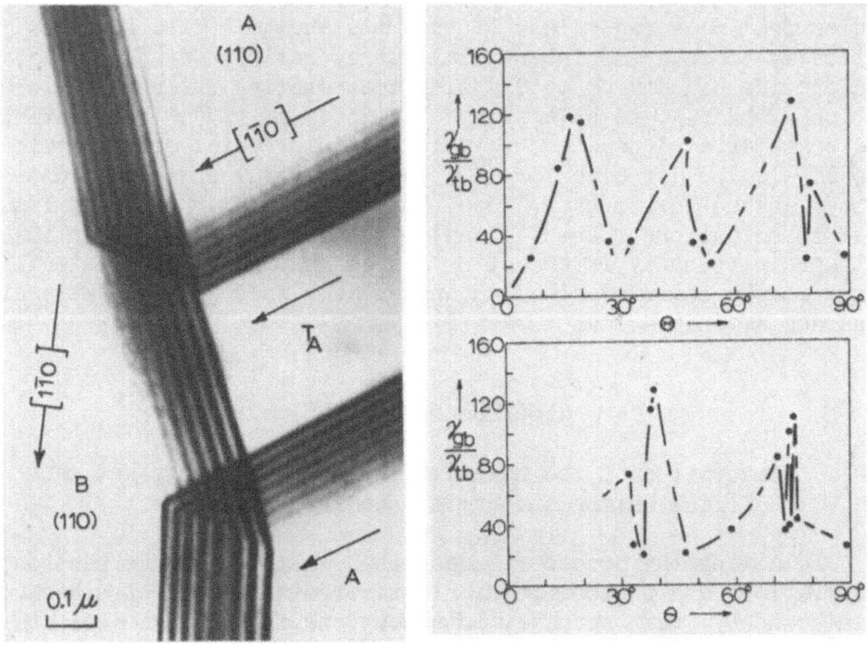

Fig. 6. (a) A twin-grain boundary intersection in a low-torque configuration. (b) Grain boundary energy fluctuations for low-torque configurations.

Low-torque configurations. A typical twin boundary-grain boundary intersection in a low-torque configuration is shown in Fig. 6(a). The extinction fringes were used following dynamical thickness calculations as a check on the thickness obtained from the measurement of the twin boundary projection width coupled with its exact inclination angle of 35.3°. The [1$\bar{1}$0] directions are drawn in for the matrix grain, on both sides of the twinned grain; in the twinned grain and in the adjacent grain. The misorientation between [1$\bar{1}$0] directions is 54°, while the mean values for the grain boundary inclinations and asymmetries are 34.4° and 81.4°, respectively. The error in rotation for [1$\bar{1}$0] was 2°, and the average geometrical error for the sum of the dihedral angles at each junction was 0.2%.

The fourteen low-torque configuration twin systems had an average geometrical error of 0.25% and an average directional-rotational error of 5°. The mean twin boundary-grain boundary energy ratio was computed to be 0.0235 with an average net effective interfacial torque ratio of 0.00125. In addition, the torques were also calculated as a percentage of the twin boundary free energy by dividing the net effective torque ratio by the interfacial energy ratio, resulting in a mean torque value equal to 4.2% of the twin boundary free energy. These torques were the result of average changes in grain boundary inclination and tilt of 2.3° and 5.2° for the 14 low-torque systems.

The inverse twin boundary-grain boundary energy ratio, which is essentially equivalent to a constant times the grain boundary energy was plotted against the corresponding parameters describing the grain boundary. Curves for γ_{gb}/γ_{tb} versus Θ and for γ_{gb}/γ_{tb} versus θ are shown in Fig. 6(b). No discernible relationship was observed to exist between γ_{gb}/γ_{tb} and the grain boundary asymmetry, Φ. In Fig. 6(b), the grain boundary energy definitely shows cusping when plotted against misorientation. Considerably more scatter is seen in the γ_{gb}/γ_{tb} versus θ curve; indicating that θ is not dominant in determining the fluctuations in grain boundary energy. However, it should be noted, because of the nature of the geometrical corrections made from measurements on the bright-field transmission images, individual points corresponding to specific γ_{gb}/γ_{tb} ratios are by themselves not to be considered significant. It is the trends and means of ratios and parameters which are meaningful.

<u>High-torque configurations</u>. Figure 7 shows a representative bright-field electron transmission image of a twin boundary-grain boundary intersection having a high-torque configuration. The constructed dislocation traces, $[1\bar{1}0]$, are shown in both the twinned band and the matrix grain, by dashed lines and labeled (111) to indicate the projection of the slip plane on which the dislocations lie. In some cases dislocation traces were also able to be constructed in the B grain; insuring a more accurate thickness calculation. As in the low-torque case, the same dynamical thickness calculations were made using extinction fringes. The arrows indicate the corrected $[1\bar{1}0]$ directions in the A, T_A and B grains. The misorientation, Θ_{AB}, is 11°, while $\Theta_{T_A B}$ is 60°. Mean values for the grain boundary inclination, θ, and asymmetry, Φ, are 27.9° and 27.2°, respectively. The rotational error for the $[1\bar{1}0]$ direction in grain A and the selected-area diffraction pattern was roughly 0°, with the average geometrical error at each junction computed to be 0.04%. Fifteen twin-grain boundary systems in the high-torque configuration had an average rotational error deviation of 3° and an average geometrical error of 0.16%. The mean values of the energy ratios and torque ratios were 0.05294 for γ_{tb}/γ_{gb}, 0.08806 for $\Sigma M/\gamma_{gb}$, and 84.8 for $\Sigma M/\gamma_{tb}$%; which corresponded to mean variations in the grain boundary parameters of 3.3°, 26.0° and 32.1° for $\overline{\Delta\theta}$, $\overline{\Delta\Phi}$, and $\overline{\Delta\Theta}$, respectively.

<u>Anomalous high-torque configurations</u>. A total of ten twin boundary-grain boundary intersection systems in a low-torque configuration ($\Theta_{AB} = \Theta_{T_A B}$) were observed to possess high-torque values. Figure 8(a) is an electron transmission micrograph showing one of the twin intersection systems in an anomalous high-torque

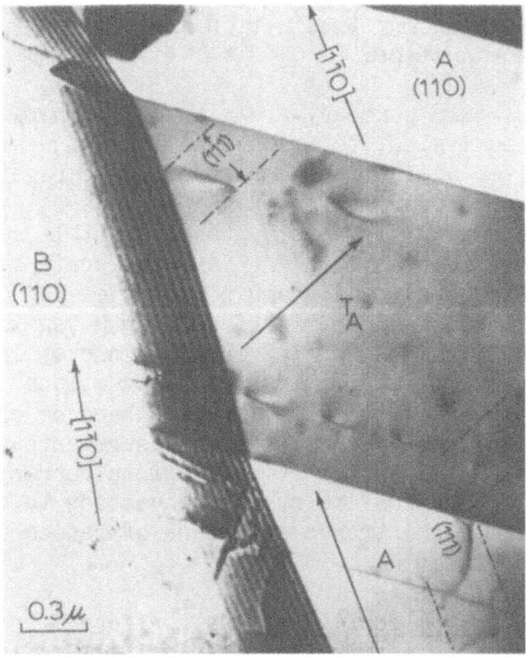

Fig. 7. A twin-grain boundary intersection in a high-torque configuration.

configuration. The twin boundary is assumed to have an inclination angle of 35.3° with the A grain surface and it possesses a [$1\bar{1}0$] trace direction, as is the case for the normal low-torque configuration. Note, however, the extreme changes in grain boundary inclination at the twin intersections. Extinction contours can be used as a thickness check as in the low-torque and high-torque twin systems. The arrows indicate the [$1\bar{1}0$] crystallographic directions. The high-torque arising for this anomalous situation is due to the change in grain boundary inclination across the junctions, $\Delta\theta$, and is exhibited by a large change in asymmetry, $\Delta\Phi$, at the junctions.

The twin system of Fig. 8(a) was observed to have an extremely large net effective interfacial torque equal to 0.29967, and an absolute torque of 152.4% of the twin boundary energy. The average change in grain boundary inclination for the two twin boundary intersections was 29.2°, with a visible change in asymmetry equal to 53.6°. As is the case for general high-torque systems, the twin boundary-grain boundary energy ratio and its inverse have no significance due to the torque terms. The error in [$1\bar{1}0$] rotation was 9.5°, with an average geometric analysis error of 0.18%. The rotation criterion was also applied to the B grain which resulted in a 1° error.

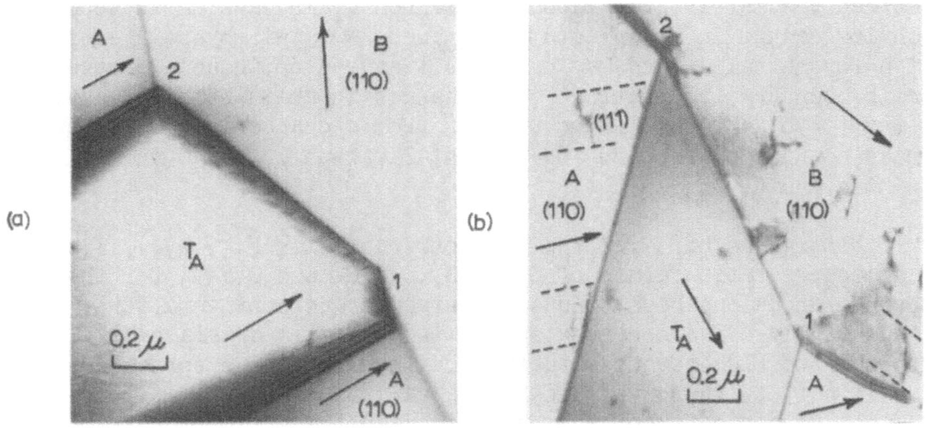

Fig. 8. Bright-field transmission images. (a) Low-torque configuration, anomalous high-torque case. (b) High-torque configuration.

Ten anomalous high-torque twin boundary-grain boundary intersection systems produced mean values of 0.12922 for $\Sigma M/\gamma_{gb}$, and 179.8% for $\Sigma M/\gamma_{tb}$ with an average change in $\Delta\theta$ and $\Delta\Phi$ equal to 8.6° and 9.4°, respectively. The average dihedral angle error and $[1\bar{1}0]$ rotational error were 0.19% and 7°, respectively.

Figure 8(b) represents a high-torque configuration which also has a noticeable change in grain boundary inclination. The results for this example, which is slightly off orientation and therefore not included in the high-torque configuration data, will be discussed in Section 4.3.

Comparison of Low-Torque, High-Torque and Anomalous High-Torque Configurations

The initial comparisons among the three configurations can be made by observing Figs. 6(a), 7, and 8. The low-torque and anomalous high-torque twins have the same crystallography; the twin plane inclinations and trace directions are the same (35° twins with $[1\bar{1}0]$ traces). The difference between these two situations arises due to a change in grain boundary inclination on either side of the twin plane junctions for the anomalous high-torque twins which causes a greater change in grain boundary asymmetry, i.e., the anomalous high-torque configurations exhibit large changes in $\Delta\theta$ and $\Delta\Phi$ relative to the low-torque configurations. Numerically, the anomalous high-torque twin has a large interfacial torque value compared to that of the low-torque twin. The high-torque configuration has its twin plane perpendicular to the film surface, either a $[1\bar{1}2]$ or $[\bar{1}12]$ trace direction, and a change in $[1\bar{1}0]$ direction between the matrix grain and the twinned grain causing an accompanying change in misorientation. These 90° twin systems exhibit high-torque values and may or may not have a noticeable change in grain boundary inclination. The changes in grain boundary asymmetry for high-torque twins results from a change in the absolute grain boundary direction and a crystallographic change in the twinned grain.

Histograms of γ_{tb}/γ_{gb} are constructed in Fig. 9(a), with mean energy ratio values of 0.02350, 0.8564 and 0.5294 for the low-torque, anomalous high-torque and high-torque configurations, respectively. The observed spread in values is considerably greater for the two high-torque situations. The low torque spread could be due in part to the crystallographic variations in γ_{gb} with respect to the degrees of freedom associated with the grain boundary. The additional spread for the anomalous high-torque and high-torque values is due to the actual torque terms. This is another way of saying that energy ratios obtained by assuming the torque to be negligible are not accurate unless there is some assurance that the ratios do not contain significant torque values.

Figures 9(b) and 10(a) show the relative net effective interfacial torque ratio distributions and absolute torque distributions for the three twin intersection configurations. The absolute torque values are given as a percentage of the twin boundary energy, so they may be considered as weighted torque values. The mean values for Fig. 9(b) are 0.00125, 0.12922 and 0.08806 for the low-torque, anomalous high-torque and high-torque situations, respectively. These values correspond to mean absolute torque values of 4.2%, 179.8% and 84.8% of the respective twin boundary energies. It should be cautioned that $\Sigma M/\gamma_{gb}$ and $\Sigma M/\gamma_{tb}$ for the high-torque cases were evaluated by assuming $\gamma_{AB} \cong \gamma_{T_A B}$, which is not as accurate an assumption as for the low-torque case, but does allow for general trends in torque values to be observed.

The average change in grain boundary inclinations at either side of the two twin boundary-grain boundary junctions is given in the histograms of Fig. 10(b). The mean values of the distributions for the low-, high- and anomalous high-torque situations are 2.3°, 3.3° and 8.6°, respectively. The lesser importance of the changes in grain boundary inclination in distinguishing low-torque configurations from high-torque configurations can be clearly seen by comparing their respective histograms. Furthermore, the predominance of the θ contribution to the torque value for the anomalous high-torque configurations relative to the low-torque configurations is also quite easily seen from their respective histograms.

Figure 11(a) reproduces the display of the average variations in the grain boundary asymmetry at each set of junctions for the three configurations, and gives the frequency distribution of the

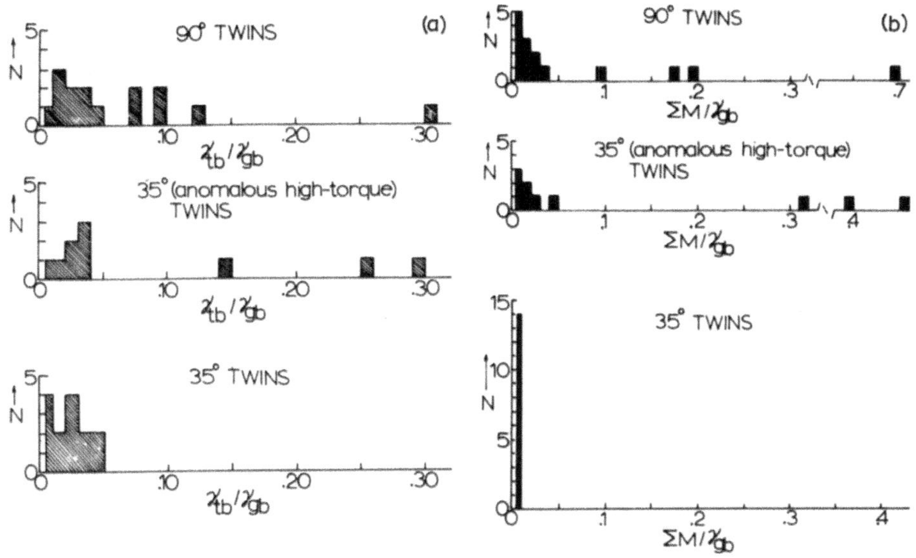

Fig. 9. (a) Distribution of relative interfacial energy ratios. (b) Net effective torque distributions.

changes in misorientation for the high-torque configurations. The mean values for $\Delta\Phi$ are 5.2°, 26.0° and 9.4° for the low-torque, high-torque and anomalous high-torque twins, respectively. The distributions of changes in $\Delta\Phi$ for the high-torque 35° twins and the low-torque 35° twins are seen to parallel their respective distributions in $\Delta\theta$, showing the correspondence between $\Delta\Phi$ and $\Delta\theta$ when the only torque producing variable is $\Delta\theta$.

The changes in misorientation seem to be weighted slightly to the lower half of the scale and there seems to be a correspondence between $\Delta\Theta$ and $\Delta\Phi$ for the 90° twins. This correspondence is probably due to the fact that Φ is defined in terms of Θ and the variations in actual boundary directions are mostly covered up by the larger variations in Θ ; that is, a variation in crystallographic direction is being measured, as opposed to measuring only a variation in absolute grain boundary direction. The mean values of 32.1° and 26.0° for $\Delta\Theta$ and $\Delta\Phi$ lend support to this proposal. In addition, the effect of torques on the boundary can be seen from the values of the variations in the dihedral angles, Ω_{tb} and $\Omega_{tb'}$, at the junctions opposite the twin plane intersections. The mean $\Delta\Omega$ values ($\Delta\Omega = |\Omega_{tb'} - \Omega_{tb}|$) for the low-torque, high-torque and anomalous high-torque configuration twins are 6.1°, 19.9° and 23.8°, respectively.

Figure 11(b) shows the results of plotting the changes in grain boundary inclination against changes in grain boundary asymmetry at the twin junctions. From the data points given for the anomalous

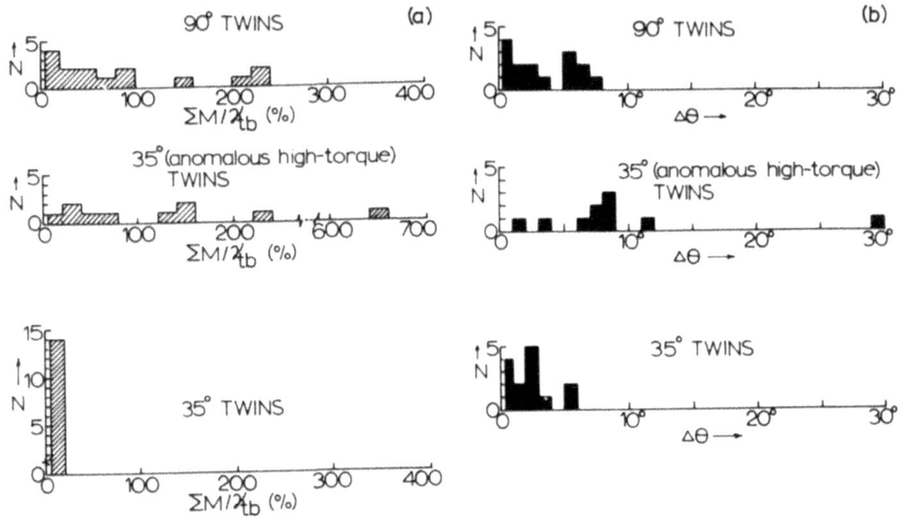

Fig. 10. (a) Spread in absolute torque values. (b) Changes in grain boundary inclinations.

high-torque twins, it would appear that $\Delta\Phi$, which shows the effect of the torque values on the boundary increases with an increasing $\Delta\theta$, which produces the torque values. This is reasonable, in light of the fact that the anomalous high-torque configuration has no $\Delta\Theta$ torque contribution and changes in $\Delta\theta$ would tend to cause changes in $\Delta\Phi$. For the high-torque situation, the curve shows that a decreasing $\Delta\theta$ will result in a higher $\Delta\Phi$ for $\Delta\Phi > 40^{\circ}$. The changes in $\Delta\Phi$ for the high-torque configuration, however, are also affected by changes in $\Delta\Theta$, and this influences the relationship between $\Delta\theta$ and $\Delta\Phi$.

Summary of Results and Configurational Test Cases

The qualitative summary of results for the three sets of twin boundary-grain boundary intersection configurations is best reviewed by looking back at Figs. 6(a), 7 and 8. The [1$\bar{1}$0] directions, represented by arrows (used for determining misorientations, constructing dislocation plane traces, and as error checks), were obtained from selected-area electron diffraction patterns. All surfaces were (110), also determined by selected-area electron diffraction patterns. Film thicknesses were obtained from 35° twin plane projections, or from constructed dislocation or slip plane projections, and checked by using the extinction fringes visible in the boundaries. Grain boundary inclinations were clearly seen from the bright-field electron transmission micrographs, as were the changes in grain boundary direction and tilt at the twin junctions. The basic differences between the 90° high-torque and 35° low-torque

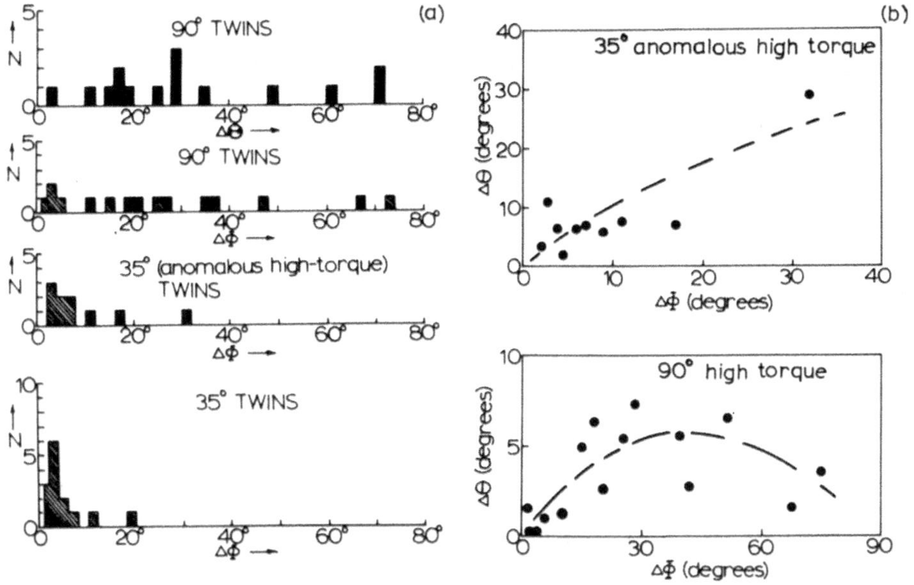

Fig. 11. (a) Changes in grain boundary asymmetry and misorientation. (b) Variation in grain boundary inclination as a function of asymmetry changes.

configurations were seen to result from an intrinsic change in misorientation at the junctions. Low-torque 35° twins and anomalous high-torque 35° twins were observed to differ mainly in the extent of the grain boundary inclination changes at the junctions. Finally, the effects of the torques pulling on the grain boundary at the junctions, as depicted theoretically in Fig. 3, could easily be differentiated between the high-torque and low-torque situations. The quantitative mean values for the three twin boundary-grain boundary intersection configurations are summarized in Table I.

Table I. Summary of mean values

Mean values	Low-torque	Anomalous High-torque	High-torque
γ_{tb}/γ_{gb}	0.02350	0.08564	0.05294
$\Sigma M/\gamma_{gb}$	0.00125	0.12922	0.08806
$\Sigma M/\gamma_{tb}$%	4.2	179.8	84.8
$\Delta\theta°$	2.3	8.6	3.3
$\Delta\Phi°$	5.2	9.4	26.0
$\Delta\Theta°$	---	---	32.1
$\Delta\Omega°$	6.1	23.8	19.9

Several twin systems which did not meet all the criteria for acceptance were selected at random to be used as test cases. They were compared with the general qualitative and quantitative results obtained from the more precisely analyzed systems in order to determine whether the results of the three configurations described above could be applied to general cases.

Figure 12 shows two examples of double-junction twins, originally postulated in Figs. 3(d) and (e). As mentioned previously, none of the surviving twin systems was a double-junction twin. However, merely a visual inspection of the micrographs in Figs. 12(a) and (b) will confirm that double-junction twins behave in a similar fashion as normal twin systems. Based solely on the intrinsic misorientation concept, the twin intersection system in Fig. 12(b) should be in a high-torque configuration. Also, from the $\Delta\theta$ concept, Fig. 12 (a) would not be an anomalous high-torque case. Moreover, visual inspection of the changes in grain boundary direction shows the twin system in Fig. 12(a) to be in a low-torque situation and the twin system in Fig. 12(b) to be in a high-torque situation. Also, the assumed balance of torques represented by arrows in Fig. 3(e) can be seen to exist in an actual case as evidenced by the pulling of the boundary junctions in Fig. 12 (b).

Another case investigated involved a twin boundary-grain boundary intersection system for which both surface crystallographies were (100). The transmission electron micrograph with superposed selected-area electron diffraction patterns, corrected for absolute rotation, is shown in Fig. 13. The twin plane has an inclination of 55° with the (100) surface and a trace direction of $[0\bar{1}1]$. Thus, Fig. 13 should represent a basic low-torque configuration, insofar as $\Delta\Theta = 0^{o}$. However, a distinct change in grain boundary inclination can be observed at the twin plane junctions; predicting an anomalous high-torque configuration. After analyzing the bright-field image and selected-area electron diffraction patterns, the crystallographic and geometric data for this twin intersection were evaluated in the same manner as all previous twin boundary-grain boundary intersection systems, and yielded the following results: $\gamma_{tb}/\gamma_{gb} = 0.18862$, $\Sigma M/\gamma_{gb} = 0.15156$, $\Sigma M/\gamma_{tb}\% = 80.4$, $\Delta\theta^{o} = 10.4$, $\Delta\Phi^{o} = 9.2$, and $\Delta\Omega^{o} = 27.8$, with an average dihedral angle error of 0.1% and a $[1\bar{1}0]$ rotational error of roughly 0°. The values for the relative net effective torque, $\Delta\theta$, $\Delta\Phi$ and $\Delta\Omega$ agree remarkably well with the mean values in Table 1 for the anomalous high-torque configurations. What is even more significant is the fact that the computed analysis is in complete agreement with the preliminary visual observations.

A high-torque configuration twin system which was not near exact orientation, as previously shown in Fig. 8(b), was used as a final test case to see if a basic high-torque configuration would

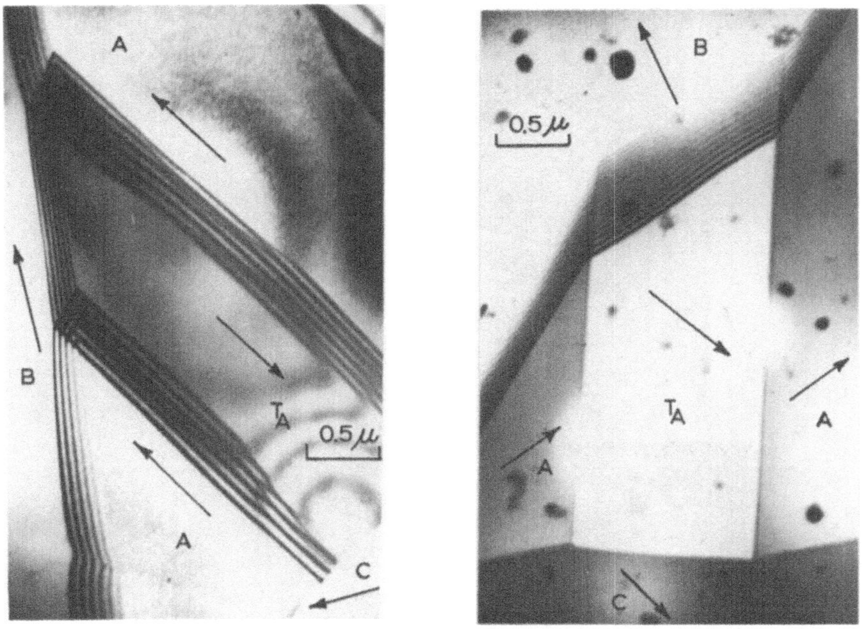

Fig. 12. Double-junction twins. (a) Low-torque configuration. (b) High-torque configuration. Arrows indicate ⟨110⟩ directions.

still exhibit calculable high-torque characteristics even though the system was not as accurately defined as the cases analyzed in the main of this investigation. The inclination of the twinning plane with the surface as determined from the micrograph was 84.3°, as opposed to a 90° inclination with the crystallographic (110) surface, and the error in $[1\bar{1}0]$ rotation was 15°. Nevertheless, both surfaces index as (110) from their respective selected-area electron diffraction patterns. After normal analysis and computation, the final results for the twin-grain intersection system shown in Fig. 8(b) were: $\gamma_{tb}/\gamma_{gb} = 0.12563$, $\Sigma M/\gamma_{gb} = 0.09726$, $\Sigma M/\gamma_{tb}\ \% = 77.4$, $\Delta\theta^{\circ} = 6.8$, $\Delta\Phi^{\circ} = 37.3$, $\Delta\Omega^{\circ} = 54.8$, and $\Delta\Theta^{\circ} = 25$, with an average dihedral angle error of 0.2%. The results compare very favorably with the values listed in Table 1, and the consistency of the theory, even for slightly off-orientation systems, is clearly evident.

In this investigation, the overall spread of the total γ_{tb}/γ_{gb} energy ratio data from the sum of the low-torque, high-torque, and anomalous high-torque results was 0.30. This spread is reduced by almost a factor of 10 to 0.035 by considering only the low-torque data. The mean torque values for the three configurations range from 0.1% to 12.9% of the average grain boundary energy with an individual range from 0.002% to 69% of the average

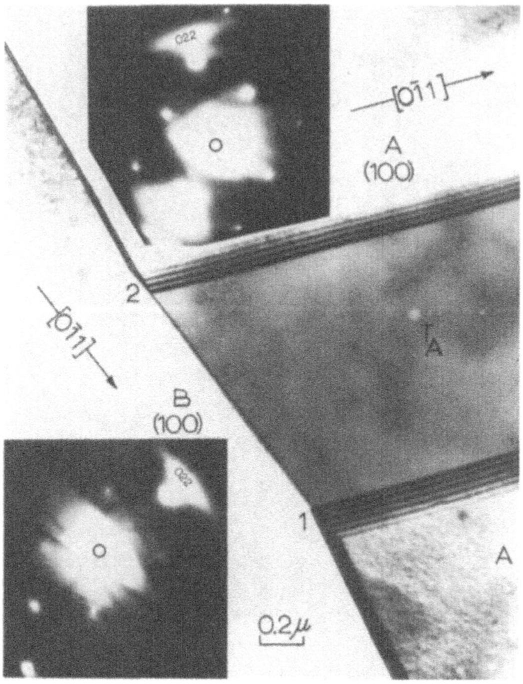

Fig. 13. Low-torque configuration for (100) surfaces.

grain boundary energy.

More specifically, it has been shown that the major contribution to the net effective torque is the result of a variation in misorientation, which can be designated as $\tau(\Theta)$. Another torque contribution arises because of changes in grain boundary inclination and can be represented by $\tau(\theta)$. In general, there should also be a third torque component for a three-degrees of freedom system which results from changes in asymmetry $\tau(\Phi)$. Both $\tau(\Theta)$ (high-torque configuration) and $\tau(\theta)$ (anomalous high-torque configuration) have been shown to exert a moment on the boundary tending to pull it into a lower energy configuration, as originally postulated by Herring.[2] The effect on the boundary can be observed both crystallography and geometrically. Moreover, for high-torque configurations the resultant torque can be made up of $\tau(\Theta)$, $\tau(\theta)$ and $\tau(\Phi)$, since in general $\Sigma M = f(\Theta, \theta, \Phi) = f[\tau(\Theta), \tau(\theta), \tau(\Phi)]$. Therefore, the configuration for a twin boundary-grain boundary intersection will reflect the resultant torque associated with it.

Because it has been established from the high-torque results that intrinsic configurational torques arise at grain boundary-twin boundary intersections as a result of variations in the misorientation of the neighbor grains, Θ_{AB}, and the misorientation across the twin grain and the neighbor grain, Θ_{T_AB}, a significant variation of grain boundary free energy with misorientation is inevitable. This variation of grain boundary energy with respect to misorientation was further confirmed in Fig. 6(b). It was stated in the low-torque results that γ_{gb}/γ_{tb} was actually a constant times the grain boundary energy. The assumption that γ_{tb} is a constant is based on the fact that a twin boundary is crystallographically defined, that all measurements were carried out at a constant temperature of 1060°C, and the additional fact that no precipitation was observed to occur at the twin boundaries. Moreover, for the low-torque configurations, all twin composition planes are (111) or ($\bar{1}\bar{1}1$) planes, having a fixed inclination angle of 35.3° to the (110) crystallographic surfaces; and they all have [$1\bar{1}0$] surface trace directions. Therefore, the inverse twin boundary-grain boundary energy ratio is in fact equal to $(1/\gamma_{tb})\gamma_{gb}$, a constant times the grain boundary free energy. This variation of γ_{gb} with misorientation in turn accounts for part of the residual spread in the distribution of relative energy ratios for the low-torque configurations.

At this point it is worth reiterating the fact that only twin boundary-grain boundary energy ratios for low-torque configuration twin systems should be considered in any energy analysis. Equation (2), which is used to determine the twin boundary-grain boundary energy ratios was developed by assuming $\gamma_{AB} \cong \gamma_{T_AB}$, and by assuming the torque terms to be negligible. Thus, the twin boundary-grain boundary energy ratios are inherently inaccurate when obtained for any system containing appreciable torque terms. This inaccuracy is reflected in the scatter of γ_{tb}/γ_{gb} data for the high-torque and anomalous high-torque configurations in Fig. 9(a). Therefore, in any energy study for fcc metals based on the analysis of twin boundary-grain boundary intersection systems, a preliminary visual examination of the transmission image should be performed in order to make a determination based on the configurational theory developed herein, as to which intersection systems are in high-torque configurations and should be discarded, and which systems are in low-torque configurations and should be further analyzed. Herring's equation in approximate form, Eq. (2), may then be used for these low-torque systems.

Finally, it may be stated that even though the nature of this investigation is statistical in conception and execution, the results show very convincingly that where a twin-grain boundary intersection system can be characteristically described in terms of structure and crystallography its' energy and torque values are distinct.

CONCLUSIONS

1. For the first time, the origin and existence of effective torques associated with configurational equilibria at twin boundary-grain boundary intersections have been theoretically predicted based upon crystallographic conventions; and verified experimentally by direct observations and measurements.

2. High-torque configurations are defined by an intrinsic change in misorientation across the twin plane, while low-torque configurations exhibit no intrinsic variation in misorientation. Anomalous high-torque configurations are basic low-torque configurations displaying high-torque values as a result of changes in grain boundary inclination at the twin plane intersections.

3. Herring's general equation for interfacial equilibrium has been experimentally verified by direct observation for the first time in a metallic solid; and it has been shown that the associated torque terms can be neglected only for low-torque situations.

4. More accurate estimates of the twin boundary-grain boundary energy ratio, γ_{tb}/γ_{gb}, may be obtained by systematically excluding high-torque configurations.

5. The optimum-mean value of γ_{tb}/γ_{gb} in 304 stainless steel has been computed from low-torque configuration systems to be 0.0235.

6. It has been shown that a residual spread in γ_{tb}/γ_{gb} values exists for low-torque configuration twin-grain boundary intersections. This statistical variation of γ_{tb}/γ_{gb} must reflect the variation in γ_{gb} with changes in grain boundary parameters.

7. Grain boundary energy versus grain boundary misorientation curves show definite cusping and reflect the predominance of misorientation in determining grain boundary free energy variations.

8. A functional relationship is seen to exist between $\Delta\theta$ and $\Delta\Phi$ for both high-torque occurrences.

9. The combination of bright-field transmission electron microscopy and selected-area electron diffraction microscopy can be used to completely characterize grain boundaries in thin metal films.

10. Net effective torques, and absolute torques associated with configurational equilibria at twin boundary-grain boundary intersections can be calculated for low-torque configurations and approximately calculated for high-torque configurations.

ACKNOWLEDGMENTS

This work was supported by the Metallurgy Branch of the Office of Naval Research under Contract N00014-67-A-0269-0010, NR 031-735.

REFERENCES

1. C.S. Smith, Trans. AIME, 175, 15, (1948).

2. C. Herring, in The Physics of Powder Metallurgy, ed. W.E. Kingston, McGraw-Hill Book Co., New York, 1951.

3. R.L. Fullman, J. Appl. Phys., 22, 448, (1951).

4. M.C. Inman and A.R. Khan, Phil. Mag., 6, 937, (1961).

5. M.C. Inman and H.R. Tipler, Met. Rev., 8, 105, (1963).

6. L.E. Murr, Acta. Met., 16, 1127, (1968).

7. L.E. Murr, J. Appl. Phys., 39, 5557, (1968).

8. L.E. Murr, R.J. Horylev and W.N. Lin, Phil. Mag., 22, 515, (1970).

9. H. Brooks, in Metal Interfaces, ASM, Cleveland, Ohio, 1952, p. 20.

10. W.T. Read, Dislocations in Crystals, McGraw-Hill Book Co., New York, 1953.

11. L.E. Murr, Phys. Stat. Sol., 19, 7, (1967).

12. L.E. Murr, R.J. Horylev and W.N. Lin, Phil. Mag., 20, 1245, (1969).

13. L.E. Murr, Electron Optical Applications in Materials Science, McGraw-Hill Book Co., New York, 1970.

14. R.J. Horylev and L.E. Murr, in Proc. Electron Microscopy Soc., ed. C.J. Arceneaux, Claitor's Publishing Division, Baton Rouge, 1970, p. 436.

GRAIN BOUNDARY CURVATURES IN ANNEALED BETA BRASS

John P. Nielsen* and Louis P. Stone**

*New York University, New York, New York

** Anaconda American Brass, Waturbury, Connecticut

The polycrystalline structure experiences two mechanisms of grain boundary area reduction in the process of grain growth on annealing. One is the readjustment of junction angles on the occurrence of a grain encounter, i.e. the first time "meeting" of two grains. As soon as the threshold of a grain encounter is crossed a rather sharp but localized drop in boundary area ensues. This correction is relatively rapid at first, gradually slowing down as the equilibrium angles for the local junctions are approached. The various junction angles do not necessarily satisfy the angles required by the faces of various polyhedra involved and this discrepancy is satisfied energetically by producing a net spherical curvature of the boundaries. The actual curvature may not always be spherical because there are other geometrical requirements when the polyhedra are not regular or not symmetrical. But the net curvature tends toward spherical curvature because this yields the minimum surface for a given volume, which is dictated by the tendency toward minimization of boundary free energy. The second mechanism takes place as the grains with a net convex curvature diminish in size, the net diffusion favoring atomic escape from such grains. This is a relatively slow process in general but a necessary one, since it leads to further encounters, which repeat the cycle just described. The process may be somewhat rapid for those grains that are approaching extinction.

This paper describes the measurement of the boundary curvatures that are present in a series of beta brass specimens subjected to grain growth, and gives an analysis of the results obtained. Beta brass was chosen primarily because annealing twins appeared to be a minor part of the structure, and because the high diffusion rate of BCC structures yields fairly rapid grain growth even for large grain size structures.

MATERIALS AND PROCEDURES.

Beta brass ingots of 52.9 atomic percent zinc, using starting materials of 99.999% purity, were cast after induction melting in a tightly capped graphite crucible and homogenized at above 750°C for at least 40 hours. The ingots were rolled at 200°C and then swaged to 18mm round bars. A recrystallized structure was obtained on annealing at 466°C for one hour. The rods were cut into discs which were then rolled down to a 50% reduction and again annealed for one hour. The resulting initial recrystallized structure was approximately 0.10mm, but clearly nonuniform in grain size distribution. A uniform distribution was apparently obtained throughout the specimen upon annealing to a grain size of about 0.25mm.

These specimens were then subjected to grain growth anneals in a 50% sodium nitrite - 50% potassium nitrate salt bath, except for the 700°C anneal, which was done in a chloride bath. A series of temperatures from 395°C to 700°C were used with annealing times ranging from less than a minute to 250 hours. Grain size vs time for different temperatures were plotted on log-log graphs. The slopes of the lines were quite constant at 0.31 ± 0.03, yielding an $\underline{n}$ value of 0.31 for the grain size-time relationship equation $\bar{d} = kt^n$, $\bar{d}$ being the average grain size and t the time of anneal at a given temperature, and k a constant.

Eight specimens were selected from these grain growth studies for grain boundary curvature measurements, as listed in Table I.

The specimens were etched to bring out the structure with dilute chromic acid spiked with hydrochloric acid. The grain size measurements were made by the line intercept method.

CURVATURE MEASUREMENTS.

Fine line pencil tracings were made of the grain boundaries from photomicrographs enlarged 30 times for the fine grain structure, and 4.75X for the coarse grain structures. These magnifications brought the fine and coarse grain sizes about equal in size for the tracings. The grains on the tracings were numbered.

Starting with grain No. 1 on the tracing sheet, the length of a boundary in mm was measured with a steel ruler and the curvature was matched to a series of graded curvatures on a transparent celluloid strip. The curvature was called positive if the radius of curvature was directed toward the grain interior, and

TABLE I

Annealing Schedules and Resulting Grain Size

Specimen	Temp. (°C)	Time (Min.)	$\bar{d}$ (mm.)
A_o	395	133	0.39
B_o	447	5	0.30
D_o	497	1	0.36
C_o	700	34	0.89
		(weighted average)	(0.368)
A_1	395	16,785	1.53
B_1	447	3,125	1.84
D_1	497	3,125	2.45
C_1	700	24	2.70
		(weighted average)	(2.054)

negative for the reverse case. The next boundary was likewise measured and so on for all the boundaries of the grain. The process was repeated for grain No. 2 and so on. Only whole grains in the microstructure were considered. It became obvious almost immediately that there were numerous arcs with very large radii, more than a meter, the limit of the arc scale used. Also, the precision for the large radii was rather low. However, inasmuch as the objective was to obtain curvature data reflecting driving force for migration, it was decided that the low precision for the long radii mattered little as such boundaries had negligible migration rates. In other words the driving force for migration is proportional to the curvature (the reciprocal of the radius), and for radii more than 100 times the average, the error in driving force estimation is necessarily less than 1% for such a boundary. A similar argument applies to very short boundaries. Thus the reliability of the mean curvature measured was acceptable.

Incidentally throughout this paper the migration of atoms and boundaries in the metallographic plane refers to the migration component in the plane.

The next significant observation made in the course of these measurements was the variety of boundary types which could not be classified as simple circular arcs, although a large fraction, approaching half, could safely be called circular arcs. These different boundaries were labeled and characterized as follows:

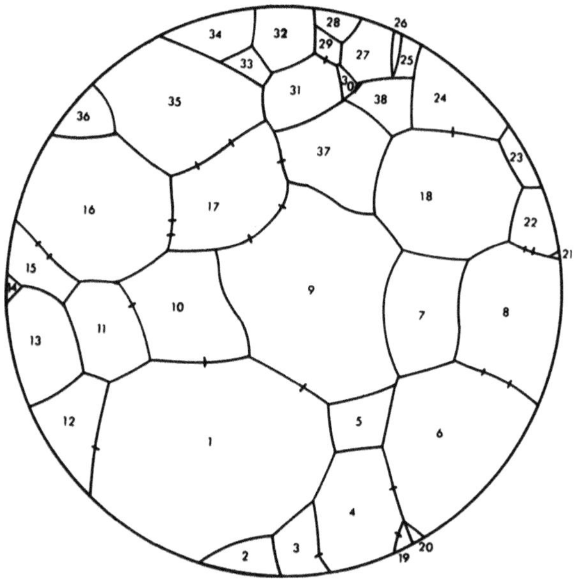

Figure 1. Boundary Tracing for Specimen D_1, 7.5X.

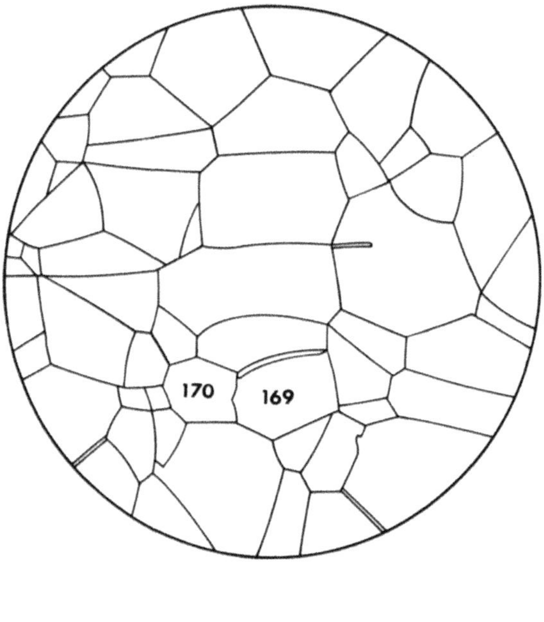

Figure 2. Boundary tracing for Specimen A_o, 20X.

S - Straight boundaries but with a barely perceptible although unmeasurable curvature. See boundary 5-6 in Figure 1. Many short boundaries were thus classified.

E - Straight edge boundaries suggesting a twin-like boundary, especially when angles at the extremities appear to approach 180°. (See boundary 32-33 in Figure 1) The eye is a good judge of such boundaries when looking at the microstructure plane at a small angle.

M - Multiarced boundaries. See boundaries 9-37, 9-17 and 9-10 in Figure 1.

M-E - Multiarced with one section straight-edged. See boundaries 1-9 and 3-4 in Figure 1.

D - Discontinuous in slope. See boundary 169-170 in Figure 2.

RESULTS OF MEASUREMENTS.

Table II gives the count and percentages of occurrence of the boundary characterizations.

TABLE II

Grain Boundary Characterizations

Spec.	Grns.	Bdies.	$\bar{d}$ (mm)	S %	E %	M %	M-E %	D %
A_o	185	1129	0.39	24.5	14.0	5.3	1.1	1.8
B_o	25	148	0.29	33.8	2.7	11.5	1.4	3.7
D_o	49	271	0.27	36.2	6.3	7.5	1.8	2.2
C_o	50	304	0.41	39.4	2.6	8.9	1.3	1.0
Averages -		(1852)	0.368	29.4	10.1	6.7	1.3	1.9
A_1	50	280	1.53	23.2	6.8	29.6	1.4	2.9
B_1	25	139	1.58	18.0	7.9	30.2	1.4	0
D_1	25	139	2.67	16.5	12.9	18.7	7.2	1.4
C_1	18	94	2.96	12.8	23.4	19.1	8.5	0
Averages -		(652)	2.054	19.2	10.7	25.9	3.7	1.5

(Averages are weighted by corresponding number of boundaries. Numbers in parentheses are totals.)

In Table II the abbreviations Grns. and Bdies are for grains and boundaries. Linear measurements are millimeters. The percentages refer to total boundaries.

TABLE III

Angles Subtended by Circular Arc Boundaries and by Net Curvature of Positive and Negative Grains

	$\bar{d}$ (mm.)	$\overline{\theta\lambda}$ (rns.)	$\bar{\lambda}$ (mm.)	$\bar{\theta}p$ (rns.)	Np	$\overline{\Sigma\lambda p}$ (mm)	$\bar{\theta}n$ (rns.)	Nn	$\overline{\Sigma\lambda n}$ (mm.)
A_o	0.39	0.10	.36	.40	103		-.35	76	
B_o	0.29	0.10	.28	.39	10		-.80	14	
D_o	0.27	0.17	.24	.78	24		-.63	24	
C_o	0.41	0.15	.26	.87	29		-.44	17	
Ave.		0.117	.320	.534	(166)	.294 x5.4	-.462	(131)	.366 x6.3
A_1	1.53	.13	1.45	.35	24		-.47	18	
B_1	1.88	.13	1.31	.54	16		-.32	9	
D_1	2.67	.09	2.00	.31	11		-.27	12	
C_1	3.78	.11	2.96	.39	4		-.38	11	
Ave.		.115	1.755	.399	(55)	1.82 x5.5	-.373	(50)	1.87 x6.6

(θ, λ, p, and n represent circular arc, boundary length, and positive and negative grains, respectively. Σ refers to summation over the boundaries of a grain, where the multiplicand is the mean boundary length and the multiplier is the mean number of boundaries. The angles are in radians, (rns.). The whole numbers in parenthesis are totals.)

In Table III under column $\overline{\theta\lambda}$ is given the mean angles subtended by circular arc boundaries, where $\theta\lambda$ represents the circular arc of a boundary of length $\bar{\lambda}$. An angle subtended required two measurements on a circular arc boundary, the radius of curvature and the length. The reciprocal of the radius times the arc length gave the subtended angle in radians. For the net angle subtended by all the boundaries of a grain, $\theta\lambda$ values were added algebraically, and these are found under columns θp and θn. Column $\bar{\lambda}$ gives the mean length of the boundaries. The mean of the summations, $\overline{\Sigma\lambda p}$ and $\overline{\Sigma\lambda n}$, is for the average boundary length for all positive grains multiplied by the mean number of boundaries or sides for a grain. Note that this product is shown explicitly in the $\overline{\Sigma\lambda}$ columns. Columns Np and Nn are the counts of positive and negative grains, respectively.

ANALYSIS OF DATA.

Referring to Table II the outstanding characterization to be noted distinguishing the large grain structure from the small grain one is the preponderance of multiple-arced boundaries. Indeed in looking at the microstructures with this new "insight"

TABLE IV

Mean Radii of Curvature for Boundaries and for Positive and Negative Grains

	$\bar{d}$ (mm.)	$\bar{r}$ (mm.)	$\bar{r}$±St.d. (x $\bar{d}$)	$\bar{r}p$ (x $\bar{d}$)	$\bar{r}n$ (x $\bar{d}$)	$\bar{R}p$ (x $\bar{d}$)	$\bar{R}n$ (x $\bar{d}$)
A_o	0.39	3.60	9.2				
B_o	0.29	2.80	9.7				
D_o	0.27	1.41	5.2				
C_o	0.41	1.73	4.2				
		2.735	7.4 ±2.1	8.1	13.6	10.4	17.1
A_1	1.53	11.15	7.3				
B_1	1.88	10.08	5.4				
D_1	2.67	22.22	8.3				
C_1	3.68	26.91	7.1				
		15.27	7.4 ±1.0	12.2	16.1	15.6	20.6

(R and r represent radii of spherical surfaces in space and of circular arcs on the metallographic plane. The last five columns give the value of the radii in units of $\bar{d}$, the mean intercept.)

one finds that it is hard to find single arc boundaries in the coarse grain structure as against in the fine grain ones. However, in the fine grain specimens one has to hunt to find a boundary with the multiple curvature. (There are enough multi-arced boundaries in the coarse structure with an inflection point to contribute to an error in the intercept method for grain size.) Likewise, although not so abundant, is the marked contrast in the number of curved and straight edge combinations in a boundary. There appear to be more flat boundaries, i.e. the S boundaries, (which may or may not be slightly curved) in the fine grain specimens than in the coarse grain ones. This may not be significant in that the fine grain structure appears to have a wider grain size dispersion, and the relatively numerous little grain had boundaries which were too small to exhibit a measurable curvature. There appear to be two trends. As the annealing temperature increases for the fine grain structures, the percentage of flat boundaries tends to increase. The trend is reversed in the large grain structures.

The data of Table III permit the calculation of the mean radii of curvature. This is obtained simply from the definition of curvature, $K = \theta/L$, where θ is the angle subtended by the circular arc length, L. The reciprocal gives these radii in the L units of length. In Table IV these mean radii lengths are given also relative to the mean grain size, $\bar{d}$. The mean radii for the positive and negative grains were calculated from the mean (net) radians listed for such grains in Table III.

As an indication of precision, the standard deviation is given for the two values of 7.4 radians, which appear to indicate that the mean radius for small grains in terms of average grain size is the same for both the fine and coarse grain structures. This may be so, but the precision is not good enough to make this assertion.

All curvatures thus far measured and calculated refer to curvatures in the metallographic plane of the specimens used. Presumably the preferred orientation in these specimens was small so that the circular arcs measured are for random cuts in space for spherical surfaces of the grain boundaries. These circular arcs are then a series of small circle arcs of various spherical surface sections. The mean radius of curvature for all the spherical surfaces in the structure can be obtained by multiplying the mean radius of the measured circular arcs by $4/\pi$[1]. These values in units of $\bar{d}$ are given in columns $\bar{R}p$ and $\bar{R}n$ in Table IV.

By extending the analysis to the grains themselves rather than to the boundaries we obtain the mean curvature per grain. In a fair number of cases a grain has both negative and positive sides, so that part of the curvature effect is cancelled out, i.e. some sides have atoms diffusing into the grain and some to the outside. Such a grain grows only to the extent of the net flow of diffusion. Therefore grains in general have a larger mean radius per grain than the mean radius per boundary. Table IV shows this where it is to be noted that the mean boundary radius for the small grain structure increased from 7.4 to $8.1\bar{d}$ for the positive grains, and to $13.6\bar{d}$ for the negative grains. The large grain structures revealed a much larger rate of increase, indicating that the grains are losing their growth capacity not only by the decrease in curvature due to size magnification, but also due to the emergence of a greater mixture of positive and negative sides per grain. In effect, as growth progresses, the distinction between a positive and a negative grain becomes less definite.

The mean radius of curvature per grain in the metallographic plane is on the average shorter than that of the grain in three dimensions because grains are in general cut by the plane a la truncating rather than through the grain centroid, thus giving the small circle effect as in the case of the single boundary cuts. Using the same factor, $4/\pi$, we obtain R values, i.e. mean radius of curvature of the spherical surfaces which comprise the surface forming a grain. (One must be careful in distinguishing the various spheres one might use to represent a grain. There is the equal volume sphere, the mean sphere of the distribution of spheres representing the individual net grain curvature, and the sphere having a mean radius of curvature on averaging over the representative radii of the net grain boundary curvature, and still others.)

Incidentally, the positive grains in the microstructure view come from positive grains in the bulk polycrystalline structure, and vice versa. This comes from the fact that all cuts of a positive curved surface polyhedron give positive curvilinear polygons, and vice versa. This is an important insight in that the apparent "small" grains in a metallographic view when yielding a net negative curvature come from a negative space grain, and in addition probably from a relatively large grain.

A further step can be taken in the analysis with respect to the question of total angle of arc subtended for an average grain. Let us assume that the metallographic tracings of Figures 1 and 2 are for two-dimensional structures, i.e. assume that the grains are prisms with the boundary tracings perpendicular to the sheet. In this case, every grain is a fraction of a circular grain, the fraction being the sum of the circular arcs subtended. Furthermore, every such grain is a definite circular fraction corresponding to the number of sides as follows:

TABLE V

Circular Fractions of (Regular) Curvilinear Polygons

Sides	$\Sigma\theta$ radians	C. F.	r / L Units	r / $\bar{d}_x$ Units
1 (Circle)	$6\pi/3$	1	.50	$.64\bar{d}_1$
2 (Lens)	$4\pi/3$	2/3	.56	$1.03\bar{d}_2$
3	$3\pi/3$	1/2	1.0	$1.42\bar{d}_3$
4	$2\pi/3$	1/3	1.93	$2.23\bar{d}_4$
5	$\pi/3$	1/6	4.78	$4.38\bar{d}_5$
6	$0\pi/3$	0	∞	$\infty\ \bar{d}_6$
7	$-\pi/3$	-1/6	6.72	$3.90\bar{d}_7$
8	$-2\pi/3$	-1/3	3.83	$2.00\bar{d}_8$

(C.F., the circular fraction, r the radius of curvature in both L, the chord length of a side, or in $\bar{d}_x$, the mean intercept for random lines cutting the x-sided polygon.)

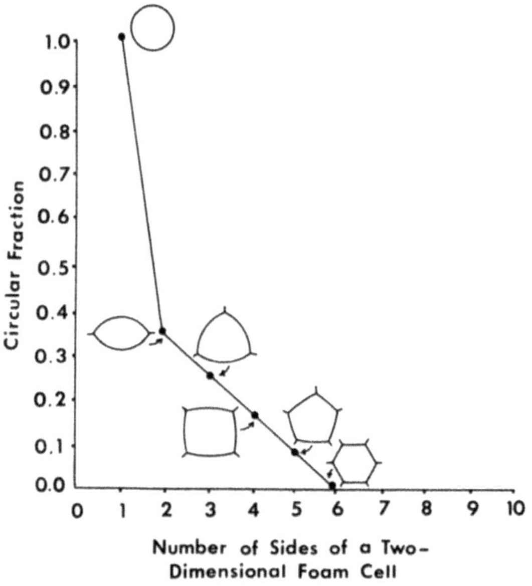

Figure 3. Circular Fractions for 2-Dimensional Grains or Foam Cells of Regular Curvilinear Polygons.

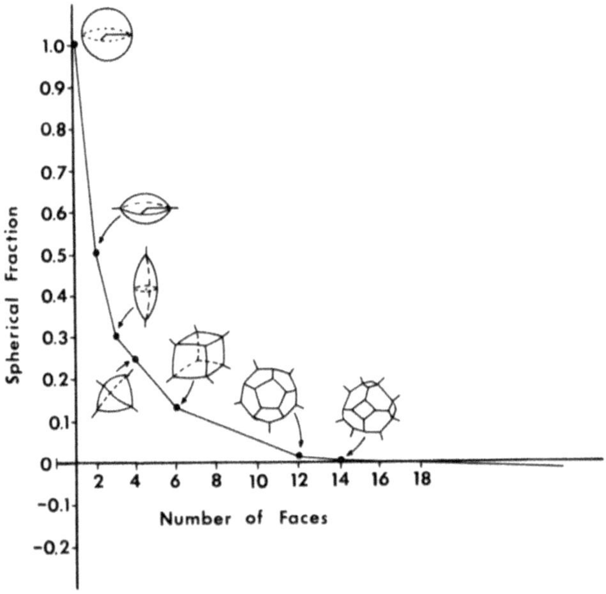

Figure 4. Spherical Fractions for 3-Dimensional Grains or Foam Cells for Regular or Semiregular Polyhedrons with Curved Surfaces.

The circular fractions of Table V are plotted in Figure 3. Except for the circle, the points all fall in a straight line following the equation:

$$C.F. = 1 - x/6, \quad \text{where}$$

x is the number of sides of a curvilinear polygon in which the boundary junctions of the arcs at the vertices make three 120° angles. Thus a 2-dimensional grain (or bubble in a foam) loses or gains atoms at the same rate per unit of circumference as does the full circle of the same radius of curvature. The circular fraction then gives the fractional rate a grain gains or loses atoms relative to a complete circular grain. This simplifies the analysis of grain growth, since only positive grains need be considered, the zero and negative grains serving only as the matrix for the positive grains. In the two-dimensional system each positive grain loses atoms at the rate proportional to the net sum of its circular arcs. If we know the density distribution of the few types of positive grains (the 3,4 and 5-sided ones, only), and label these f_3, f_4, and f_5, respectively, we have a mass transfer rate equation:

$$\frac{dm}{dt} = f(\bar{\gamma})\ (f_3 \cdot 3/6 + f_4 \cdot 2/6 + f_5 \cdot 1/6)$$

where $f(\bar{\gamma})$ is a function containing $\bar{\gamma}$, the mean surface tension of the boundaries and where the f_x's are the density, or frequency, distribution coefficients for the corresponding circular fractions. This form of the mass transfer rate equation adds a certain insight to the grain growth process. If γ, as well as the f_x terms, are constant relative to the $\bar{d}p$ variable, then one can show by integrating the simple equation $d(\bar{d}p^2) = Kdt$ (the mass term is proportional to the mean diameter squared, $\bar{d}p^2$), that $d = Kt^{1/2}$. However, the data on beta brass show that $\bar{\gamma}$ as well as the f_x quantities change rather strongly as $\bar{d}$ increases. Thus the assumed constants are actually significant functions of $\bar{d}$. (The symbol $\bar{d}p$ represents the mean diameter of positive grains.)

For example, suppose that the f_x's and $f(\bar{\gamma})$ decrease with increase in $\bar{d}$, as they appear to do, so that $f(\bar{\gamma}) \cdot \bar{f}_x = K/\bar{d}$, K being a constant. If no other factors affecting growth are involved (coalescences, grain boundary disintegration, second phase particles, etc.), then we arrive at $\bar{d} = Kt^{1/3}$, as was found for beta brass. These variables, $f(\bar{\gamma})$ and the f_x's can be investigated experimentally as a function of grain growth, and so in due course the mystery of the low slopes in the log $\bar{d}$ vs log t curves for grain growth at low temperature anneals might be resolved. The fact that these slopes gradually increase and appear to reach 1/2 near the melting point as shown by Hu[2] might be partially explained by the fact that the grain boundary energies as a function of grain orientation mismatch at grain boundaries decreases with increasing temperature according to McLean[3]. Thus the $f(\bar{\gamma})$ term in the above equations decreases its dependence on $\bar{d}$, which in turn yields more curvature in the continually created boundaries, making the f_x's terms somewhat less dependent on $\bar{d}$.

In the spatial polycrystalline case, which is of course the real case, grains are doubly-curved surface polyhedrons. The curved surface of each symmetrical polyhedron on calculation yields a specific spherical fraction. The calculations are not by any means as simple as in the 2-dimensional case. The spherical fractions as a function of number of faces for each symmetrical polyhedron is given in Figure 4. The 14-sided Kelvin cell (the cubo-octahedron, semi-regular polyhedron) has zero spherical fraction, as the hexagon has zero circular fraction in the 2-dimensional case. Each face of the Kelvin cell has zero net curvature. The quadrilateral faces of this cell are flat, while the hexagonal faces exhibit complex curvatures. Because the Kelvin cell has curved surfaces it is not the minimum surface per unit volume cell in the structure. The minimum surface is probably near the 12.5 to 13 faces per cell. (This may explain the data McNutt [4] assembled that foam cells and grains tend to average somewhat less than 14 faces per grain on actual specimens.)

What is important in Figure 4 is that we can interpolate the spherical fraction for any "equiaxed" grain, if we know the number of faces and vice versa. For the beta brass of this study the mean radius of spherical curvature for the positive grain was found to be $8.1\bar{d}$ for the fine grain state, and $12.2\bar{d}$ for the coarse grain state. The relative dispersion for the grain size distribution narrows as the grains grow, thus reducing the driving force for boundary migration. This is evident from the fact that the curvature decreases faster than $\bar{d}$ increases.

Figure 4 permits the estimation of the number of faces for the average positive and negative grains. They are found to be as follows:

Fine grain		
positive	-	12.0
negative	-	12.8
Coarse grain		
positive	-	17.5
negative	-	16.3

It is of interest to analyze the course of change in boundary type as grain growth progresses. The fine grain structure is characterized by numerous straight and single curvature boundaries, while the coarse grain structure has numerous multiarced and combination straight edge and curved boundaries. If we note that in the present case the growth was about 5.5 times in going from the fine to the coarse, then each grain in the coarse structure represents a consumption of $(5.5)^3$ grains or about 175 grains. There are about 6.5 encounters, or new boundaries, created for each grain that disappears, or about 1135 new boundaries were created and later disappeared and about the same number of old boundaries were destroyed. Inherent in this process of creating and destroying grain boundaries is a bias toward lower energy boundaries. The low energy boundaries tend to be longer in length and also have a smaller driving force to tend to eliminate them. The process has some aspects of the survival of the fittest rule. The multiarced and the straight-edge, plus the curved boundaries, are low energy boundaries. The multiarced are of low energy type because the low angle boundaries have low attractive forces holding the dislocations together that comprise the low angle boundary. The straight edge sections are very likely coincidence boundaries, some being twin boundaries, which of course are also low in surface energy

SUMMARY

There is a definite effect on the nature of the boundaries during grain growth in beta brass (and presumably other metals). In general the boundaries tend to more complex curvature with mixture of coincidence-type straight-edge boundaries. The radius of net spherical curvature for the positive grains in the fine grain structure is 10 times the average grain size, while the negative grains is about 17 times. For the coarse grain structure the net radius of curvature per positive grain averages about 16 times and the negative grains about 21 times the average grain size. Each grain (or foam cell) can be considered a spherical fraction in which the growth or decrease of the grain is proportional to its spherical fraction. The 14-faced grain neither grows or decreases because its spherical fraction is zero.

These observations and the new insight of the grain growth process furnished by the spherical fraction concept simplifies the analysis of grain growth mechanisms and grain growth behavior in general.

REFERENCES

1. E.E. Underwood, Surface Area and Length in Volume, in Quantitative Microscopy, R.T. DeHoff and F.N.Rhines eds., McGraw-Hill (1968).

2. H. Hu and B.B.Rath, Met. Trans., 1, 3181 (1971).

3. M. McLean, Acta Met. 19, 1387 (1971).

4. J.E.McNutt, The Shape of Equilibrium Cells in Nature, in Quantitative Microscopy, R.T. DeHoff and F.N.Rhines eds., McGraw-Hill (1968).

This paper is in partial fulfillment for the degree of Doctor of Engineering Science by Louis P. Stone.

ACKNOWLEDGMENT

The drawings and laboratory work were partly carried out by Peter Lillienthal, graduate student, and David Boehm and Edward Thompson, undergraduates.

The work was conducted under National Science Foundation Grant, GK-2172.

GRAIN BOUNDARY MOTION AND RELATED PHENOMENA

ON THE THEORY OF GRAIN BOUNDARY MOTION

K. Lücke, R. Rixen, and F. W. Rosenbaum

Institut für Allgemeine Metallkunde und
Metallphysik, Technische Hochschule
Aachen, Germany

ABSTRACT

In order to predict the exact mechanism of grain boundary motion one has to know the atomistic structure of the grain boundary. Although some progress in the understanding of grain boundary structure has been made in recent years, a generally accepted model which is widely applicable is not yet available. In the present paper several alternatives will be treated. First, models will be considered where the boundary is assumed to consist of a distorted zone having a large width compared to the atomic distance. Here thermodynamic formulas can be applied. Secondly, grain boundaries are assumed to have a width of only atomistic dimensions. Here, in particular, the ledge model will be considered. Special emphasis will be given to the impurity drag model and to the orientation dependence of grain boundary motion which is considered largely responsible for the formation of recrystallization textures. Finally a general comparison between theoretical predictions and the various experimental observations of grain boundary motion will be made.

INTRODUCTION

Theoretical prediction of the mechanism and laws of grain boundary motion requires the knowledge of the grain boundary structure. Although some progress in the understanding of the atomic structure of a grain boundary has recently been made (see, for example, the survey article of Gleiter[1]), many important questions remain unanswered. Consequently, existing models for grain boundary motion are rather fragmentary.

In solid solutions, segregation at the grain boundary occurs to various degrees, depending upon the structure of the boundary. Because of this segregation the grain boundary motion is often strongly influenced by small amounts of impurities even in so-called "high purity metals". This presents a further difficulty in the study of the process.

In the present paper, current ideas on the structure of the boundaries and the mechanisms of their motion in pure metals will first be discussed. Our present understandings of the effect of foreign atoms on grain boundary motion (impurity drag theory) will then be summarized. Following these, an application of the impurity drag theory to some of the experimental results will be given; and finally, examples showing the role of grain boundary motion in the formation of annealing textures will be presented.

GRAIN BOUNDARY MOTION IN PURE METALS

Structure of Grain Boundaries

Since our knowledge about the grain boundary structure has been reviewed rather recently[1], only a few remarks shall be made here concerning mainly the question of the width of the boundary.

1. It has been proposed repeatedly that a grain boundary can be considered as a region of disorder having a width w, which is large compared to the atomic distance b (Fig. 1b). The structure of this region is often thought to be similar to that of an under-cooled melt. It has been shown, however, that this model of a "wide" boundary is unrealistic[1]. It is mentioned here only because several treatments of grain boundary motion are (at least implicitly) based upon this model.

2. There are boundaries where the range of disorder is certainly not much larger than b, e.g. twin boundaries or boundaries with short periodic structures (also called coincidence boundaries*). Since these boundaries are made up of identical units of relatively small size, the range of distortion must also be small. These are shown schematically in Figs. 2a and 3a. One must imagine that here only those atoms in the immediate vicinity of the boundary surface

* Originally lattice coincidences instead of the boundary coincidences had often been discussed in connection with grain boundary motion. It has been shown, however, that these lattice coincidences are meaningless for this process[1,2].

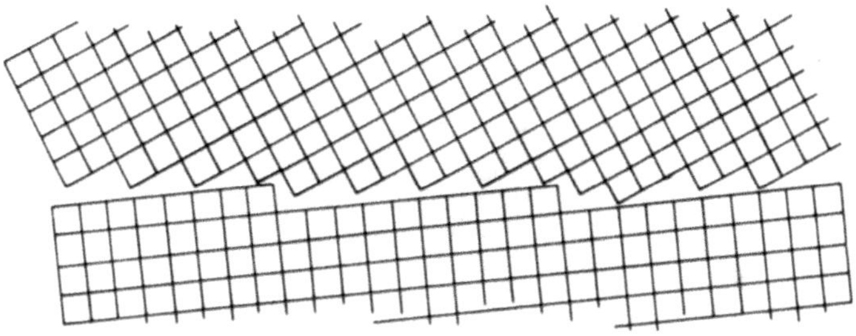

Fig. 1a. Schematic representation of a narrow grain boundary

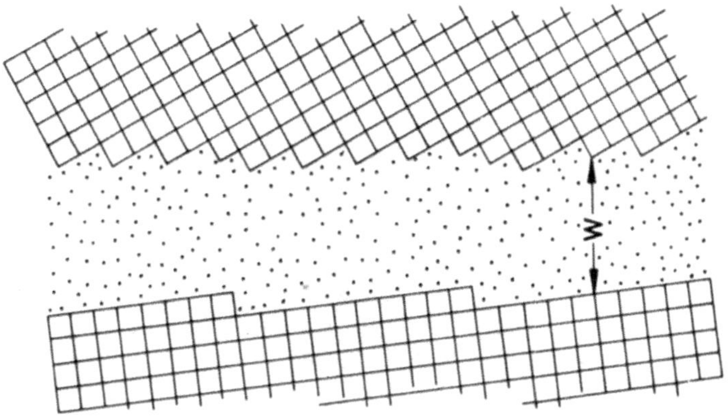

Fig. 1b. Schematic representation of a wide grain boundary

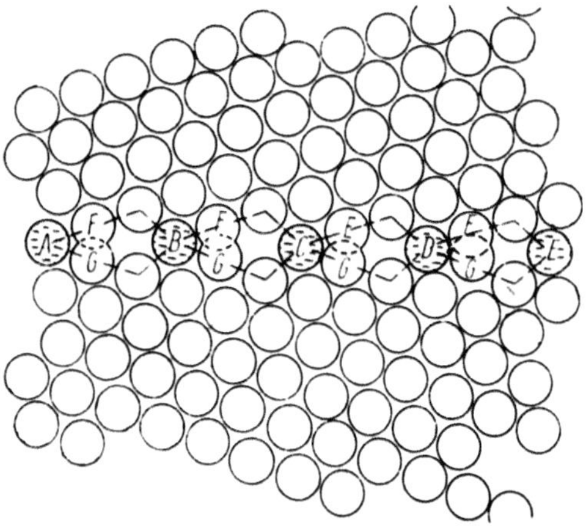

Fig. 2a. Boundary between two crystals according to the coincidence model [1]. (38° <111>)

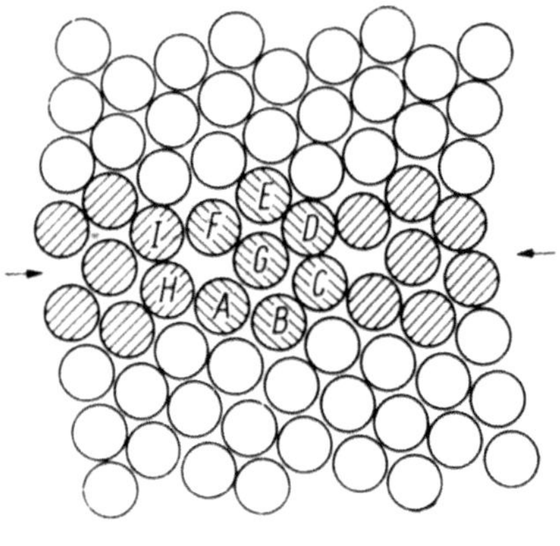

Fig. 2b. Calculated structure of a grain boundary [3]. (38° <111>)

relax into somewhat different positions in order to decrease the energy. Figure 2b shows the computer calculated lowest energy configuration of the atoms of the periodic 38° <111> -boundary in f.c.c. metals[3].

3. If one assumes that such coincidence boundaries are preferred because of their low energies, a boundary with a somewhat deviating orientation can be obtained by superimposing upon this boundary a dislocation array corresponding to a low angle grain boundary. Since such a dislocation boundary has an elastic strain field of the range of the distance between the dislocations*, a boundary with an orientation very close to a coincidence boundary may have a rather far-reaching strain field, and may therefore be considered as a "wide" boundary. On the other hand, if one considers, for example, that the elastic energy of this strain field decreases very rapidly with decreasing misorientation, it can thus be neglected just as in the case of the very far-reaching strain. This would mean that if such a boundary possesses a strain energy at all worth mentioning, it is concentrated in a narrow range along the boundary. The same aspect would be true also for the segregation of foreign atoms. Since the interaction energy with foreign atoms decreases with increasing distance from the dislocation, hence the boundary, and since the concentration of foreign atoms decreases exponentially with decreasing interaction energy, the great majority of segregated atoms is situated very close to the boundary. This means that from the viewpoint of energy or segregation, a boundary with a long range strain field must be considered as a "narrow" boundary.

4. As shown in Fig. 1a, a general boundary can be described in a similar way as a periodic boundary which is illustrated in Fig. 2a. According to the ledge model, the boundary surfaces of the two joining crystals are thought to consist of ledges limiting the closed packed planes (Fig. 4). The two surfaces must then be thought to have been packed together (Fig. 1a), with the atoms near the surfaces relaxed into better positions. For the same reasons as for periodic boundaries, these boundaries are "narrow" boundaries.

Mechanism of Grain Boundary Motion

Here the models of a "wide" and a "narrow" boundary shall be treated separately.

---

* If such a boundary is made up by mixing structural units of different coincidence boundaries[1], the strain field, in general, must be rather similar.

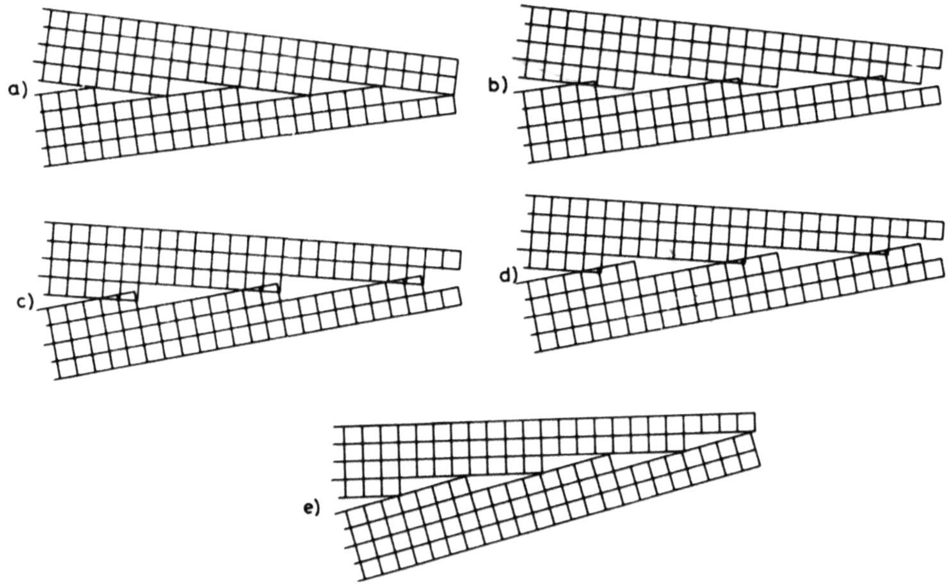

Fig. 3. Different stages (during the migration) of a narrow stepped grain boundary

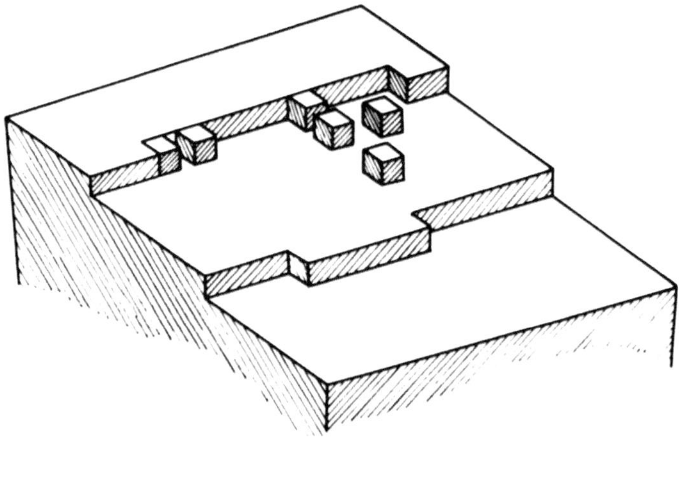

Fig. 4. The sequence of different states of an atom during grain boundary migration in Gleiter's model[5].

Wide boundary. In the simplest case corresponding to Fig. 1b, one has the free energy chart given in Fig. 5. Since the free energy of an atom in the boundary is higher by an amount g_B than that of an atom in the crystal, the activation energy g_1 for jumping into the boundary is larger than g_2 for leaving the boundary and joining a crystal. Since in this model the width w of the boundary must have a predetermined constant value, each jump out of the boundary must somehow be coupled with a jump into the boundary so that the larger value g_1 acts as effective activation energy. If there is a driving force (here Δg), the activation energies for forward and backward jumps become different, and one has a net motion of the boundary.

Bolling[4] used such a scheme based on the undercooled liquid model. Entrance into the boundary would correspond to melting, and exit, to crystallization; and g_B would be equal to the free energy of melting. However, if this undercooled liquid model cannot be applied to describe the boundary structure, it also may not be applied to explain grain boundary motion.

In a rather detailed paper, Gleiter[5] applied quantitatively the theory of evaporation and condensation for crystals to the process of grain boundary motion. He considered a ledge model for the grain boundary as shown in Figs. 1, 3a, and 4, and treated the entrance into the boundary as evaporation from the crystal, and the exit, as condensation. Since, however, it is essential for the mathematical treatment in this theory that single atoms can freely diffuse on the close packed faces to and from the ledges (Fig. 4), and since this is possible only in a gaseous phase and not in the rather densely packed grain boundary (Fig. 3), it is felt that this treatment is also rather unrealistic.

Recently the scheme of Fig. 5 has been modified[6,7] by also considering atomic motion within the boundary region. This leads to the question whether this motion, or the entrance into the boundary region, is rate determining for the boundary migration. The difficulty with these models is that they do not specify either the physical meaning of the different atomic jumps, or the energetical and geometrical factors which control their sequence and rate. Therefore, practically no conclusions could yet be drawn by the application of these models.

For the simple version of this model (Fig. 5) it is typical that part of the resulting activation energy is given by the grain boundary energy $g_B \approx 0.3$ to 0.6 eV per atom.

Narrow boundary. As illustrated schematically in Fig. 1a, one has in the case of a narrow boundary $w \approx 2b$. Here, in the limiting case, each atomic jump causes simultaneously a separation of an

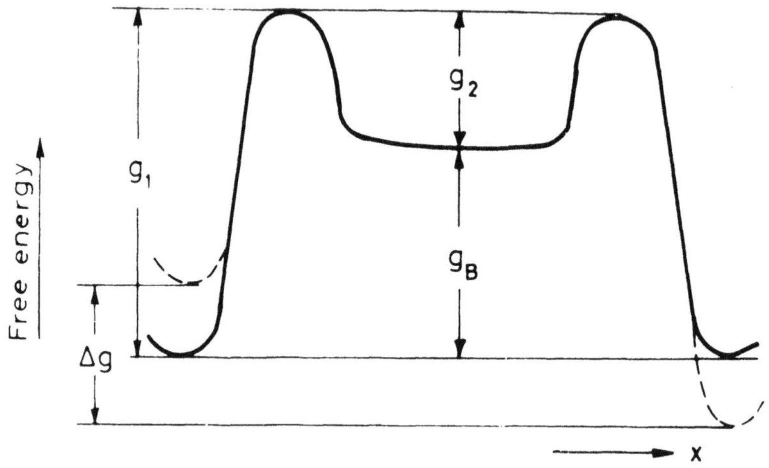

Fig. 5. The free energy of an atom crossing a wide grain boundary.

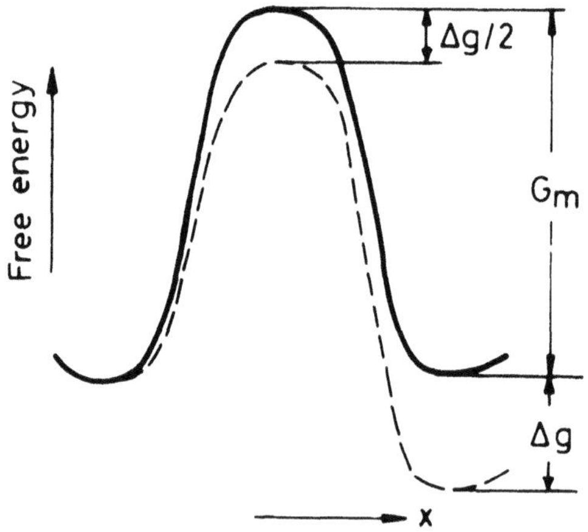

Fig. 6. The free energy of an atom crossing a narrow grain boundary.

atom from the one crystal, and a crystallization onto the other. Since by such a jump the number of atoms in the boundary is not changed, the activation energy does not contain terms resulting from the grain boundary energy, but is purely a diffusion type quantity.

The free energy diagram for this process is shown in Fig. 6. If each jump leads to a change in free energy, $\Delta g = pb^3$ (p is the driving force), and if the activation energy in one direction is increased and in the other decreased by $\Delta g/2$, one obtains the boundary velocity,

$$v = \alpha \nu_m b \left[\exp\left\{ - \frac{G_m - \frac{1}{2}\Delta g}{kT} \right\} - \exp\left\{ - \frac{G_m + \frac{1}{2}\Delta g}{kT} \right\} \right] \qquad (1)$$

or, with $\Delta g \ll kT$ and the diffusion constant

$$D_m = \alpha \nu_m b^2 \exp\left\{ \frac{S_m}{k} \right\} \exp\left\{ - \frac{H_m}{kT} \right\}, \qquad (2)$$

the boundary velocity can be expressed as

$$v = \frac{D_m \Delta g}{b kT} = \frac{D_{mo} \Delta g}{b kT} \exp\left\{ - \frac{H_m}{kT} \right\} \qquad (3)$$

where ν_m is the attack frequency, $G_m = H_m - TS_m$ is the free energy of activation, and $\alpha < 1$ is a geometrical factor. The condition $\Delta g \ll kT$ is always fulfilled in the case of recrystallization. For example, a dislocation density of 10^{12} cm^{-2} leads in the case of copper to a stored energy $\Delta g \approx 2 \cdot 10^8$ erg/cm^3, or $\Delta g/kT \approx 3 \cdot 10^{-2}$ (for T = 500°K). Only in the case of a phase transformation can the difference in free energy be much larger so that an exponential dependence of v upon the driving force might then be obtained.

If v is measured as a function of T, and if the driving force is known, the quantities H_m and $D_{mo} = \alpha \nu_m b^2 \exp\{S_m/k\}$ can be determined from the Arrhenius plots of the experimental results. With $\nu_m \approx 10^{13}$ to 10^{14} and $S_m \approx 2k$, values of $D_{mo} \lesssim 0.05$ to 1 should be expected. In many cases, however, values larger by several powers of ten were obtained. Also for H_m rather large values were reported, sometimes being larger than the activation energy H_D for self diffusion, even though diffusion steps in the boundary zone should be easier than in the lattice.

Lücke and Detert[8] first pointed out that large values for H_m and D_{mo} are mostly caused by foreign atoms and that for this effect only very small concentrations are necessary, since the foreign atoms segregate in the boundary. Therefore, for checking Eq. (3), only metals of extreme purity are suited. Table I shows some of the results given in literature. One recognizes that the values for H_m are always much smaller than the self diffusion value H_D, and that for D_{mo} values both larger and smaller than the theoretical range are found. If one considers that the materials may not yet be pure enough, that the driving force Δg is often not well known, and that the D_{mo} values are obtained by extrapolation in the Arrhenius-plot only within an accuracy of several powers of ten, it can be concluded that the values in Table I are not in disagreement with Eq. (3). Vandermere and Gordon[9] came to the same conclusion.

In the past, proposals were made that during boundary motion not single atoms but larger groups of atoms change from one crystal to the other by a single activation event[10,11,12]. This would lead to an increase in the pre-exponential factor for v. Very recently, Haessner and Hofmann[6] proposed a similar process, namely, the activation of a single atom causes a larger group of atoms to change sides. They came to this conclusion since they felt that the observed D_{mo} values are too large to be explained by Eq. (2). The present authors, however, feel that in view of the above discussion this conclusion is not justified*. Moreover, it seems difficult to understand how by the activation of a single atom groups of the order of 10^3 atoms can reorder themselves (this is the magnitude of the factors Haessner and Hofmann wanted to explain).

The model underlying Eq. (3) is rather undetailed. In reality the factor α might depend still upon the orientation difference and the boundary structure. Details are not known. Gleiter[5] assumes that grain boundary motion is achieved by atoms moving from ledges at one crystal surface to the ledges on the other crystal surface. This is shown schematically in Fig. 3a-e and would require diffusion parallel to the grain boundary. Unclear is whether, or to what extent, vacancies are necessary for this diffusion and, therefore, for boundary motion[2].

Figure 3 also shows that for the understanding of boundary motion it is not sufficient to know the equilibrium (i.e. the lowest energy) structure of the boundary. One recognizes that during building and unbuilding of the ledges, a stage is reached

---

* A closer look at the data collected by Haessner and Hofmann shows that, except for some errors in calculation, one has in many cases agreement between experimental and theoretical D_{mo} values, and in some cases, indications for insufficient purity of the material.

TABLE I

Data Collected from Literature on Grain Boundary Motion

No.	Ref.	Material	D_{mo} $cm^2/sec.$	H_m eV	D_o $cm^2/sec.$	H eV	U_o eV
1	18	Al + Mg	1.6	0.6	50 - 80	1.05 - 1.23	0.22 - 0.52
2	18	Al + Cu	1.6	0.6	800 - 1100	1.02 - 1.1	0.38 - 0.56
3	9	Al + Cu	16.0	0.68	≈ 0.23	0.87 - 0.91	0.36 - 0.45
4	21	Au + Fe	$4.1 \cdot 10^{-4}$	0.88	$\approx 3 \cdot 10^{-3}$	≈ 1.04	0.46 - 0.57
5	22	Cu + ?	-	0.67	$\approx 10^{-4}$	0.76 - 0.87	0.47 - 0.57
6	23	Al + ?(I)	-	-	-	-	0.42 - 0.49
7	23	Al + ?(II)	-	-	-	-	0.067 - 0.12

(e.g. Fig. 3c) where the energy is certainly higher than in the best fitting structure (Fig. 3a and e). This would mean that during the motion of a grain boundary its energy changes periodically, i.e., for the boundary motion an energy with a meaning similar to that of the Peierls-energy for dislocation motion must also be assumed.

THE IMPURITY DRAG THEORY

General Considerations

The impurity drag theory describes the influence of dissolved foreign atoms upon the motion of a grain boundary. Its principles have first quantitatively been formulated by Lücke and Detert[8]. They postulated: (i) that the grain boundary represents for the foreign atoms a region of different (e.g. of lower) potential energy so that a segregation of foreign atoms in the boundary will take place; (ii) that in the case of a moving grain boundary this atmosphere of foreign atoms lags behind the boundary thus exerting a dragging force on the boundary, hence diminishing its velocity; and (iii) that at sufficiently high velocities the boundary breaks away from this atmosphere, and moves with a velocity comparable to that in a pure metal.

Cahn[13] and Lücke and Stüwe[2] refined this theory by a more accurate consideration of the diffusion of the foreign atoms with respect to the moving boundary. Assuming (implicitly) "wide" boundaries, small concentrations and some additional simplifications, the final formulas of both papers came out to be identical.

More recently, Lücke and Stüwe in a second paper[14] and Rosenbaum and Lücke[15] gave a more transparent formulation of the theory and a further extension by treating also "narrow" grain boundaries and larger concentrations ("approximated atomistic theory"). In this section a survey over the theoretical situation in this area shall be given.

Figure 7a shows schematically the potential $U(x)$ felt by the foreign atoms near the boundary. U_o is the maximum free energy of interaction. Figure 7b shows schematically the concentration of foreign atoms $C(x)$ near a moving boundary; the foreign atoms are mainly concentrated behind the boundary (moving to the right). The atomic planes, having the distance b from each other, are indicated by the thin vertical lines and numbered by the subscripts ξ, $\xi + 1$. The boundary motion is then controlled by two rate equations:

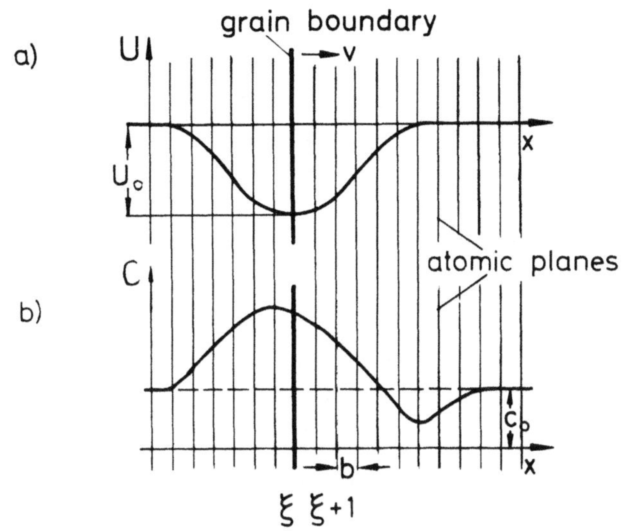

Fig. 7. (a) The potential U(x) between a solute atom and a grain boundary.

(b) The distribution of solute atoms near a moving grain boundary.

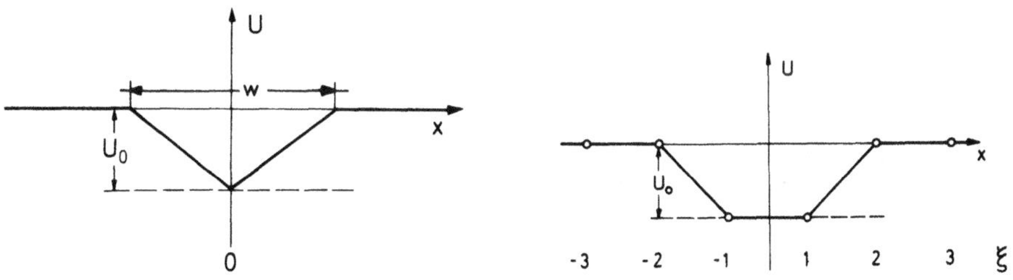

(a) for the continuum model (b) for the atomistic model

Fig. 8. Potential of foreign atoms near the grain boundary.

1. The first equation describes the rate of jumping of the atoms (matrix and foreign) across the grain boundary by which the shift of the boundary is achieved. It leads to the boundary velocity (cf. Eq. (11)),

$$v = \alpha\, b\, \nu_m \left[\exp\left\{ - \frac{G_m - \frac{1}{2}\Delta g - \frac{1}{2}\Sigma_\xi C_{\xi+1}\Delta U_\xi}{kT} \right\} - \exp\left\{ - \frac{G_m + \frac{1}{2}\Delta g + \frac{1}{2}\Sigma_\xi C_{\xi+1}\Delta U_\xi}{kT} \right\} \right] \tag{4}$$

The original activation energy G_m for such a jump is here modified for two reasons: Firstly, each jump leads to an additional change in energy Δg, since it causes a motion of the boundary with respect to the driving force. Secondly, each jump leads to an additional change in energy $\Sigma_\xi C_{\xi+1} \Delta U_\xi$, since by the motion of the boundary its potential field is shifted with respect to the foreign atoms. Here, C_ξ is the concentration in the atomic plane ξ, and ΔU_ξ, the change in the free energy for an atom moving from plane ξ to $\xi + 1$.

2. The second rate equation describes the rate of diffusion of the foreign atoms. It leads to the diffusion current,

$$j = \frac{\nu_\xi}{b^2} \left[C_\xi (1 - C_{\xi+1}) \exp\left\{ - \frac{G_\xi - \frac{1}{2}\Delta U_\xi}{kT} \right\} - C_{\xi+1}(1 - C_\xi) \exp\left\{ - \frac{G_\xi + \frac{1}{2}\Delta U_\xi}{kT} \right\} - \frac{v}{b^3} (C_{\xi+1} - c) \right] \tag{5}$$

The first term gives the number of jumps from any plane ξ to $\xi + 1$, and the second, the number of backward jumps. Being similar as in Fig. 6, the activation energy G_ξ for a jump is decreased or increased, respectively, by half of the potential difference ΔU_ξ between the two planes. The activation energy G_ξ and the attack frequency ν_ξ are written with a subscript ξ since the corresponding diffusion constant $D_\xi = \nu_\xi b^2 \exp\{-G_\xi/kT\}$ may still depend somewhat upon the distance from the boundary. The very last term of Eq. (5) is a convection term. It must be introduced in order to relate the diffusion to the moving coordinate system connected with the boundary, which is assumed to move with the constant velocity v. ξ (and later x) describes then the distance of the considered atomic plane from the boundary, c is the macroscopic concentration (in contrast to the local concentration C_ξ).

The Approximated Continuum Theory

In the case of the approximated continuum theory the following principal approximations which will be discussed later are introduced into Eqs. (4) and (5)

$$\Delta g = pb^3 \ll kT \tag{6a}$$

$$C_\xi \ll 1 \tag{6b}$$

$$\exp\left\{\frac{\Delta U_\xi}{kT}\right\} = 1 + \frac{\Delta U_\xi}{kT} \tag{6c}$$

$$\Delta U_\xi = b\,\frac{dU}{dx} \tag{6d}$$

$$C_{\xi+1} = C_\xi + b\,\frac{dc}{d\xi} \tag{6e}$$

This leads to

$$v = \frac{D_m b^2}{kT}\,(p - p_F); \qquad p_F = \int_{-\infty}^{+\infty} C(x)\,\frac{dU(x)}{dx}\,dx \tag{7}$$

$$j = \frac{D}{b^4}\left(\frac{dC}{dx} - \frac{C}{kT}\frac{dU}{dx}\right) - \frac{v}{b^3}\,(C - c) \tag{8}$$

Since here U, D and C are continuous functions of x, this treatment will be named continuum theory.

By assuming for U(x) the simple triangular potential shown in Fig. 8a and setting within the boundary region $|x| < w$ for the diffusion constant $D(x) = \text{const} = D$ (with the free energy of activation $G = H - TS$), Cahn[13] and Lücke and Stüwe[2] obtained (with an additional minor approximation) for v a cubic equation. Using reduced coordinates[14] this equation reads

$$y^3 - y^2 + \frac{1+\rho}{\sigma^2}\,y - \frac{1}{\sigma^2} = 0 \tag{9}$$

where

$$y = \frac{v}{v_{Free}} = \frac{v}{m\,p} \tag{10}$$

is the reduced velocity,

$$\sigma = K \frac{D_m}{D} \frac{b^3 p}{U_o} \tag{11}$$

is the reduced driving force, and

$$\rho = 2K^2 \frac{D_m}{D} C \tag{12}$$

is the reduced concentration. Furthermore, v_{Free} is the velocity of the grain boundary without foreign atoms, given by Eq. (3), m its mobility and

$$K = \left[\frac{kT}{U_o}\left(\exp\left\{\frac{U_o}{kT}\right\} - \exp\left\{-\frac{U_o}{kT}\right\} - 2\frac{U_o}{kT}\right)\right]^{\frac{1}{2}} \tag{13}$$

The solution of Eq. (9), $y = y(\rho, \sigma)$ is given in Fig. 9 in a contour-line representation with logarithmic scales. One sees that the plane is divided by a wedge-shaped area. Above and to the left, practically no segregation at the boundary occurs ("free boundary"), and one has the solution

$$y \approx 1/\rho \tag{14}$$

In the wedge-shaped region three solutions exist: The one corresponding to the free boundary, the one corresponding to the loaded boundary, and an unstable solution which has no physical meaning. In this region it depends upon the pretreatment of the specimen whether a free or a loaded boundary is obtained.

Figure 10a shows cuts through Fig. 9 parallel to the ρ -axis for different σ. One recognizes that with increasing σ the critical concentration also increases above which the boundary is loaded, and the $v \propto 1/c$ law is valid. Instead of cuts parallel to the σ -axis, in Fig. 11a the p - dependence of v is shown directly (schematic). One recognizes also that for the loaded boundary $v \propto p$, only the mobility is strongly diminished compared to that of the free boundary. The length of the thin horizontal lines in Fig. 11a gives directly the magnitude of the retarding force p_F due to the foreign atoms, which according to Eq. (7) must be

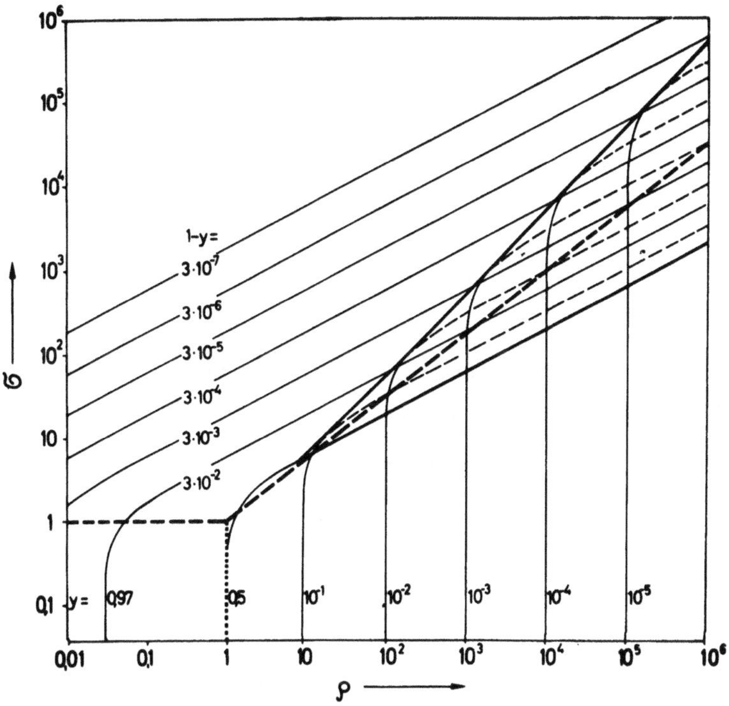

Fig. 9. Contour-line representation of the function $y = y(\rho, \sigma)$.

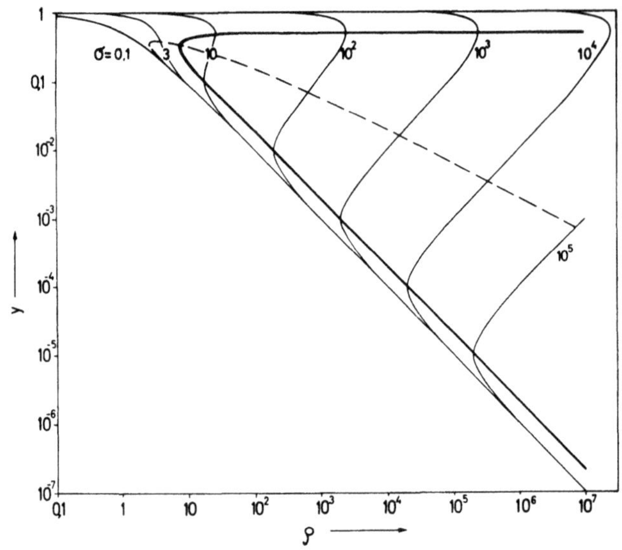

Fig. 10a. The function $y = y(\rho)$ for different values of σ (continuum model)[14].

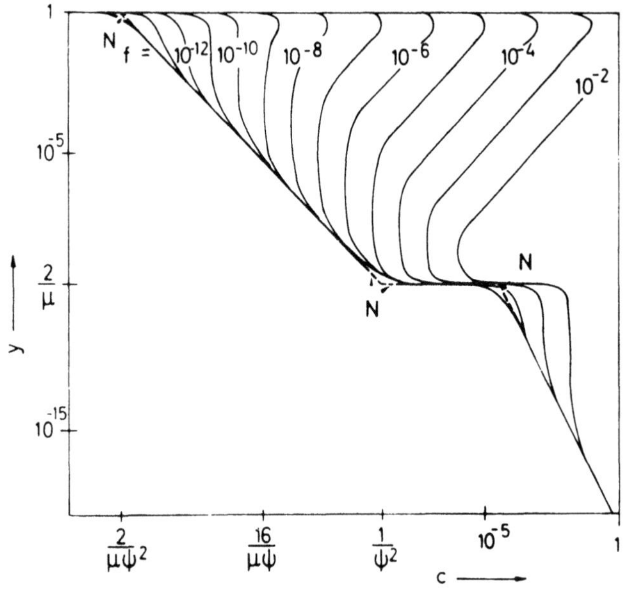

Fig. 10b. The function $y = y(c)$ for different values of the driving force (atomistic model)[15]
$\mu = (U_o/kT)\ (D_m/D)$; $\psi = \exp\{U_o/2kT\}$

deduced from the external driving force p in order to obtain the effective driving force.

Furthermore, the kind of plot in Fig. 9 has the advantage that the meaning of a temperature change can easily be recognized. Since it can be assumed that $H > H_m > U_o$ both ρ and σ increase with decreasing T in a rather similar manner. This means that a decrease in temperature corresponds to a shift in the $(\rho - \sigma)$ -diagram in the nearly 45°- direction towards larger values of ρ and σ. If ln v vs. 1/T is plotted directly one obtains from Eqs. (14), (10), (3), (12), and (13) the "activation energies" for the free and loaded bounaries:

$$H_{Free} = H_m - kT \tag{15}$$

$$H_L = H - kT + \frac{d \ln K^2}{d(1/kT)} \tag{16}$$

If by changing an external parameter p, c, or T, a path through the $(\rho - \sigma)$ -diagram is obtained which runs through the 3-valued (wedge-shaped) area, a discontinuous change in v is to be expected. This is indicated in Fig. 11a for a change in p by the vertical lines with arrow. If, however, this area is avoided but the dashed or the dotted line is cut, a curve for v with an inflection point or a sharp bend is expected (e.g. in Fig. 10a the curve for $\sigma = 10^3$). Some examples for such dependencies are given in Fig. 12.

An Approximated Atomistic Theory

In Eqs. (6a) to (6e) the suppositions for the continuum theory were listed. Their validity will now be discussed:

(a) The condition $pb^3 \ll kT$ is always fulfilled (cf. Sec. on "Mechanism of Grain Boundary Motion".)

(b) Within the boundary region the condition $C_\xi \ll 1$ is not always fulfilled, even for dilute solutions. It requires that $C_{max} = c \exp\{U_o/kT\} \ll 1$. Since for U_o values as large as 0.3 eV must be assumed (see following Sec. on "Discussion of Experimental Results"), one obtains, for example, for $c = 10^{-3}$ and $T = 500°K$, a value of $C_{max} \approx 1$.

(c) This supposition requires $\Delta U/kT \ll 1$, or, with $T = 500°K$, $\Delta U \ll 0.08$ eV. With $U_o = 0.3$ eV this condition is only fulfilled if the range w of the potential is larger than at least 10b which,

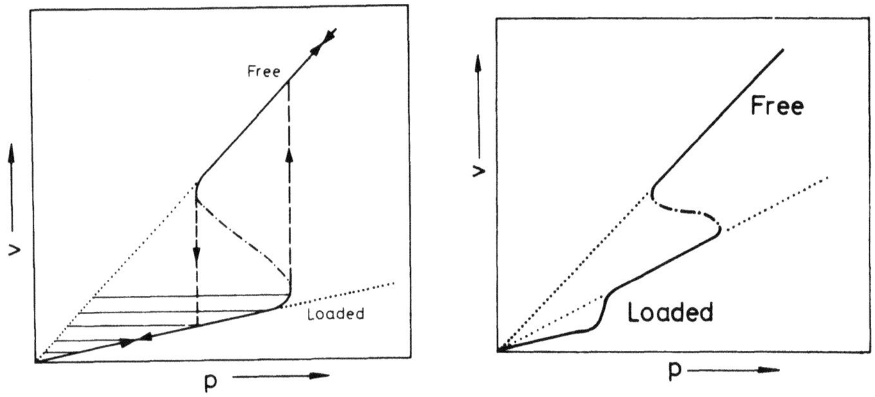

(a) continuum model (b) atomistic model

Fig. 11. Boundary velocity v as a function of the driving force p (schematic).

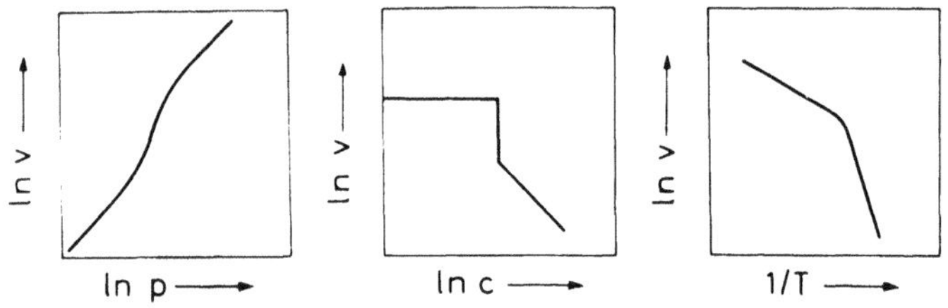

Fig. 12. Some possible types of transitions from free to loaded boundary (schematic).

in general, cannot be expected (see Sec. on "Structure of Grain Boundaries").

(d) and (e) The last two suppositions are only fulfilled for a long range potential and not for a narrow boundary with $w \approx 2$ to 3 atomic layers.

As a result of these considerations, one must conclude that, under more realistic conditions, the suppositions for the continuum theory seem to fail. For this reason Rosenbaum and Lücke[15] have calculated the grain boundary migration rate $v = v(p, c, T)$ for a model which is based on the atomistic model of Lücke and Stüwe[14] and on the following suppositions:

1. The discrete values for C_ξ and ΔU_ξ as used in Eqs. (4) and (5) are retained and not replaced by differential expressions as in Eqs. (6d),(6e), (7), and (8).

2. The boundary region, i.e. the region where the potential for foreign atoms is decreased *, is considered to have a thickness of only one, two, or three atomic layers. As an example, Fig. 8b shows the potential used in the 2-layer model.

3. It is assumed that within the boundary the concentration may reach values of the order of one. Therefore, the $(1 - C)$ terms in Eq. (5) are not neglected. This means that the Boltzman statistics has been replaced by the Fermi statistics.

Using some proper approximations, also for these much wider suppositions the calculations could be carried out analytically. The results, however, were much more complicated than in the continuous case and, therefore, shall be discussed here only qualitatively.

Here the solutions can no longer be described in terms of the 3 reduced coordinates as $y = y(\rho, \sigma)$, instead all 4 coordinates must be used leading to solutions $v = v(c, p, T)$. Instead of the universal $(\sigma - \rho)$ -diagram (Fig. 9), $(p - c)$ -diagrams for different temperatures must be drawn. Also the $(p - c)$ -diamgrams show wedge-shaped 3-valued regions similar to those in the $(\sigma - \rho)$ -diagrams, but they have additional lines along which discontinuities, inflection points, or bends can occur. This can be recognized also in the schematic $(v - p)$ -diagram (Fig. 11b), or the $(v - c)$-diagram

---

* The calculations have been carried out also for positive interaction energy, i.e. for a depletion of foreign atoms in the grain boundary. But no new points of special interest were obtained.

(Fig. 10b). These lines complicate the evaluation of experiments, since they increase the number of possibilities for interpreting experimental curves. Moreover, since they mostly indicate that the concentration in a certain atomic plane within the boundary region comes close to one, the theoretical expressions for these lines and, therefore, the constants derived from them, are less accurate.

DISCUSSION OF EXPERIMENTAL RESULTS

General

If the velocity v of the grain boundary is measured as a function of p, c, or T, a complete evaluation on the basis of the impurity drag theory can only be made if the corresponding curve exhibits a discontinuity, an inflection point, or a bend as, for instance, shown in Fig. 12. These special points must be interpreted by relating them to certain lines in the (σ - ρ) - or (p - c) -diagrams (e.g. Fig. 9). From the measured coordinates of these points and from the slopes of the curves, the parameters of the impurity drag theory can then be determined. These are mainly the interaction energy at U_o and the diffusion coefficients D_m and D. In the case that measurements are made at different temperatures, the corresponding activation energies H_m and H and the pre-exponential factors D_{mo} and D_o can be determined.

In the following, the results of such evaluations are described for some measurements taken from literature. A more detailed discussion is given by Rosenbaum and Lücke[15]. As to be seen in Table I, in many cases instead of unequivocal values only ranges of values could be given for the parameters. One of the reasons for this is that there are only few experiments with sufficient and good enough information for a thorough analysis, which also allows a cross-check of the results. In particular, often the driving force or the nature and concentration of the effective impurities is not well known. Furthermore, it is not clear which model to choose (continuum model, atomistic model with boundary width of 1b, 2b, or 3b). Moreover, it is not completely clear that to which lines of the diagrams the special points of the measured curve must be related. For instance, if a discontinuity is found it is not obvious whether it corresponds to the right or left boundary of the wedge-shaped region in Fig. 9, or to a value in between or, in the case of the atomistic model, to another line. Therefore, in the evaluation all these possibilities have been tried out and, as far as possible, the consistency of the results has been tested by cross-checking. In some cases it proved to be impossible to completely avoid contradictions. This, however, may not be taken too seriously, since the quantitative character of the theories -

especially in the case of high concentrations in the boundary - is rather approximate.

Recently, the idea was put forward that transformations of the boundary structure can take place at a given temperature[16,17]* In evaluating experimental results one should also keep in mind that these transformations - if they take place at all - lead to a change in kinetics and to a bend in the curves.

Variation of Concentration

The measurements which allow the most complete evaluation are those of Frois and Dimitrov[18] carried out on aluminum of extreme purity (impurities 0.1 ppm) with small additions of Mg or Cu. As predicted by the impurity drag theory, the migration rate as a function of concentration shows clearly an independence for small concentrations, an approximate $1/c$ dependence at large concentrations, and between the two regions (at $c_{crit} \approx 10^{-4}$) a discontinuity (Fig. 13a). Also the activation energy H shows the expected course (Fig. 13b): At $c < c_{crit}$ a small value H_m corresponding to the pure metal is obtained, and at $c > c_{crit}$ a larger value corresponding to H_L of Eq. (16).

One recognizes that the interaction energies come out to be rather high (0.25 - 0.5 eV). It should be pointed out that such large values are obtained by the application of the continuum theory which is only valid for low values of U_o/kT and for which, therefore, the evaluation is not self-consistent. A similar situation exists for the application of the atomistic model of Rosenbaum and Lücke, which is valid for $U_o > kT$. This leads us to believe that the order of these values is correct, and that the inconsistencies are not a result of wrong suppositions in the theory. Also, the order of the activation energy H for the diffusion of the foreign atoms should come out rather reliably. It is found to be somewhat smaller than that of self-diffusion (H_D = 1.32 eV), suggesting that here the diffusion occurs in the vicinity of the boundary. Less reliable, however, are the values obtained for the pre-exponential factor D_o, which due to simplifications in the theory might be falsified by a few powers of ten.

Figure 13 and Table I show also the results of Vandermeer and Gordon[19] on Al with Cu-additions which agree rather well with those of Frois and Dimitrov[14] for this alloy. Here instead of a discontinuity, an inflection point is obtained for v = v (c). The activation energy for the transition region is found to be larger than that for the loaded boundary. But this has no physical

*See also paper by Hart in this volume (Note added by Editor).

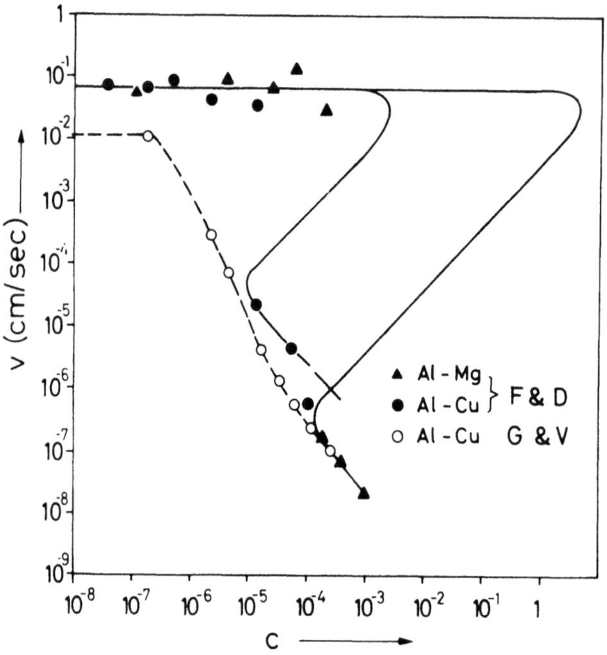

Fig. 13a. Experimental results of Frois and Dimitrov[18]. and Gordon and Vandermeer[19].

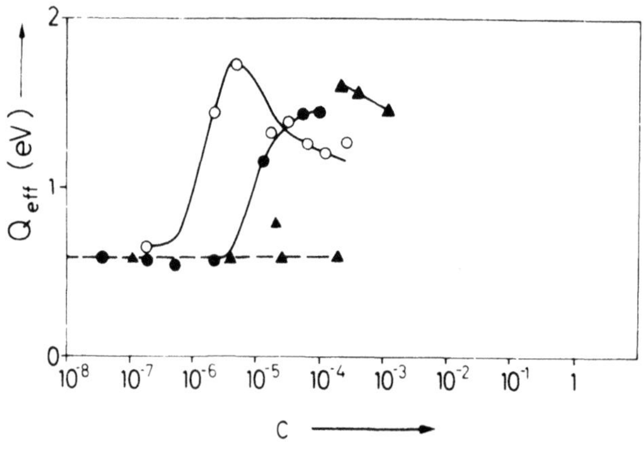

Fig. 13b. Activation energies reported by Frois and Dimitrov[18]. and by Gordon and Vandermeer[19].

meaning*. The measurement of Aust and Rutter[20] on Pb with Sn additions could not be evaluated, since neither a region with constant velocity at small concentrations nor a region with a 1/c- dependence at large concentrations has been found. It must be assumed that all their measurements are within the transition region.

Variation of Temperature

Figure 14 shows ln v as a function of 1/T for Au 99.999% pure as measured by Grünwald and Haessner[21]. The curve consists of two linear portions. These authors assumed that the high temperature portion (left of the inflection point) corresponds to a free grain boundary (small slope) whereas the low temperature portion corresponds to a boundary loaded with Fe acting as most effective impurity (large slope). The results of the quantitative evaluation (which differ somewhat from those reported by Grünwald and Haessner themselves) are also listed in Table I.

Figure 15 gives the reciprocal values of the half-life times of recrystallization of high purity copper as measured by Scheucher[22]. An evaluation can here be carried out under the assumptions that v is proportional to $1/\tau_0$, and that to the left of the bend the grain boundary is free. The results are given in Table I for the type A-copper. For the type B-copper no evaluation is possible. The high temperature part of this curve runs below that of curve A, whereas, if it would represent the free boundary, it should coincide with curve A. Therefore, one must assume that at the bend of curve B some type of impurity might evaporate, but other types of impurity remain segregated in the boundary and will evaporate only at still higher temperatures.

It should also be pointed out that the temperature change experiments lead to rather high values of U_0 and to relatively reasonable values for the other constants.

---

* This is an apparent activation energy which only indicates the increase in velocity with increasing temperature by going from the loaded to the free state. In this way, values even higher than that for self-diffusion can be obtained. Also the inflection point in Fig. 13a may not be of the type as predicted by the impurity drag theory (e.g., as shown in Fig. 12) but may be caused by macroscopic inhomogenities of the material.

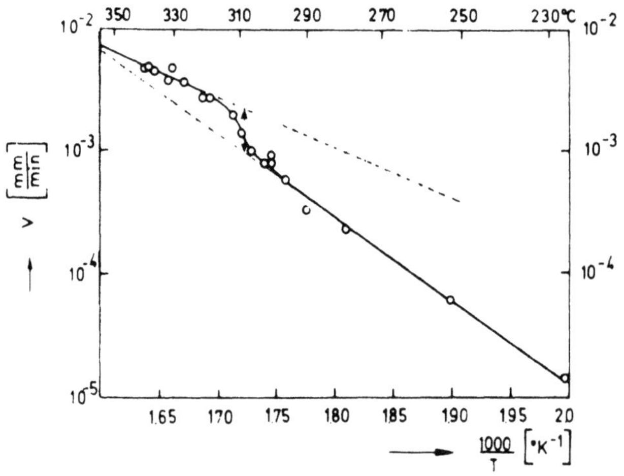

Fig. 14. Experimental results of Grünwald and Haessner[21].

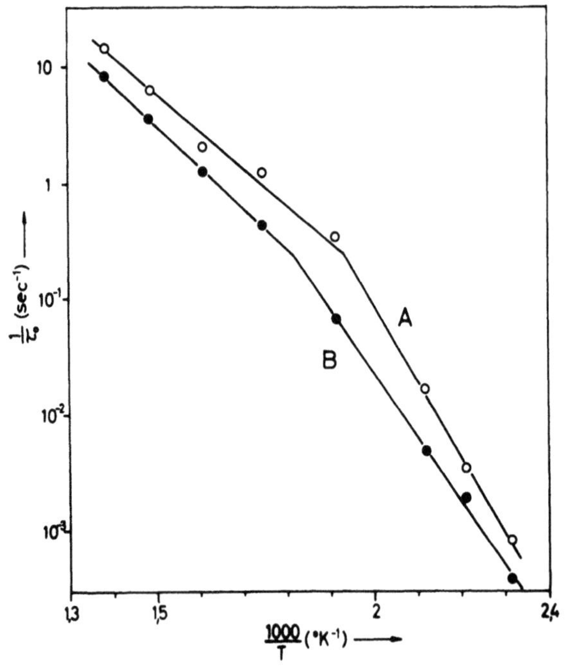

Fig. 15. Experimental results of Scheucher[22].

Variation of Driving Force

In Fig. 16 measurements of Rath and Hu[23] for high purity Al are plotted for which the driving force was varied. Additionally, some points taken from the work of other authors on Al are inserted. Since the measurements of Frois and Dimitrov are obtained for extremely pure Al and for the largest driving force, it seems reasonable to assume that their data correspond to the free boundary, hence the dashed line with the slope one running through this data point gives the driving force dependence for the free boundary. All the other measurements (except those of Vandermeer and Gordon) would then correspond to a segregated boundary. The very large slope of the curves of Rath and Hu could be taken as an indication that one is here in the transition zone. It is somewhat suspicious, however, that neither at very low nor at very high driving forces the theoretically predicted slope of one is reached, although the range of the velocity covers many powers of ten. A quantitative evaluation of these measurements could be made only with additional assumptions that the values given in Table I are not very reliable.

The data presented in Fig. 17 are calculated by the present authors from grain growth measurements in high purity Pb and Sn of Drolet and Galibois[24]. Here it looks as if the bend in the curves could be easily explained by assuming that the portion of the curves at high driving forces corresponds to the free boundary, and that at low driving forces, to the transition region. As Drolet and Galibois pointed out, however, this interpretation is not possible. They found in investigations on Pb-Sn alloys that for the high driving force branch the velocity decreases with $1/c$ which is typical for a loaded boundary. The bend might possibly be caused by the evaporation of other impurities. Because of this situation a quantitative evaluation of this data was not possible.

These authors observed additionally that the growth rate was lower after multi-anneal, i.e. after repeated heating and cooling of the specimen than after a single anneal. This is shown in Fig. 17c. The authors interpreted this effect by the assumption that during the cooling the boundary came to rest and collected additional impurities, thus bringing it from the free into the loaded state. This explanation, however, is not possible. As can be seen by discussing the path in the $(\sigma - \rho)$ -diagram of Fig. 9, it is impossible to move from the field corresponding to the free boundary into the field of the loaded boundary only by a decrease of temperature. This can be explained by the fact that the rate of diffusion of the foreign atoms to the boundary decreases faster with decreasing temperature than does the rate of grain boundary motion.

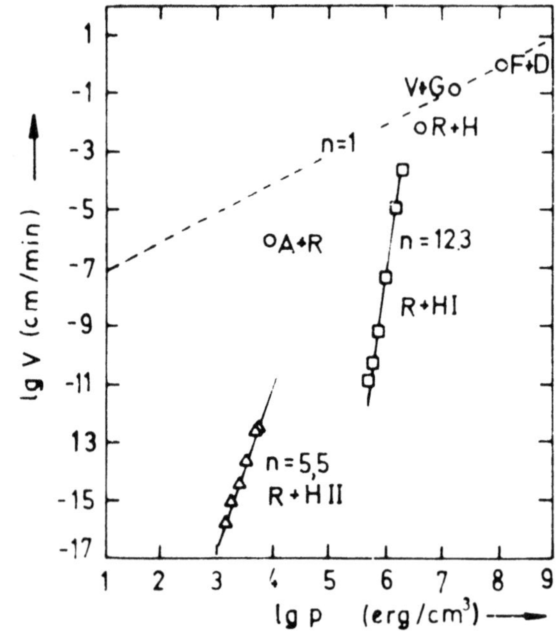

Fig. 16. Grain boundary velocity in aluminum as a function of the driving force as observed by several authors (after Rath and Hu[23]).

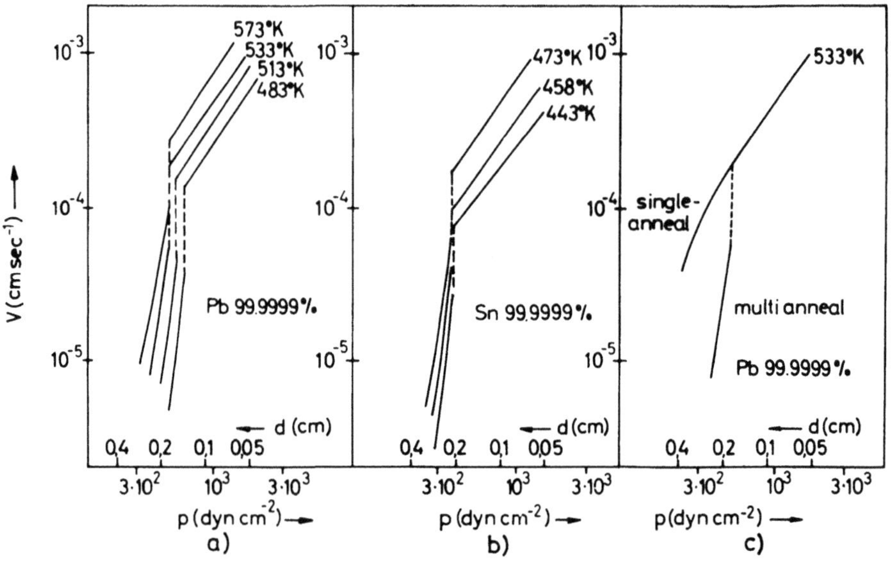

Fig. 17. Grain growth rate in lead (a, c) and tin (b) as a function of the driving force (calculated after results of Drolet and Galibois[24]).

It should be mentioned that the evaluation of grain growth experiments has the additional difficulty that at different migration rates, the impurity concentrations in the boundary, hence the grain boundary energies, are different. This means that, in contrast to the above considerations, the driving force is no longer proportional to $1/R$ (R = grain radius). Calculations for considering this effect are in progress.

ORIENTATION DEPENDENCE OF GRAIN BOUNDARY MIGRATION AND FORMATION OF TEXTURES

Investigation of the Orientation Dependence of Grain Boundary Migration

In the previous sections no consideration has been given to the orientation parameters of the moving grain boundary. This question shall be treated now. So far two types of experiments have been conducted to obtain information about this subject:

1. Direct measurements. Bicrystals of predetermined orientations are made in such a way that one crystal has a higher free energy than the other. Then the second crystal can grow into the first and the rate of boundary motion can be measured. Since such experiments are very laborious, only few investigations are carried out; e.g. those of Liebmann, Lücke and Masing[25] and, most of all, Aust and Rutter[20]. Since they are often cited, they shall not be described here again. The main findings of these works, especially those of Aust and Rutter, are that coincidence boundaries have a higher mobility than random boundaries, and that this effect exists only for metals of normal purity but not for metals of extremely high purity. This is interpreted by the authors that in normal purity metals the impurities segregate at random boundaries, thus diminish their velocity by the impurity drag effect; but no such segregation occurs at coincidence boundaries, hence they retain nearly the velocity of the high purity metal. It would be of great interest to check also for other metals whether for extreme purity the boundary mobility is orientation independent.

2. Growth selection experiments. Growth selection experiments are much simpler to carry out. Many crystals of different orientations are allowed to grow into a deformed single crystal matrix. If there is a difference in growth rate among the different crystals, a growth selection will take place, and only the fastest growing crystal with the most mobile grain boundary will finally be found[26].

In this kind of experiments, that the finally observed orientation is a consequence of growth selection and not of oriented nucleation is shown by the results[27] of a 99.999% pure aluminum in Fig. 18. Figure 18a gives the stereographic projections of the $\{111\}$ poles of the growing crystals with those of the deformed matrix in standard projections, showing the orientation distributions of the growing crystals at 3 subsequent stages of growth selection. Figure 18b shows the corresponding distributions of the rotation angles for the best fitting $\langle 111\rangle$ - rotation axes. These results indicate that growth selection starts with a rather random orientation distribution (stage I), and ends in only one type of orientation which is characterized by a 42° rotation around a common $\langle 111\rangle$ axis* (stage III).

Figure 19 shows results of growth selection in aluminum single crystals with various degrees of deformation and an alloy addition (0.5 at.% Mn). One recognizes that a higher degree of deformation leads to the same ideal orientation relationship (again $\approx 40^{\circ}$ $\langle 111\rangle$) but to a sharper orientation distribution[29]. An addition of only 0.5 at.% Mn in solution, however, changes the orientation relationship[30]. The best fitting $\langle 111\rangle$ -rotation axes cluster at a small distance from the $\langle 111\rangle$ poles of the matrix (near $\langle 335\rangle$), and the angle of rotation is 154°.

These observed orientation relationships are at present not yet understood. The frequently observed $\approx 40^{\circ}$ $\langle 111\rangle$ relationship has often been interpreted as a favouring of the 38° $\langle 111\rangle$ coincidence boundary. This explanation, however, is unsatisfactory to us for the following reasons:

1. The orientation distribution as seen from Fig. 19b is rather broad ($\pm 15^{\circ}$). It covers the range of many coincidence boundaries.

2. As can be seen clearly in the case of narrow distributions (Fig. 18, stage III; and Fig. 19, 80% rolled), the angle of rotation is definitely not 38°, but closely around 42°.

3. Both the position of the rotation axis and the angle of rotation change with the addition of solute atoms, as is shown for the Al-Mn alloy in Fig. 19. This is not in agreement with the idea of coincidence boundaries.

---

* The rotation angle of 162°, shown in Fig. 18b, is identical with 42° because of the three-fold symmetry of the $\langle 111\rangle$ axes. That the angles around 160° instead of around 40° have been chosen here is due to the method of selecting the best fitting rotation axes [26,28].

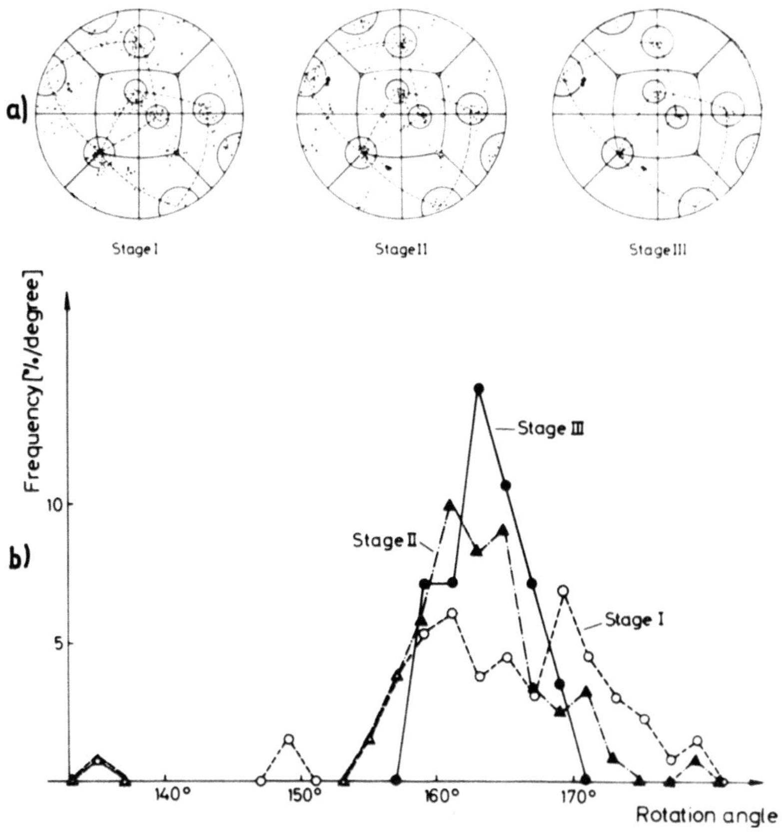

Fig. 18. Growth selection in the 20% rolled aluminum single crystals as observed at different stages.

(a) {111} poles of the new crystals (deformed matrix in standard projection);

(b) Frequency of the rotation angles around the best fitting <111> rotation axes.

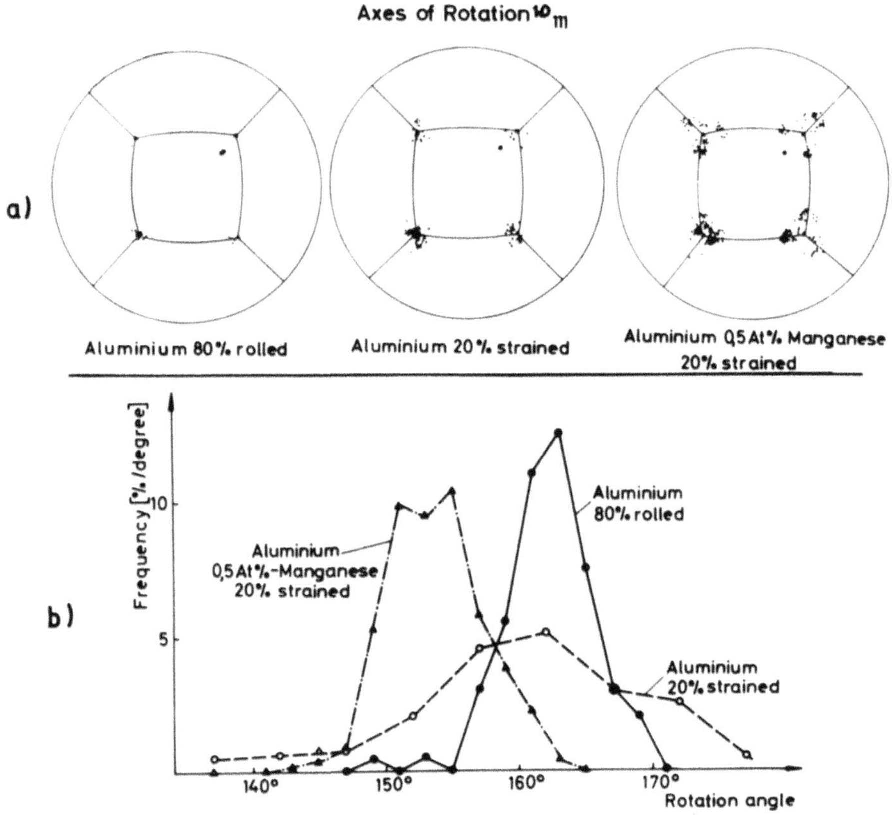

Fig. 19. Growth selection in single crystals of pure aluminum rolled 80%, or strained 20% and of Aluminum - 0.5 at.% Manganese strained 20%.

(a) Best fitting <111> rotation axes (deformed matrix in standard projection);

(b) Frequency of the rotation angles around the best fitting <111> rotation axes.

As pointed out previously, not only the orientation difference between the growing grain and the deformed matrix, but also the orientation of the boundary plane influence the migration rate. Fig. 20 shows the results of growth selection in fcc Al and in bcc Fe-3% Si (in the latter case the fastest growing orientations are 27° and 84° <110>). A comparison of the rotation axes with the position of the grain-boundary plane normal shows that most grains were in such an orientation that in aluminum the moving boundary has tilt character, and that in Fe-3% Si, twist character. This leads to the conclusion that tilt boundaries in aluminum and twist boundaries in Fe-3% Si are the fastest moving ones.

Formation of Recrystallization Textures by Growth Selection

It was due mainly to Beck[31] that in many instances recrystallization textures are formed by growth selection. Figure 18 (first pole figure) gives an example for the case that the rolled matrix consists only of a single orientation. In polycrystalline materials, it is much more difficult to conduct such growth selection studies, since the rolled matrix consists of many orientations. In many cases, however, one obtains for the rolled polycrystalline matrix an orientation distribution characterized by a rather narrow scattering around some ideal orientations. Because of the symmetry of the rolling process, one has in the simplest cases two or four of such components with crystallographically equivalent orientations.

In the single-crystal matrix, growth selection leads to the orientation having the maximum growth rate with respect to the one matrix orientation. In polycrystals, however, the new grain, in order to obtain a large volume, must be able to grow into the matrices having several quite different orientations*. Therefore, in contrast to a single-crystal matrix, in polycrystals not those grains having maximum growth rate with respect to one component of the deformed matrix will finally dominate, but those which can grow sufficiently fast in several components of the deformation texture. For this reason it is not possible to obtain information on maximum growth rate orientations from polycrystalline recrystallization experiments. In such cases, instead a "compromise texture" is obtained, which is characterized by the fact that the new orientations are as close as possible to the maximum growth rate orientations with respect to several components of the rolling texture.

---

* Of course, this is true only if the recrystallized grain has a sufficient size (see Lücke and Rixen[32]).

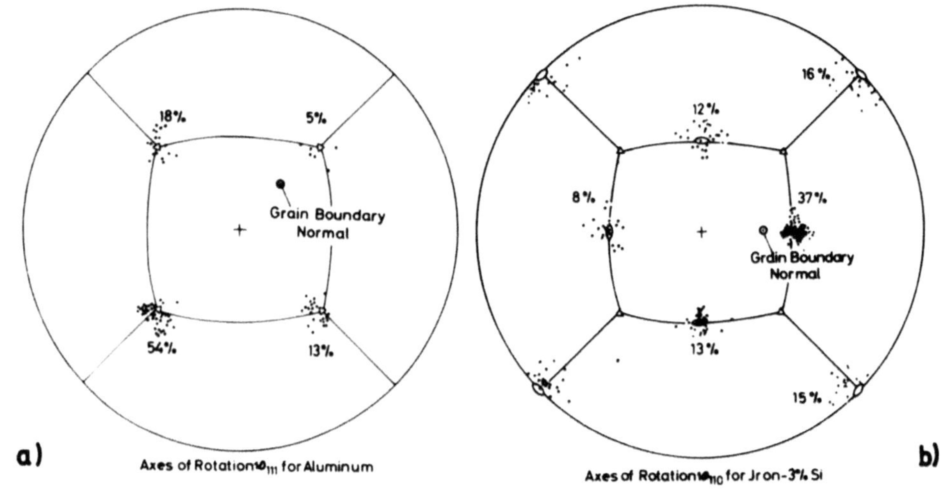

Fig. 20. Best fitting rotation axes for Aluminum <111> and for Iron - 3% Silicon <110> in relation to the grain boundary normal.

This was shown in great detail by Lücke and Rixen[32] for hexagonal metals. In their paper it was outlined that the grain-growth textures following the primary recrystallization also are a result of this mechanism. In the following, two examples for texture formation during primary recrystallization in fcc metals shall be given. They can be explained on the basis of compromise texture if one assumes those orientations given by a $\approx 40^{o}$ <111> -rotation corresponding to maximum growth rate orientations.

1. Figure 21a shows the {111} pole figure of the rolling texture of α-brass which consists of two symmetrical components[33]. Figure 21b shows two orientations which are obtained by a 40° <111> -rotation around the pole A and pole B of the two components. Figure 21c shows the observed recrystallization texture[34]. One recognizes that its ideal orientation lies between, and very close to, the two orientations shown in Fig. 21b. This means that this "compromise orientation" can grow with ease in both components of the deformation texture. Another "compromise orientation" i.e., (112) [110], can also theoretically be deduced from a $\approx 30^{o}$ rotation around the <111> axis in the transverse direction, which is common to both components of the deformation texture. However, this orientation for unknown reasons is not observed in the recrystallization texture. According to Beck[35], this (112) [110] texture is unfavorably oriented for growth in relation to the principal orientation spread of the deformed α-brass.

2. Figure 22a shows the rolling texture of an Al single crystal having initially (111) [112] orientation. One sees that due to rolling this crystal is split into two orientations. Here two good compromise orientations can be derived. Figure 22b gives the one obtained by 40° <111> rotations around the poles A_1 and B_1, and Fig. 22c the one obtained by similar rotations around A_2 and B_2. Correspondingly, the observed annealing texture (Fig. 22d) shows both compromise orientations[36].

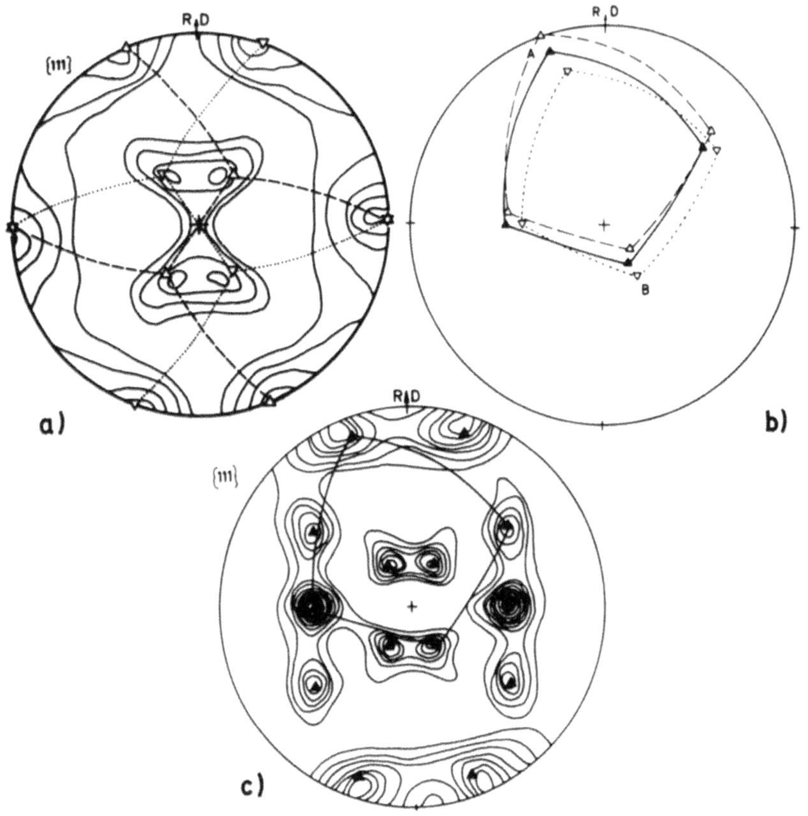

Fig. 21. Recrystallization texture formation in α-Brass

(a) Deformation texture of Brass - 20% Zinc, rolled 95% at 25°C ; △▽ (011) [112].

(b) △▽ 40° <111> -rotations about axes A and B, respectively, of (011) [112]
▲ Compromise orientation (326) [63$\bar{4}$]

(c) Annealing texture, 400°C, 30 min
▲ Compromise orientation (326) [63$\bar{4}$]

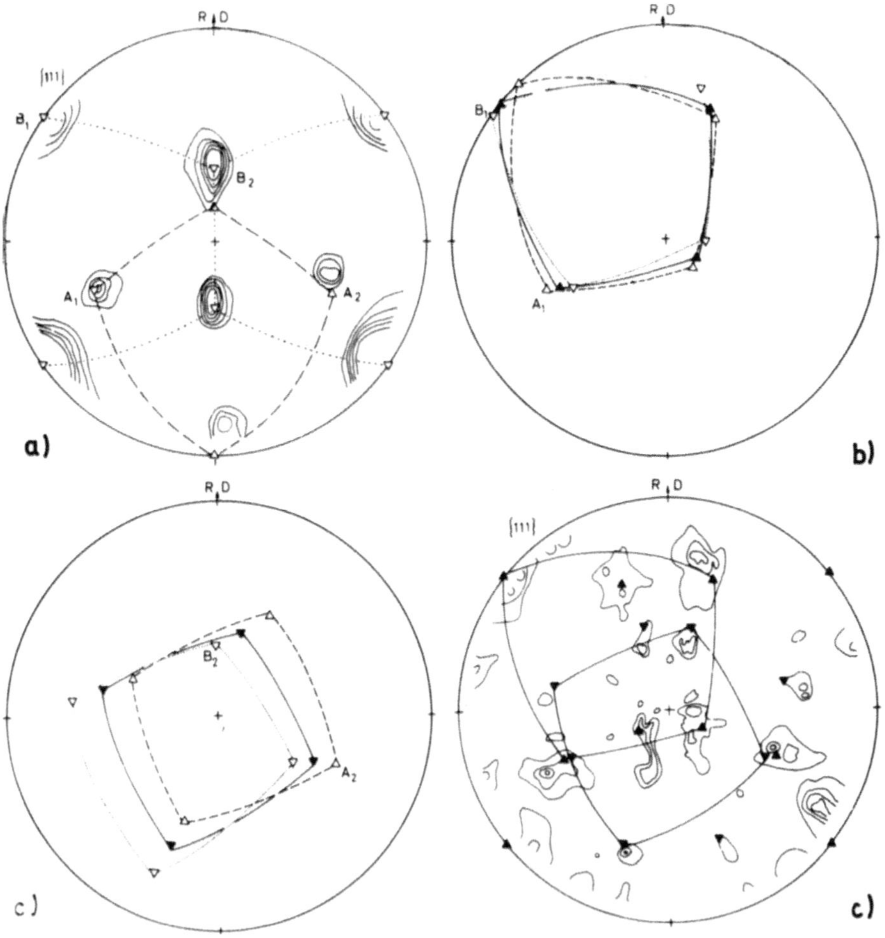

Fig. 22. Recrystallization texture formation in rolled aluminum single crystal of initial orientation (111) [11$\bar{2}$]

(a) Aluminum single crystal, rolled 80%, initial orientation (111) [11$\bar{2}$]
△ (112) [11$\bar{1}$]   ▽ (110) [001]

(b) △▽ 40° <111> -rotations about axis A_1 of (112) [11$\bar{1}$] and axis B_1 of (110) [001]
▲ Compromise orientation (112) [5$\bar{1}\bar{2}$]

(c) △▽ 40° <111> -rotations about axis A_2 of (112) [11$\bar{1}$] and axis B_2 of (110) [001]
▼ Compromise orientation (117) [52$\bar{1}$]

(d) Annealing texture, 350°C, 1000 sec ___
▲▼ Compromise orientations (112) [5$\bar{1}\bar{2}$] and (117) [52$\bar{1}$]

REFERENCES

1. H. Gleiter, Phys. Stat. Sol. (b), 45, 9-38, (1971).

2. K. Lücke and H. P. Stüwe, in Recovery and Recrystallization of Metals, ed. by L. Himmel, Interscience, New York, 1963, pp.171-210.

3. M. Weins, B. Chalmers, H. Gleiter and M. Ashby, Scripta Met., 3, 60, (1969).

4. G. F. Bolling, Ford Publication Preprint.

5. H. Gleiter, Acta Met., 17, 853-62, (1969).

6. F. Haessner and S. Hofmann, Z. Metallkde, 62, 807-10, (1971).

7. M. Feller-Kniepmeier and K. Schwartzkopf, Acta Met., 17, 497-503, (1969).

8. K. Lücke and K. Detert, Acta Met., 5, 628-37, (1957).

9. P. Gordon and R. A. Vandermeer, in Recrystallization, Grain Growth and Textures, ASM, Metals Park, Ohio, 1966, pp.205-66.

10. K. Lücke, Z. Metallkde., 41, 114-24, (1950).

11. R. Smoluchowski, Phys. Rev., 83, 69, (1951).

12. N. F. Mott, Proc. Phys. Soc. (London), 60, 391, (1948).

13. J. W. Cahn, Acta Met., 10, 789-98, (1962).

14. K. Lücke and H. P. Stüwe, Acta Met., 19, 1087-99 (1971).

15. F. W. Rosenbaum and K. Lücke, to be published, F. W. Rosenbaum, Doctor-Thesis, T. H. Aachen, (1971).

16. H. Gleiter, Z. Metallkde, 61, 282-86, (1970).

17. C. J. Simpson, K. T. Aust, and W. C. Winegard, Met. Trans., 2, 987-91 and 993-97, (1971).

18. C. Frois and O. Dimitrov, Mem. Sci. Rev. Met., 59, 643-48, (1962).

19. P. Gordon and R. A. Vandermeer, Trans. TMS-AIME, 224, 917-28, (1962).

20. K. T. Aust and J. W. Rutter, Trans. TMS-AIME, 215, 119-27, (1959).

21. W. Grünwald and F. Haessner, Acta Met., 18, 217-24, (1970).

22. E. Scheucher, Z. Metallkde, 60, 422-28 and 808-12, (1969).

23. B. B. Rath and H. Hu, Trans. TMS-AIME, 245, 1243-52 and 1577-85, (1969).

24. J. P. Drolet and A. Galibois, Acta Met., 16, 1387-99 (1968).

25. B. Liebmann, K. Lücke and G. Masing, Z. Metallkde, 47, 57-63, (1956).

26. G. Ibe and K. Lücke, in Recrystallization, Grain Growth and Textures, ASM, Metals Park, Ohio, 1966, pp. 434-447.

27. G. Sixt, Diplom-Thesis, T. H. Aachen, (1969).

28. G. Ibe, W. Dietz, A. C. Fraker and K. Lücke, Z. Metallkde, 61, 498-507, (1970).

29. R. Rixen, Diplom-Thesis, T. H. Aachen, (1966).

30. H. Perlwitz, Diplom-Thesis, T. H. Aachen, (1963).

31. P. A. Beck, (a) Acta Met., 1, 230-34, (1953); (b) Phil. Mag. Sup. (Advances in Physics), 3, 245-324, (1954).

32. K. Lücke and R. Rixen, Met. Trans., 1, 259-66, (1970); and Discussion, ibid, 2342-44, (1970).

33. R. Alam, Doctor-Thesis, T. H. Aachen, (1967).

34. U. Gebauer (now U. Schmidt), Diplom-Thesis, T. H. Aachen, (1970).

35. Ref. 31(b), p.309.

36. J. Hintermayer, Diplom-Thesis, T. H. Aachen, (1971).

THE BEHAVIOR OF GRAIN BOUNDARIES DURING RECRYSTALLIZATION OF DILUTE ALUMINUM-GOLD ALLOYS*

R. A. Vandermeer

Metals and Ceramics Division, Oak Ridge National Laboratory, Oak Ridge, Tennessee 37830 USA

ABSTRACT

The early stage of isothermal recrystallization of five polycrystalline alloys of zone-refined aluminum containing different amounts of gold and deformed 40% by rolling at 0°C was studied by quantitative metallography. All results could be interpreted in terms of a site-saturated, matrix grain edge nucleation at zero time with subsequent growth controlling the recrystallization kinetics. The length of the matrix edges producing recrystallized colonies depended on the penultimate grain size and was longer the larger that grain size. A modification of the original phenomenological model was proposed which accounted for the observed discrepancies between the measured recrystallization parameters and the ones calculated from the earlier model. The presence of $AuAl_2$ precipitates had no apparent effect on the nucleation mode. However, the apportionment of gold into a precipitated state caused an enormous (10^4) increase in boundary migration rate (recrystallization rate). With the exception of an alloy containing 3.4 ppm gold, the activation energy for recrystallization was 25,700 cal/mole, did not vary with the extent of recrystallization, and was in approximate agreement with the solute drag theory of boundary migration. The migration behavior of the 3.4 ppm gold alloy suggested the possibility of a grain boundary structure transformation. A theory of vacancy-enhanced grain boundary migration during recrystallization of dilute alloys was proposed to account for the observations.

---

*Research sponsored by the U. S. Atomic Energy Commission under contract with the Union Carbide Corporation.

INTRODUCTION

The annealing phenomenon in which comparatively strain-free grains form within a strained matrix and grow until the latter is consumed is called primary recrystallization. In polycrystalline materials, nucleation of new grains frequently takes place preferentially on sites located at the grain boundaries of the cold-worked matrix. For example, recrystallization studies[1,2] of moderately deformed, dilute zone-refined aluminum alloys clearly established that some of the ancestral grain boundary edges, i.e., triple points of the matrix grain structure on a plane of polish, were highly preferred nucleation sites, and the new grains were strung out in colonies along these edges.

The growth of new grains is accomplished through the migration of high-angle grain boundaries and thus factors affecting the rate of boundary migration play a dominant role in controlling recrystallization kinetics. Recovery processes occurring in the cold-work matrix ahead of the migrating recrystallization interface can noticeably lower the instantaneous driving force for growth, thereby, slowing down recrystallization.[3,4]

Solid solution impurities in minute quantities can profoundly reduce boundary migration rates[5-7] and retard recrystallization. The theory formulated to explain this profound effect both in its initial development by Lucke and Detert[5] and in its more vigorous treatment by Cahn[8] and Lucke and Stuwe[9] assumed it was the force exerted by a solute atom "atmosphere" that is responsible for "capturing" a moving boundary and slowing its speed. Gordon and Vandermeer[10] tested the predictions of the theory with their own experimental data[6] obtained from recrystallization studies of zone-refined aluminum containing copper and the data of Holmes and Winegard[11] from post-recrystallization grain growth studies on lead-doped zone-refined tin. Excellent correlation and agreement between theory and experiment was found. It should be obvious, therefore, that a knowledge of the nature and behavior of grain boundaries is essential to a thorough understanding of the recrystallization process. Conversely, careful studies of recrystallization in deliberately selected and well-characterized systems can provide valuable information about grain boundaries. In the present paper, recrystallization studies on zone-refined aluminum containing varying amounts of gold will be described. The experimental data will be analyzed in terms of the phenomenological grain-edge nucleated, growth-controlled model of recrystallization[1,2] which has already been shown to characterize two of the alloys.[2] The effect of penultimate grain size on certain recrystallization parameters will be emphasized in one part of this paper.

Grain boundary migration rate data were also obtained for these alloys with the aim of further testing the impurity-drag theory of boundary migration. Attention is concentrated on the activation energy, i.e., temperature dependence, of the rate. Effects of solute composition and the latter's apportionment into precipitates on the measured rates will be described. Evidence suggesting a change in grain boundary structure during migration will also be presented.

EXPERIMENTAL

The experimental details and procedures have already been described in depth[2] and will only be summarized here. The starting base material was Cominco grade 69 zone-refined aluminum with a measured resistivity ratio ($\rho_{300K}/\rho_{4.2K}$) of 2047. A 99.999% gold was melted and zone-leveled with this aluminum to compose the five alloys whose chemical analyses are listed in Table I. For some unknown reason alloy 6 had two to three times as many trace elements as the other four alloys. The directionally solidified alloys were fabricated into polycrystalline strips by a series of moderate rolling reductions alternated with recrystallization anneals.

The solid solubility of gold in aluminum has been studied by von Heimendahl.[12,13] In agreement with those findings, it was noted in this work that in all but the lowest gold containing alloy, Al_2Au precipitates (as determined by x-ray analysis) could be produced. It was concluded that at temperatures of less than 200°C, fewer than 10 atomic ppm of gold can be dissolved in aluminum at equilibrium. Because of this low solid solubility resulting in interferring solute segregation and precipitation, it was not possible to develop the identical penultimate grain size in all of the alloys. Table I gives the grain sizes, as measured by lineal analysis.[14]

For the most part, the alloys were cooled in a manner approximately equivalent to air cooling from the penultimate annealing temperature which was always at or slightly above the solubility limit. Thus the alloys were supersaturated in gold when they were given the final deformation consisting of rolling at 0°C to a 40% reduction in thickness.

Recrystallization anneals were accomplished isothermally by immersing samples directly into a boiling organic liquid bath. Metallographic measurements were made from an electropolished surface carefully prepared to avoid any surface recrystallization. The structural parameters, X_V, the volume fraction recrystallized and S_V, the grain boundary area per unit volume separating

Table I. Chemical Analysis of Alloys.

Alloy Number	Gold Content: Nominal (atomic ppm)	Gold Content: Photoelectric Analysis (atomic ppm)	Metallic Trace Elements: Heavy[1] (weight ppm)	Metallic Trace Elements: Light[2] (weight ppm)	Penultimate Grain Size (cm)
1	540	480	1.5	0.7	.060
2	103	45	2.5	1.4	.033
3	20	40	1.6	1.7	.052
4	4	3.4	1.8	0.7	.041
6	45	27	2.7	4.3	.038

[1]Includes the elements Cr, Cu, Fe, Mn, Ni, Ti, V.
[2]Includes the elements B, Ca, K, Mg, Na, P.

recrystallized regions from unrecrystallized ones were estimated using point counting analysis.[2] Annealing times were chosen so as to confine the majority of measurements to the early stages of recrystallization only, i.e., $X_V < 0.2$. For the most part, the anneals were always shorter than the times when precipitation hardening became detectable.[12]

In order to determine what effect solute segregation might have on the recrystallization characteristics of these alloys, a separate series of experiments were performed on alloy 6. After a penultimate anneal of 2 hr at 415°C one of the following procedures was executed prior to cold rolling: (1) The strip was quenched into ice water to place as much gold in solid solution as possible, or (2) the strip was given an additional 44 hr at 415°C followed by 44 hr at 300°C, 170 hr at 200°C, and seven days at 100°C with furnace cooling between steps to precipitate as much gold as possible. Procedure 2 should be more than ample to place the alloy in an overaged condition[12] with the maximum amount of the precipitated equilibrium phase Al_2Au.[13]

RESULTS

Isothermal Recrystallization Kinetics

The Avrami equation[15]

$$X_V = 1 - \exp(-Bt^n) , \tag{1}$$

where B and n are constants and t is the annealing time, is often used for describing the kinetics of recrystallization at a particular temperature. If Eq. (1) is followed, a plot of log $\ln(1/1 - X_V)$ versus log annealing time should give a straight line. Over the range of times and temperatures studied, four of the five alloys (alloy 4 is the exception) exhibited such behavior and Fig. 1 is an example.* Figure 2 demonstrates the differing behavior of alloy 4 which is revealed by the distinct change in slope at $X_V \approx 0.025$.

As previously reported,[2] temperature compensating factors capable of normalizing data such as that depicted in Fig. 1 (or 2) to a <u>single</u> curve may exist. If the temperature dependence of the

---

*Similar plots were shown in Fig. 1 of Ref. 2 for alloys 3 and 6. The compositions of these alloys as quoted in Ref. 2 were the "nominal" compositions rather than those actually determined by chemical analysis and tabulated in Table I of this paper.

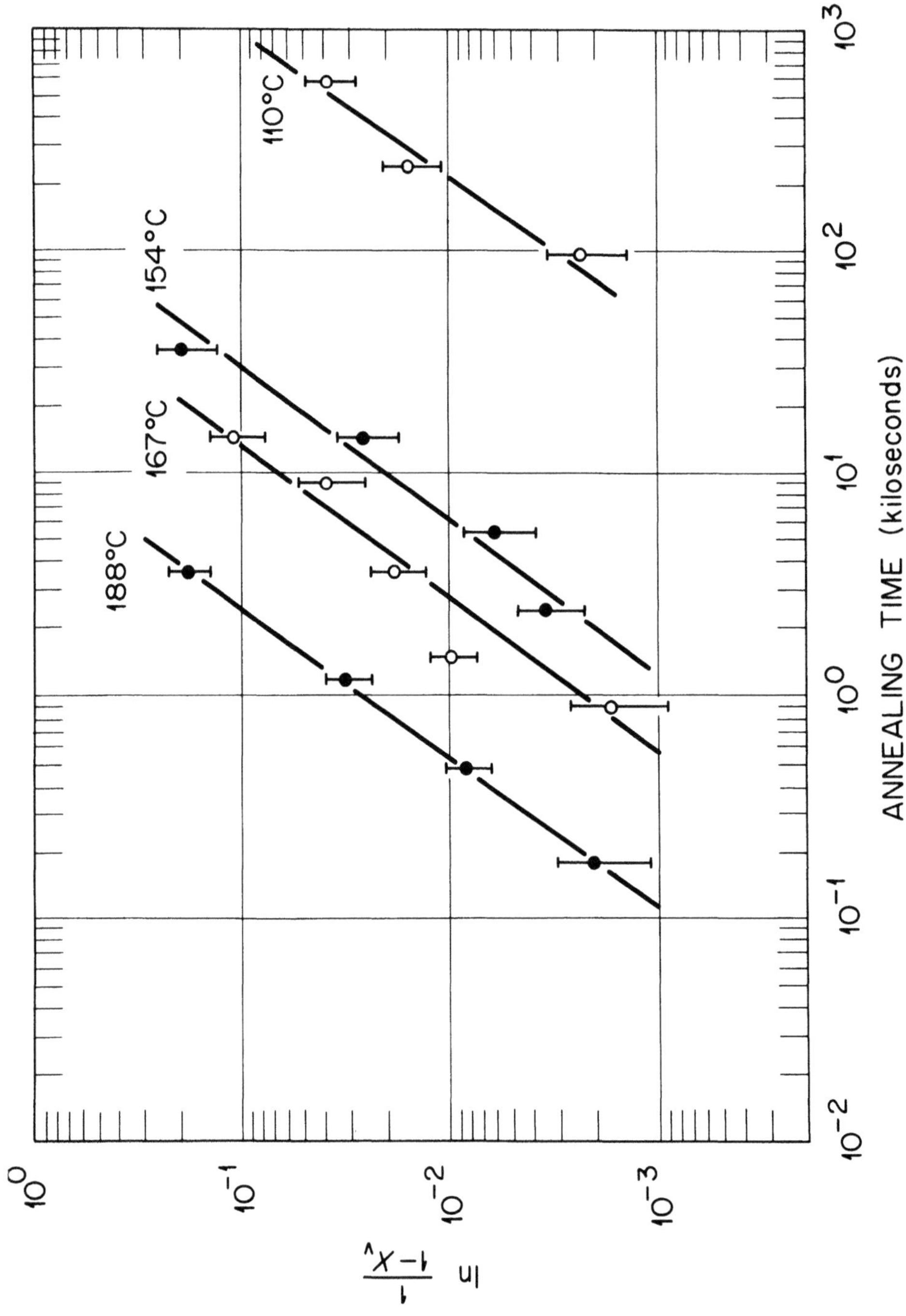

Fig. 1. Recrystallization kinetics for aluminum containing 480 ppm gold.

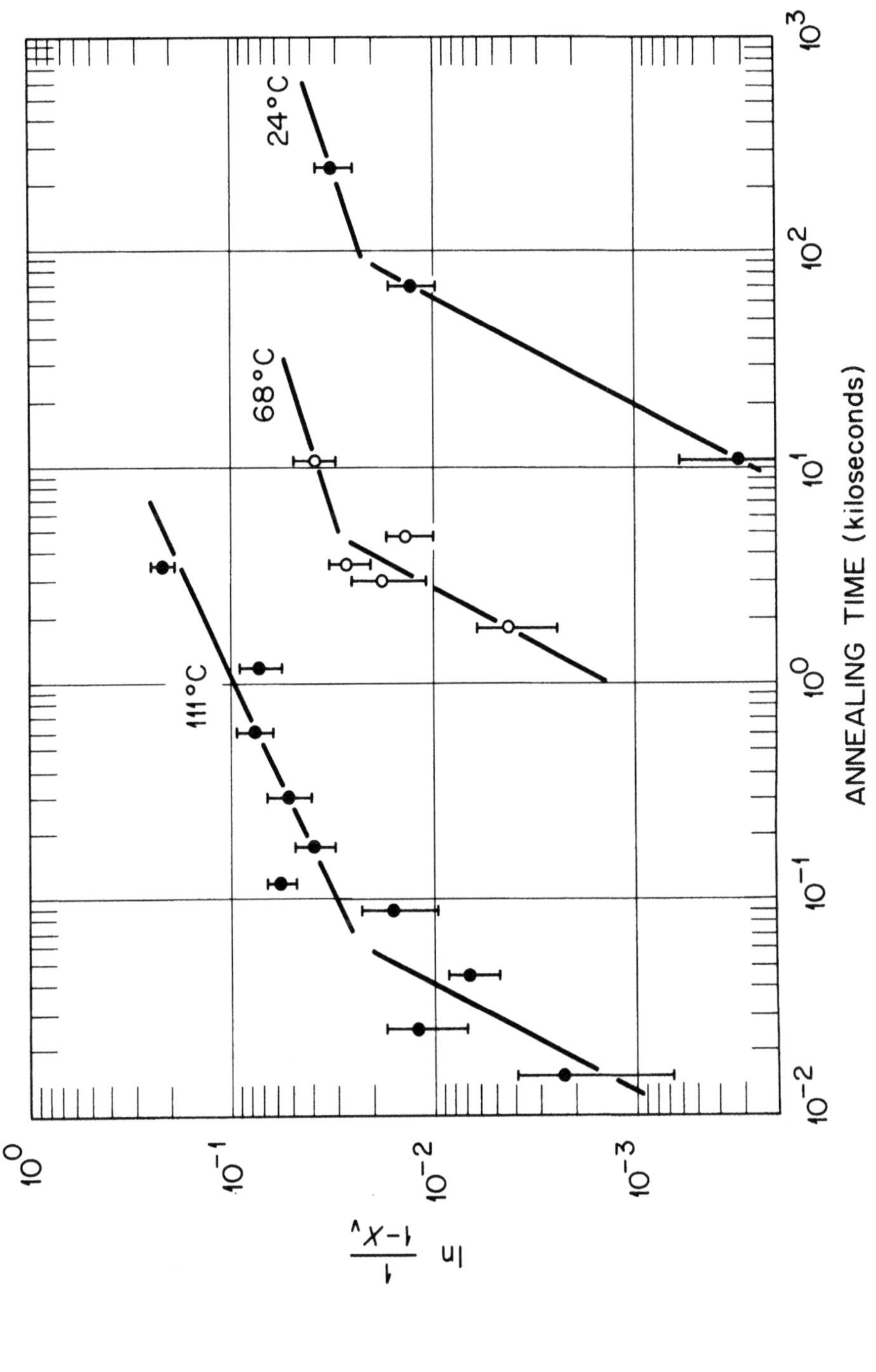

Fig. 2. Recrystallization kinetics for aluminum containing 3.4 ppm gold.

rate of recrystallization, Y, can be represented by the usual exponential (or Arrhenius-type) expression

$$Y = Y_o \exp - \left(\frac{Q_Y}{RT}\right) , \qquad (2)$$

where Y_o is the frequency factor, R the universal gas constant, T the absolute temperature, and Q_Y the activation energy for recrystallization, then a reduced time, $t(\infty)$, may be defined:

$$t(\infty) = t \exp - \left(\frac{Q_Y}{RT}\right) \qquad (3)$$

where t is the annealing time at absolute temperature, T. Substituting Eq. (3) into Eq. (1) allows the Avrami equation to be written as

$$X_v = 1 - \exp - [B(\infty)\ t(\infty)^n] \qquad (4)$$

where $B(\infty) = B \exp (nQ_Y/RT)$. The data of Figs. 1 and 2 have been replotted according to Eq. (4) in Fig. 3. Here is summarized the recrystallization kinetics for all five alloys. Since the data for alloys 3 and 6 have already been published in this form,[2] only the lines describing those data are shown here. Again four out of five alloys appear well behaved. The activation energy applied to normalize these data was Q_Y = 25,700 cal/mole.

Alloy 4, the one containing the least amount of gold did not adhere to the proposed normalization scheme. Above about $X_v = 0.03$ this alloy may be adequately represented by the suggested temperature compensating factors but for $X_v < 0.03$ this is definitely not correct. Choosing a different activation energy, Q_Y = 14,100 cal/mole, improved the situation at low X_v but worsened the fit at high X_v. It was concluded that the activation energy for recrystallization of this alloy changed rather abruptly in the vicinity of $X_v = 0.03$ from a low value of 14,100 to a much higher one approaching 25,700.

Figure 3 reveals that as gold was added to zone-refined aluminum, recrystallization was increasingly delayed, i.e., the curves are shifted to longer times. This is to be expected if impurity-dragging was operative. Alloy 6 seemed to be out of line. This may be attributed to the higher residual impurity content (two to three times as much) of this material, see the analyses of Table I.

The values of the Avrami exponent, n, assigned to the various curves in Fig. 3 are n = 2 for alloys 2, 3, and 6, and n = 3/2 for alloy 1. The curve for alloy 4 was divided into two segments. For $X_v > 0.03$, n = 1/3 and for $X_v < 0.03$, n = 2. These latter

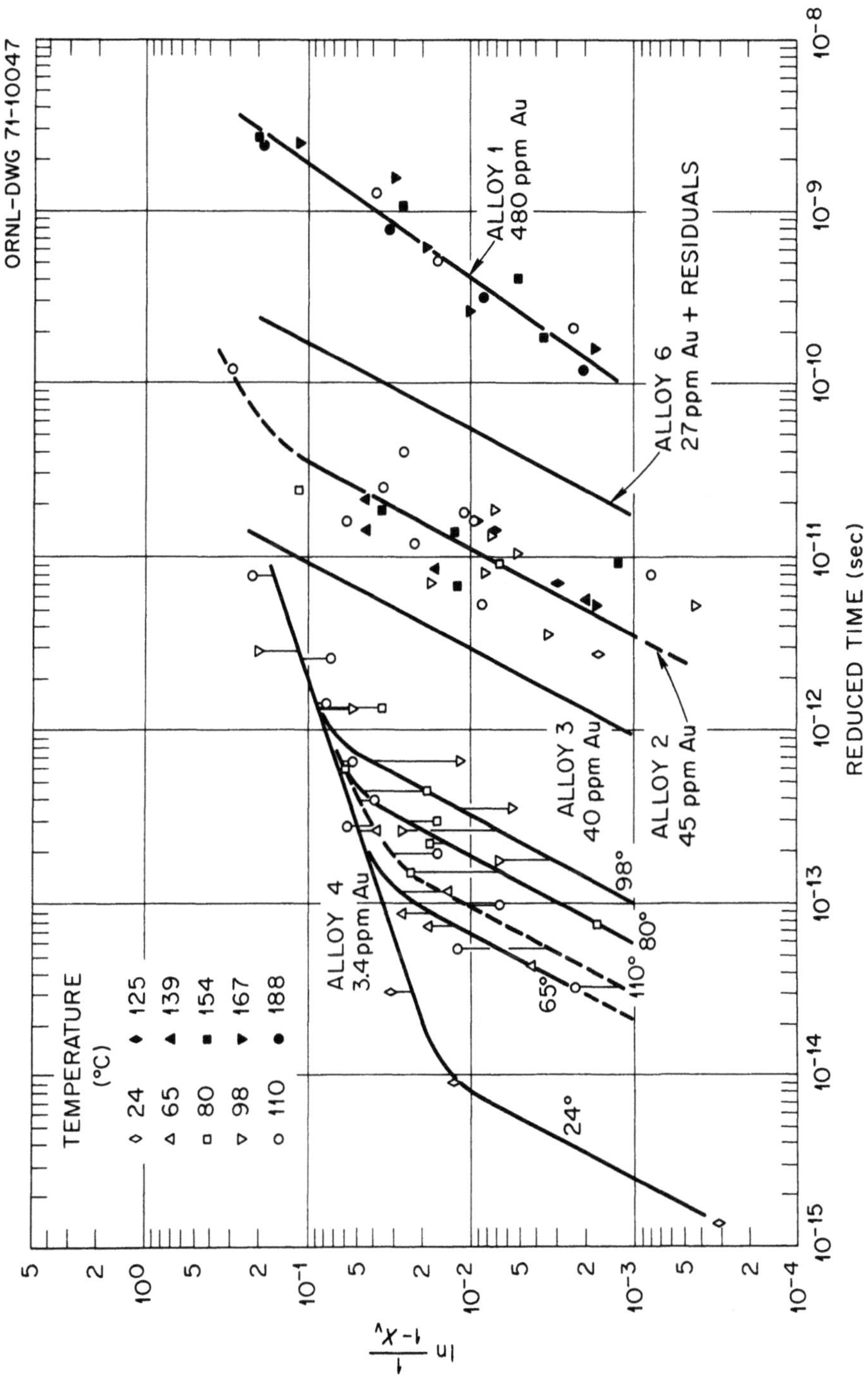

Fig. 3. Recrystallization kinetics plotted on a reduced time basis for 5 aluminum-gold alloys.

choices were somewhat arbitrary but they are reasonable representations of the data.

Relationship Between Grain Boundary Area and Volume Fraction

Figure 4 shows the interrelationship between S_v and X_v plotted on a log-log scale for four of the five alloys. (The data for alloy 6 were given in Fig. 3 of Reference 2.) Straight lines may be used to adequately describe this data except at very low values of X_v where negative deviations are evident. These deviations occurred at $X_v < 0.003$ for alloys 1, 3, and 6; they were not apparent for alloy 2 and existed to a higher X_v, i.e., $X_v \cong 0.025$, for alloy 4. These deviations are believed to arise because not all recrystallizing grains cut through by the plane of polish were resolved at low X_v by the metallographic technique. This view was supported by a previous analysis.[2] The regions of straight line behavior in Fig. 4 allowed the data to be represented mathematically by the empirical equation

$$S_v = K_s (X_v)^k \tag{5}$$

where K_s and k are experimentally determined constants for each alloy.

Average Growth Rates

$\bar{G}$, the average growth rate during recrystallization may be estimated from the experimental data using the Cahn-Hagel formula[16]

$$\bar{G} = \frac{1}{S_v} \frac{dX_v}{dt} \quad . \tag{6}$$

Combining Eq. (5) with Eq. (6) yields for the growth rate

$$\bar{G} = \frac{1}{K_s} X_v^{-k} \frac{dX_v}{dt} \quad . \tag{7}$$

Recognizing further that $X_v \cong Bt^n$ for small X_v, Eq. (7) can be written as

$$\bar{G} = K_G \, t^{n(1-k)-1} \tag{8}$$

where $K_G = \frac{n}{K_s} B^{1-k}$ and is a constant for a given alloy at a

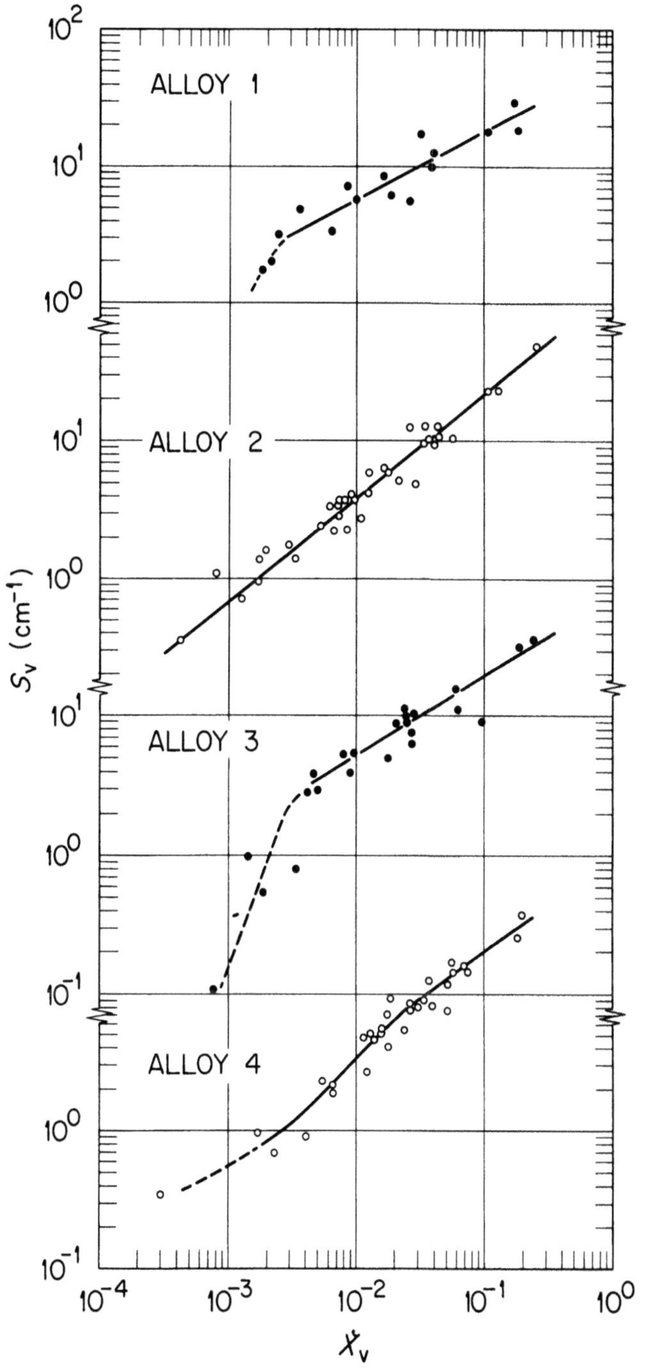

Fig. 4. Interrelationship between S_v and X_v plotted on log-log coordinates.

particular temperature. According to Eq. (8), $\bar{G}$ would remain constant during annealing if $1 - k = 1/n$. None of the alloys studied exhibited this behavior; $\bar{G}$ always decreased with time. It was assumed that recovery processes competing for the stored energy of cold work were responsible for the decrease in $\bar{G}$ with time during recrystallization.[4] Some alloys showed stronger time dependencies than others.

Table II summarizes the phenomenological parameters found in this investigation, and Table III lists the grain boundary migration rates measured at 110°C. Note that all the measured velocities in the gold-containing alloys were substantially lower than those in zone-refined aluminum with the exception of the 27 ppm alloy that had been fully aged prior to deformation and recrystallization. The rate in this aged alloy in fact appeared to have about the value expected for unalloyed, zone-refined aluminum at that temperature.

Activation Energies

Figure 5 portrays the temperature dependence of the measured migration rates for the 27 ppm and 3.4 ppm gold alloys and compares it to zone-refined aluminum.[6] Both Tables II and III tabulate the activation energies calculated from the data. Alloys containing more than 3.4 atomic ppm gold exhibited the same activation energy for boundary migration within experimental error. As alluded to before, a transition in activation energy seemed to take place in the 3.4 ppm alloy. Below $X_v \approx 0.025$ and for temperatures less than 110°C an activation energy of 14,100 cal/mole was evident. Beyond $X_v = 0.03$, the activation energy changed and approached a value of 25,700 cal/mole.

Effect of Solute Atom Segregation

Figure 6 illustrates the enormous effect of solute disposition on the recrystallization kinetics (on a reduced time basis) of the alloy containing 27 ppm gold. Water quenching from the penultimate annealing temperature (at the solubility limit), thereby allocating a substantial amount of gold to solid solution, caused recrystallization after cold-rolling to be sluggish (i.e., requiring long times). It was slower (by almost a factor of 2) than it was in specimens given the usual air cooling from the penultimate anneal temperature (dotted line in Fig. 6). In the fully segregated state on the other hand recrystallization happened extremely fast; so rapidly that at comparable temperatures it was in its later stages (i.e., high X_v) already after the shortest feasible experimental times. The recrystallization rate

Table II. Summary of Recrystallization Parameters

Alloy Number	Gold Content (ppm)	Penultimate Grain Size (cm)	Avrami Exponent n	Constant k in Eq. (5)	Time Dependency of $\bar{G}$	Activation Energy, Q_Y (cal/mole)	X_v at $t = t_c$
1	480	0.060	3/2	1/2	$t^{-1/4}$	25,700	~ 1.0
2	45	0.033	2	3/4	$t^{-1/2}$	25,700	< 0.001
3	40	0.052	2	4/7	$t^{-1/7}$	25,700	0.20
4	3.4	0.041	2 $X_v < 0.03$	---	---	14,100 $X_v < 0.03$	
			1/3 $X_v > 0.03$	2/3 $X_v > 0.03$	$t^{-8/9}$	25,700 $X_v > 0.03$	
6	27	0.038	2	3/5	$t^{-1/5}$	25,700	0.045

Table III. Boundary Migration Rates and Activation Energies

Alloy	Migration Rate, $\bar{G}$ (cm/sec) $X_v = 0.01$	Migration Rate, $\bar{G}$ (cm/sec) $X_v = 0.03$	Activation Energy (cal/mole) $Q_{meas.}$	Activation Energy (cal/mole) Q_{GB}	Activation Energy (cal/mole) Q_L^i
Zone-refined Al[a]	1.6×10^{-3} (extrapolated)		15,000	13,100[b]	
3.4 ppm Au	1.5×10^{-4}		14,100		
		5×10^{-6}	25,700		
40 ppm Au	2.7×10^{-6}		25,700		
45 ppm Au	6.9×10^{-7}		25,700	11,000 to	27,000 to
480 ppm Au	1.2×10^{-8}		25,700	14,000[c]	28,000[d]
27 ppm Au					
Air cooled	1.8×10^{-7}		25,700		
Water quenched	1.1×10^{-7}		25,700		
Aged	1.3×10^{-3}		17,000		
Al-Cu alloys[a]			29,000		32,000[a,d]

[a]Reference 6.

[b]$Q_{GB}^{Al} = 0.47\ Q_L^{Al}$ with $Q_L^{Al} = 28,000$. See Reference 27.

[c]See text.

[d]References 27 and 28.

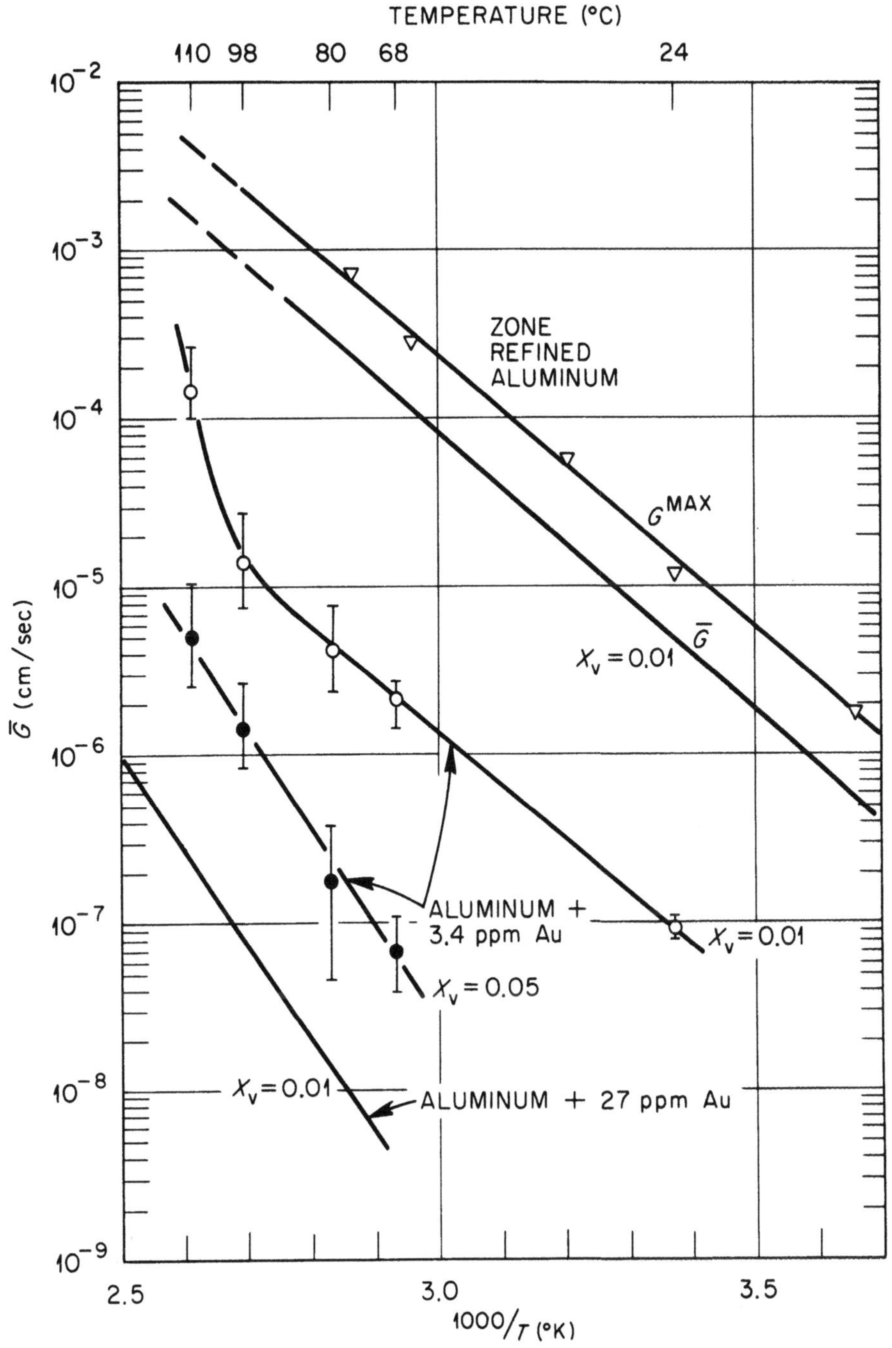

Fig. 5. Temperature dependence of grain boundary migration rate.

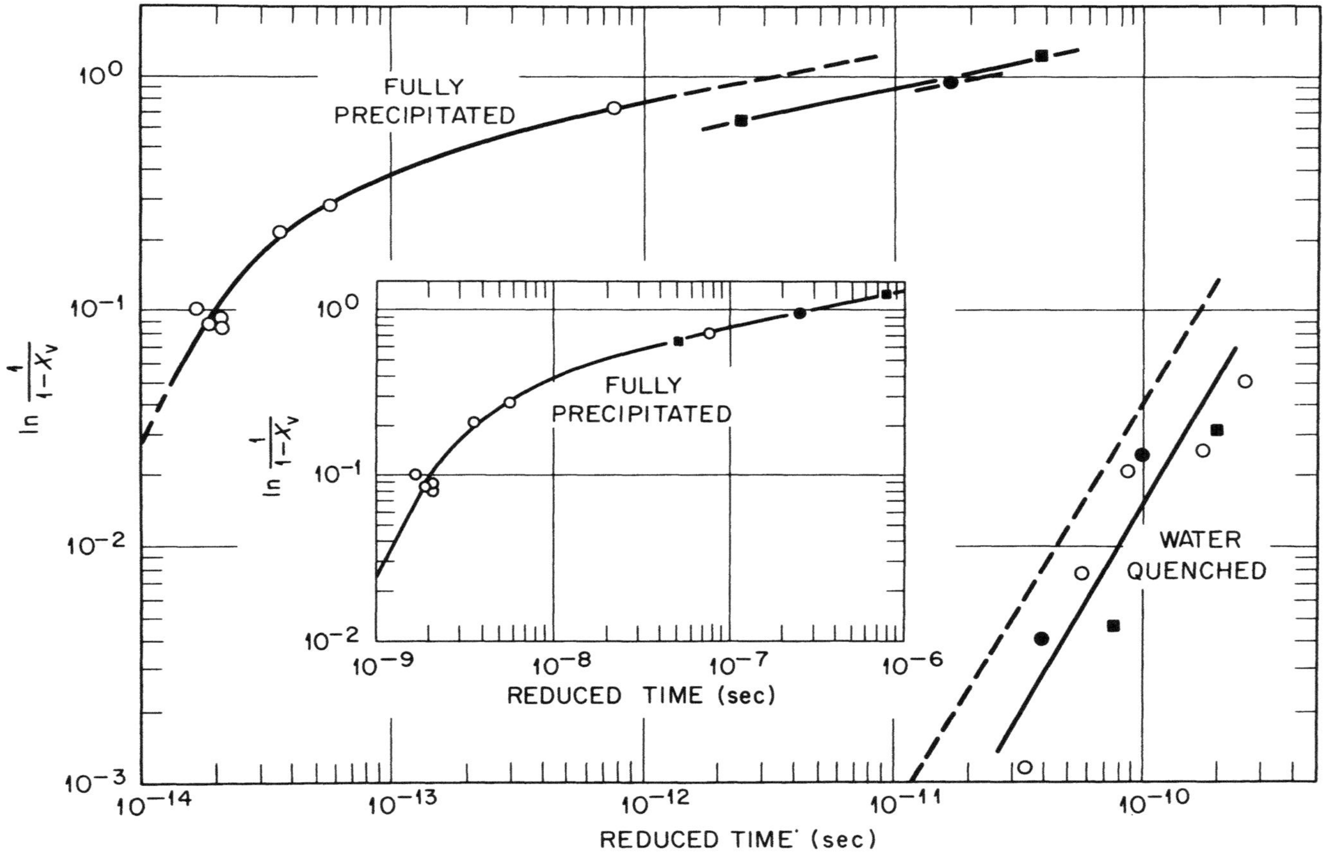

Fig. 6. Effect of state of solute segregation on the recrystallization kinetics of an aluminum-27 ppm gold alloy.

was better than 10^4 times faster (at $X_v = 0.1$) in these specimens. Since the nature of nucleation of recrystallized grains was essentially unchanged by the presence of precipitates, the explanation of this drastic effect must be sought in the mechanism of grain boundary migration.

The temperature compensating factors used in plotting Fig. 6 assumed $Q_Y = 25{,}700$ cal/mole. For the precipitated specimens an activation energy of $\sim 17{,}000$ cal/mole allowed a better normalization of the data as the insert to Fig. 6 reveals.

DISCUSSION

Phenomenology

Recrystallization Model. The important characteristics of recrystallization in these aluminum-gold alloys may be rationalized on the basis that recrystallization took place by site-saturated, matrix grain boundary edge nucleation at zero time with subsequent growth of new grains.[1,2] Further discussion and calculations will be within the framework of a previously proposed idealized, phenomenological model[1,2] based on the following assumptions: The recrystallized grains strung out in a "string of beads" fashion along a matrix grain boundary edge (called a colony of grains) may be considered to be a uniform cylinder. To a first approximation there are a constant number, N_v, of these per unit volume and each has the same length, ℓ. Recrystallization proceeds by the two-dimensional thickening of the cylinders but lengthening has been ignored. Since it was observed that grains grew at different rates,[2] a distribution of cylinder sizes from a maximum radius, R_m, to a minimum, R_o, was postulated such that the fraction having radii in the interval r to r + dr was

$$f(r)dr = Ar^{-m}\,dr \tag{9}$$

where m and A are constants.

On the basis of these postulates the volume fraction recrystallized (for small X_v) can be written as

$$X_v \cong Bt^n = \pi N_v \ell \bar{R}^2 \; F(z,m) \tag{10}$$

where $\bar{R}$ is the radius of the average colony (cylinder) in the distribution. F (z,m) is given by

$$F(z,m) = \frac{(2-m)^2}{(1-m)(3-m)} \cdot \frac{(1-z^{1-m})(1-z^{3-m})}{(1-z^{2-m})^2} \tag{11}$$

where z is the ratio, R_o/R_m. For typical values of m and z, F(z,m) will not be greatly different from unity and in a rough calculation could be ignored.

The model also predicts that

$$\frac{S_v}{X_v^{1/2}} = I_c(z,m) + \frac{2}{\ell} X_v^{1/2} \tag{12}$$

where $I_c(z,m)$ is a constant if the colony distribution does not change and ℓ is the length of the nucleating edge. The data of Fig. 4 were replotted according to Eq. (12) with the results shown in Fig. 7. The solid lines were drawn according to a least-squares analysis of the data and ℓ was calculated from the slopes with the results listed in Table IV. The deviations from the expected behavior of low X_v may be rationalized, as previously stated, in terms of a lack of resolution of the metallographic technique for small colonies. The data for alloy 4 suggest a change in the colony distribution parameters, m and z, as recrystallization proceeded. The value of ℓ for alloy 1 seemed to be exceptionally large. This was undoubtedly due to the large scatter in the data and the commensurate uncertainties arising because the slope was also very small.

Table IV. Calculated Colony Lengths

Alloy Number	Gold Content (ppm)	Penultimate Grain Size (cm)	Nucleating Edge Length (cm)
1	480	0.060	0.737(?)
2	45	0.033	0.014
3	40	0.052	0.042
4	3.4	0.041	0.024
6	27	0.038	0.023
0.008 wt % Cu[1]		0.050	0.036

From these results it was concluded that ℓ did not depend on alloy content. Rather, as Fig. 8 suggests, it appeared to be definitely related to the penultimate grain size, i.e., the matrix grain size prior to cold rolling. If grain boundary edges were the only nucleating sites, this should be expected since the

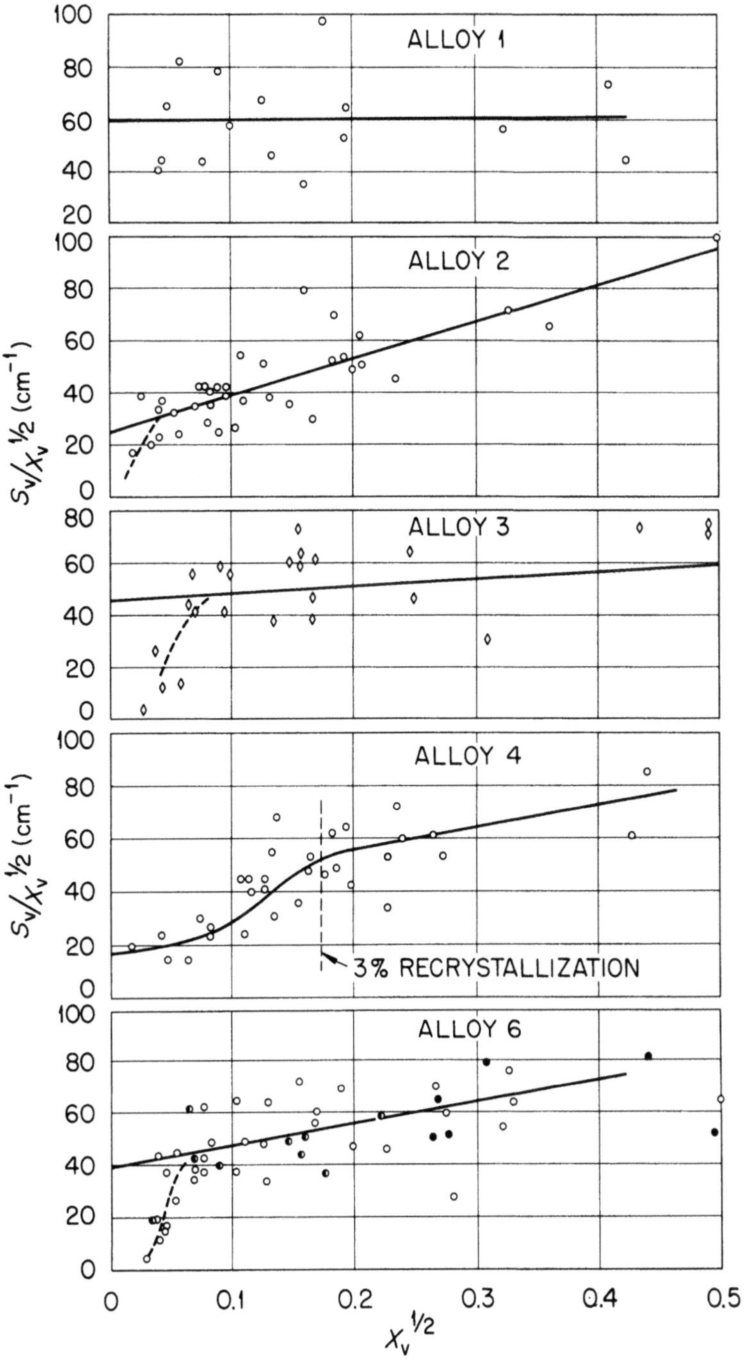

Fig. 7. Interrelationship between S_v and X_v plotted according to Eq. (12) and the cylinder model.

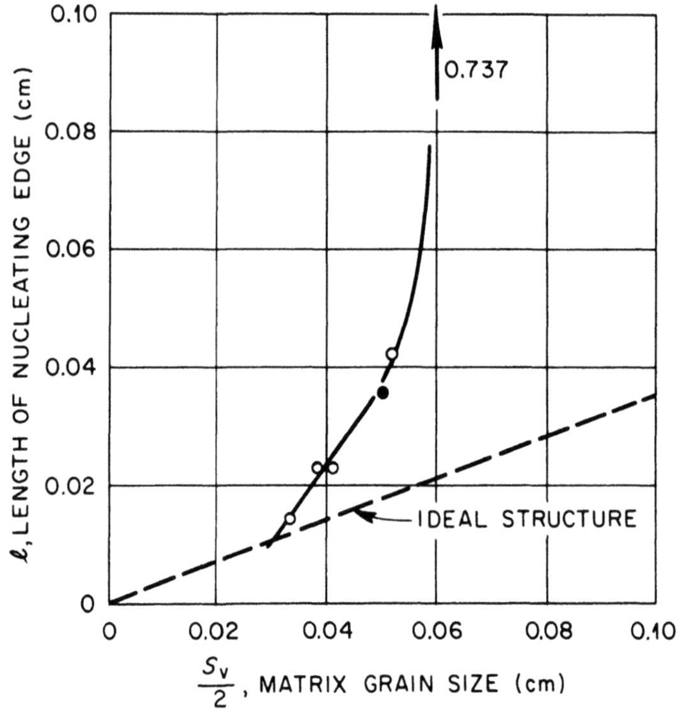

Fig. 8. Length of nucleating grain boundary edges in recrystallization versus matrix grain size in aluminum.

penultimate grain size would determine what matrix edge lengths existed in the structure and hence were available for the recrystallization nuclei. For example, in an ideal structure with matrix grains thought of as equally large tetrakaidecahedra arranged to fill space

$$\ell_T = \frac{1}{2\sqrt{2}} d_T \tag{13}$$

where ℓ_T is the edge length (all of which are equal) and d_T is the distance between square faces—a measure of the polyhedron's size.[17] The dashed line in Fig. 8 corresponds to Eq. (13). In real structures, however, the grain boundary edges do not all have the same length and Eq. (13) should only be approximate with ℓ_T and d_T being replaced by average values. The difference between observations and Eq. (13) could mean that in the larger grained structures only longer than average matrix grain boundary edges nucleate new grains.

In a recent study, Eylon, Rosen, and Niedzwiedz[18] examined recrystallization in point-deformed, high-purity aluminum single crystals. They found that point deformations introduced by a sharp-edged indentor, but not with a ball-shaped indentor, caused the appearance of a recrystallized grain after annealing. They assumed recrystallization was initiated because of a discontinuity in plastic flow around the sharp-edged indentation. A relatively dense dislocation entanglement was thought to have been created when different slip systems met under the sharp-edged indentor. Larson and Kocks[19] suggested that because macroscopic compatibility conditions impose more stringent limitations on plastic deformation at grain edges than at grain boundaries or in the grain interior, one might well expect more slip systems to be operative in these regions, i.e., they are "built-in" discontinuities. Thus, if the number and combination of previously active slip systems is important in the nucleation of a recrystallized grain, then grain edges should be highly favored nucleation sites. Furthermore Weissmann[20] observed that in deformed aluminum the subgrains with the largest disorientation angles were associated with lattice domains found at the grain edges of the deformed matrix. These subgrains also grew the most rapidly during annealing—another factor contributing to the auspiciousness of grain edge nucleation tendencies.

The state of solute segregation had no measureable effect on the mode of nucleation in these alloys. Figure 7 plots the data for alloy 6 irregardless of precipitation state. The solid points are for specimens in which as much gold as possible was precipitated. These points overlap and fall within the scatterband of points for specimens in which the gold was retained in solid solution and there were no precipitates. Such behavior would not be expected if precipitates were preferential sites for nucleation of new grains as was found by Leslie et al.[21] in iron alloys. Many more nuclei would have been created in the precipitated condition were that the case.

The Avrami Equation

All quantities in Eq. (10) were assumed by the recrystallization model to be constants except $\bar{R}$. Recognizing that $d\bar{R}/dt = \bar{G}$ and if growth starts at time t = 0, then

$$\bar{R} = \int_0^t \bar{G}\, dt \,. \tag{14}$$

Replacing $\bar{G}$ with Eq. (8) and integrating gives

$$\bar{R} = \frac{K_G}{n(1-k)}\, t^{n(1-k)} \tag{15}$$

except when $n(1-k) = 0$. Substitution of Eq. (15) into Eq. (10) yields

$$X_V = \pi N_V \ell\, F(z,m) \left[\frac{K_G}{n(1-k)}\right]^2 t^{2n(1-k)} = Bt^n. \tag{16}$$

Equation (16) reveals that k should be equal to 1/2 if the recrystallization product can in fact be represented as ideal cylinders that do not lengthen.

Table II lists the experimental constant, k, which may be compared with the expected one, 1/2 for cylinders. Only for alloy 1 was there no discrepancy between the two. This finding suggested a deficiency in the model. The discrepancies between the k's appear to be larger, the smaller was ℓ and therein was the difficulty.

Throughout previous calculations,[1,2] it was assumed for the purpose of simplification that the recrystallizing colonies (cylinders) did not lengthen, they only thickened. Thus the colony length was always taken as the nucleating edge length. As long as the colonies were long compared to their thickness, i.e., $\ell \gg R$, the approximation resulting from this assumption did not give serious error. When only a few, i.e., several percent,[1,2] of the edges present in the structure are capable of producing colonies, it would of course be more realistic to suppose that the colonies would also grow in the edge length direction, i.e., that the recrystallized grains at the grain corners, forming the ends of the colony string would grow out in all directions and not just normal to the edge. The following calculation, therefore, relaxes the restriction that ℓ be a constant. According to the previous procedure,[2] we can write

$$dX_v = vdN_v = \pi r^2 \ell_o N_v f(r)dr + \frac{4}{3}\pi r^3 N_v f(r)dr \tag{17}$$

where v is the volume of one cylinder, ℓ_o is the length of the nucleating edge and r is the radius of the cylinder. The second term in Eq. (17) takes into account the lengthening of the colony. In effect hemispherical end caps have been placed upon the cylinders of our original model. The length of the colony is now $\ell = \ell_o + 2r$ where previously $\ell = \ell_o$. We are assuming that the rate of boundary migration is the same for all the colony surfaces. Carrying out the integration of Eq. (17) over the entire distribution of cylinders gives

$$X_v = \pi N_v \ell_o F(z,m)\bar{R}^2 + \frac{4}{3}\pi N_v F'(z,m)\bar{R}^3 \tag{18}$$

where F(z,m) is still defined by Eq. (11) and

$$F'(z,m) = \frac{(2-m)^3}{(4-m)(1-m)^2} \cdot \frac{(1-z^{1-m})^2(1-z^{4-m})}{(1-z^{2-m})^3} . \tag{19}$$

Then Eq. (16) which ignored lengthening, should be replaced with

$$X_v = \pi N_v \ell_o F(z,m) \left[\frac{K_G}{n(1-k)}\right]^2 t^{2n(1-k)}$$
$$+ \frac{4}{3}\pi N_v F'(m,z) \left[\frac{K_G}{n(1-k)}\right]^3 t^{3n(1-k)} \tag{20}$$

which takes lengthening into account. According to Eqs. (18) and (20) if $\ell_o >> \bar{R}$, the second term will be very small relative to the first and thus negligible. This corresponds to the case of long, thin cylinders and the predicted k would be 1/2. In the other extreme if $\bar{R} >> \ell_o$ the first term contributes only very little to X_v and may be ignored. In this case the recrystallized colonies may be approximated as spheres rather than cylinders and the predicted k would be 2/3.

Alloy 1 with its long nucleating edges fell into the first category over the entire experimentally observed range and Eq. (16) was sufficient. For the other alloys $\bar{R} \approx \ell_o$ very quickly (because of their smaller ℓ_o's) and the second term in Eqs. (18)

and (20) could not be dropped. This caused the experimental k's for these alloys to be greater than 1/2. Alloy 3 was a borderline situation but alloys 2, 4, and 6 all required the end cap correction.

This can also be shown in another way. If, t_c, is defined as the time when the first and second terms in Eqs. (18) and (20) made equal contributions to X_v, the following equation can be obtained:

$$t_c = \left\{\frac{3}{4}\,\ell_o\,\frac{F(z,m)}{F'(z,m)}\cdot\frac{n(1-k)}{K_G}\right\}^{\frac{1}{n(1-k)}} . \tag{21}$$

The volume fraction corresponding to t_c calculated for each alloy from the experimental data has been tabulated in Table II for all alloys except alloy 4. A correlation between when the end cap contribution became sizeable and the deviation of the experimental k from 1/2 was apparent. The smaller X_v at $t = t_c$ was (also the smaller was ℓ_o), the larger was that deviation (see Table II). Alloy 2, the one with the smallest nucleating edges, showed the largest deviation and the smallest X_v at $t = t_c$. In fact the colonies in this alloy may be regarded as spherical rather than cylindrical very soon after the start of recrystallization.

Finally, if the end cap contribution is subtracted from the experimental curve as has been done in Fig. 9 for alloy 6, the resulting curve would have a slope (Avrami constant) of approximately 1.7. This compared very favorably with the value of 1.6 calculated for this alloy from the cylinder model if the growth rate was not constant. Thus it was concluded that colony lengthening cannot be considered unimportant when $\ell_o \leqslant 0.030$ and its inclusion in the model accounts quantitatively for the experimental k's being larger than the 1/2 predicted from the simplified model [Eq. (16)].

Migration Rates and Activation Energies

Impurity Drag Theory. In view of previous success[10] at explaining boundary migration rate behavior by impurity-drag phenomena, interpretation of the present results within the scope of that theory was attempted. Boundary migration rate, G, driving force, ΔF, and solute composition, C_s, may be related through the equation

$$\Delta F = \lambda G + \alpha C_s G/(1 + \beta^2 G^2) , \tag{22}$$

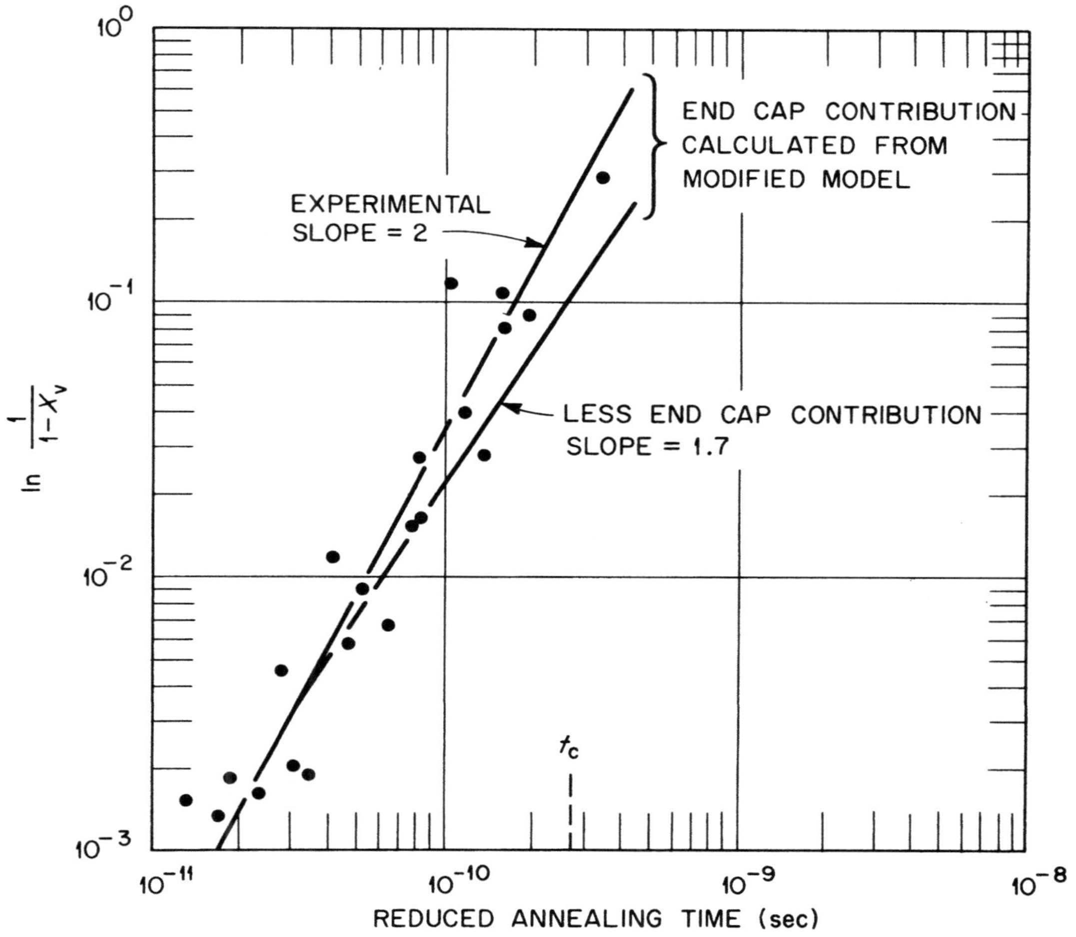

Fig. 9. End cap contribution to recrystallization kinetics in an aluminum–27 ppm gold alloy.

where λ is the reciprocal of the boundary mobility in the complete absence of solute and α and β are impurity drag parameters whose magnitudes depend on the solute atom's diffusivity in the boundary region and its interaction with the potential field of the boundary. The behavior predicted by Eq. (22) may be approximately regarded in terms of two limiting types of boundary migration and a transition between these. At high velocities the boundary sweeps through so rapidly that the slight "frictional effect" of the solute atoms causes only a minor alteration in velocity. This behavior has been referred to as solute independent migration which strictly is not correct, but practically it reasonably describes the situation. The other limiting case is the low velocity region where the moving boundary has been "captured" by a solute atmosphere which must diffuse along with the boundary. The atmosphere may be either dragged along behind or pushed ahead of the migrating boundary depending on the sign of the interaction potential. Low velocity behavior may be represented by the equation

$$G = \Delta F/(\lambda + \alpha C_s), \tag{23}$$

or if $\alpha C_s >> \lambda$ as

$$G = \Delta F/\alpha C_s \tag{24}$$

which is equivalent to the formula derived in the first theoretical work on this problem by Lucke and Detert.[5]

The impurity-drag theory cannot as yet tell from first principles whether a measured velocity is a low one or a high one. However, migration rates measured in zone-refined aluminum, i.e., $C_s \cong 0$ with a history similar to the present alloys[6] give a clue as to what magnitude to expect for a high velocity. With reference to Table III and Fig. 5, it was concluded that in the alloys the measured velocities were suggestive of low (or in one case perhaps transition region) velocity behavior except for the 27 ppm gold alloy in the fully aged condition.

The remarkable increase in migration rate achieved in alloy 6 by merely altering the solute atom arrangement from a state of supersaturated solid solution to one of complete precipitation can be rationalized by the impurity drag theory. With most of the gold in solid solution, i.e., $C_s \approx 27$ ppm, low velocity behavior was apparent. Note that a low velocity and high activation energy were measured (Table III).

When long time aging followed the penultimate anneal but preceded cold-working and recrystallization, the matrix was effectively drained of gold atoms which then mostly resided in the precipitate particles. During subsequent recrystallization this lack of gold in solution could give the migrating boundaries the impression they were traveling through a pure material, provided the second phase inclusions did not strongly interact with the moving boundaries in the manner proposed by Zener.[22] Whether they could or not depended on the magnitude of the inclusion restraining force relative to the driving force. The restraining force may be estimated by[22]

$$\Delta F_I = \frac{3}{2}\frac{f\sigma}{r} \tag{25}$$

where f is the volume fraction of precipitate present; σ is the particle-matrix interfacial energy and r is the particle radius. For the alloy containing 27 ppm gold, $f = 0.89 \times 10^{-4}$. Taking $r = 500$ Å[13] and $\sigma = 600$ ergs per cm^2 gave $\Delta F_I = 1.6 \times 10^4$ erg/cm^3. The driving force for boundary migration during recrystallization of cold-worked aluminum has been found to be about 1 cal/mole[4] or 4×10^6 erg/cm^3 which is more than two orders of magnitude larger than ΔF_I. Essentially this means the moving boundaries were not held up by the second phase particles. The C_s going into the impurity drag equation in this case then is the solubility limit.

This amount of solute in solution was insufficient, however, to "capture" the moving boundary for the driving force present in the material and high velocity behavior was, therefore, the result. On the basis of this interpretation and since the boundaries in this aged alloy traveled faster than those in the 3.4 ppm alloy at 110°C, it was concluded that the solubility limit at 110°C was less than 3.4 ppm Au-extremely small.

Activation Energies. Generally activation energies near that for lattice diffusion of the solute in the solvent have been observed for solute-controlled migration.[6,11,23,24] It will be shown later, however, that this need not always be the case. On the other hand, it can be argued that the activation energy for a high velocity, solute-independent migration ought to be equal to the activation energy for grain boundary diffusion.[10] Activation energies of this magnitude have been observed in very pure aluminum by a number of different investigators.[6,7,25,26] Aside from the early stage recrystallization behavior of the 3.4 ppm alloy, the present data fall into this expected pattern of either low-velocity, high-activation energy or high-velocity, low-activation energy.

For solute-controlled migration behavior, the impurity drag theory predicts a constant activation energy for migration provided the grain boundary structure does not change. The experimental data for $C_s > 3.4$ ppm Au seemed to manifest this behavior.

From Eq. (24) it follows that the activation energy for migration should be

$$\frac{-\,d\,\ell n\,G}{d\,\frac{1}{RT}} = \frac{d\,\ell n\,\alpha}{d\,\frac{1}{RT}} = Q_Y\ . \qquad (26)$$

The quantity α depends on the nature of the solute-grain boundary interaction energy, $E(x)$, coupled with the solute diffusivity, $D(x)$, where x is the distance normal to an arbitrarily chosen center plane of the boundary. From theory[8]

$$\alpha = \frac{4\,RT}{V_m}\int_{-\infty}^{+\infty}\frac{\sinh^2\frac{E(x)}{2\,RT}}{D(x)}\,dx\ , \qquad (27)$$

where V_m is the molar volume and R is the universal gas constant. For the assumed profile shown in Fig. 10 it can be deduced from Eq. (27) that

$$\alpha = \frac{4\,RT}{V_m D_L^i}\Bigg[\frac{m'\delta D_L^i}{D_{GB}^i}\cdot\frac{RT}{E_o}\left\{\sinh\frac{E_o}{RT}\left(\frac{1}{m'}-1\right)-\frac{E_o}{RT}\left(\frac{1}{m'}-1\right)\right. \qquad (28)$$

$$\left.+\sinh\frac{E_o}{RT}-\frac{E_o}{RT}\right\} - m'\delta\cdot\frac{RT}{E_o}\left\{\sinh\frac{E_o}{RT}\left(\frac{1}{m'}-1\right)-\frac{E_o}{RT}\left(\frac{1}{m'}-1\right)\right\}\Bigg]\ ,$$

which is valid for $m' \geq 1$. In Eq. (28), δ represents the half-width of the high-diffusivity (high-porosity) region of the boundary, $m'\delta$ is the distance from the boundary center a solute atom begins to interact with the boundary potential well, E_o is the maximum depth (or height) of the well at $x = 0$, and D_{GB}^i and D_L^i are the solute diffusion coefficients at the grain boundary center and the bulk lattice, respectively. The parameter m' is useful for discussions about boundaries with different relative ranges over which $E(x)$ is altered compared to $D(x)$. Two extreme cases should be considered here; Case 1, "tight" or good fit boundaries, and Case 2, "loose" or porous boundaries (see Fig. 11). A tight boundary is one in which $m' > 1$ and the region of high diffusivity is narrow. Because $D_{GB}^i \gg D_L^i$ the first term in the square brackets of Eq. (28) may be neglected and it can be derived that

$$Q_Y = Q_L^i + E_o\left(1-\frac{1}{m'}\right)\left[\frac{\cosh\frac{E_o}{RT}\left(1-\frac{1}{m'}\right)-1}{\sinh\frac{E_o}{RT}\left(1-\frac{1}{m'}\right)-\frac{E_o}{RT}\left(1-\frac{1}{m'}\right)}\right] - 2\,RT\ , \qquad (29)$$

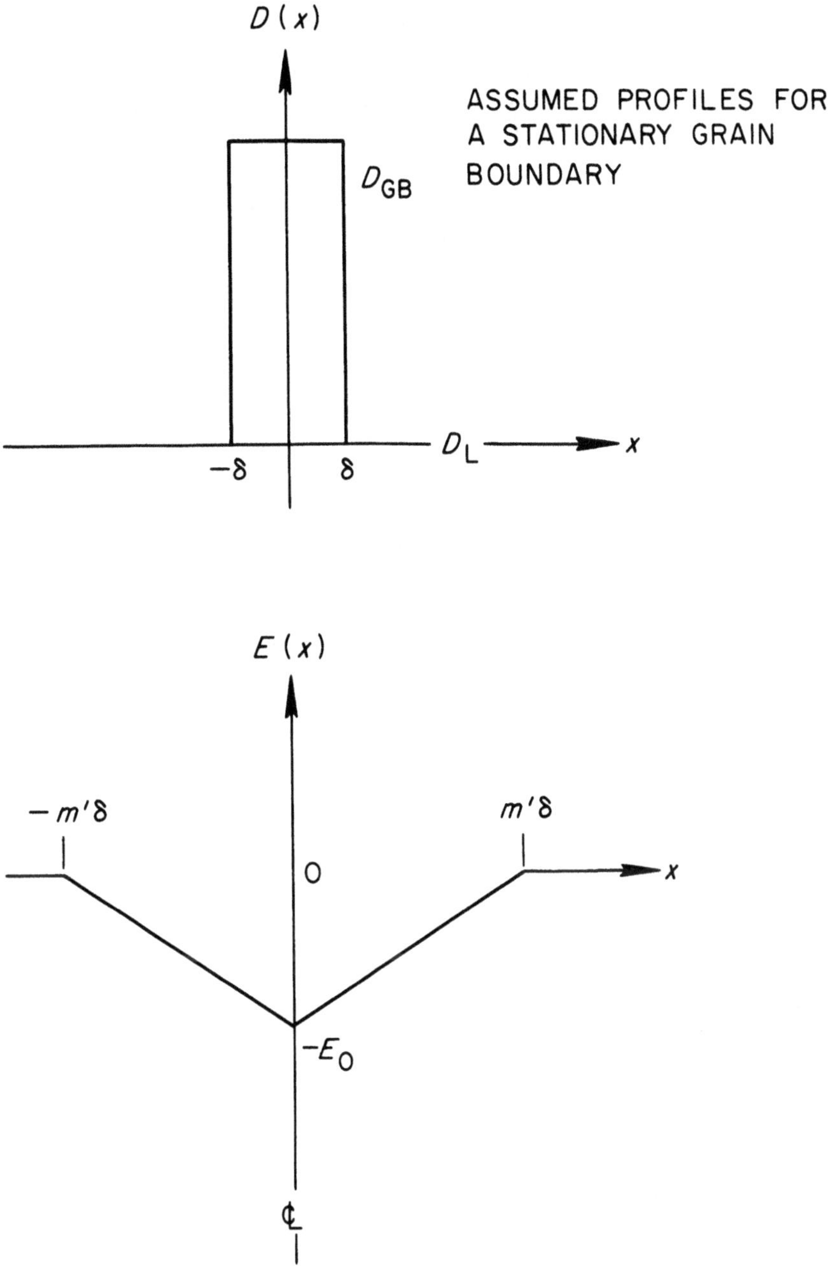

Fig. 10. The interaction energy profile and diffusivity profile assumed for calculating the impurity drag parameter α.

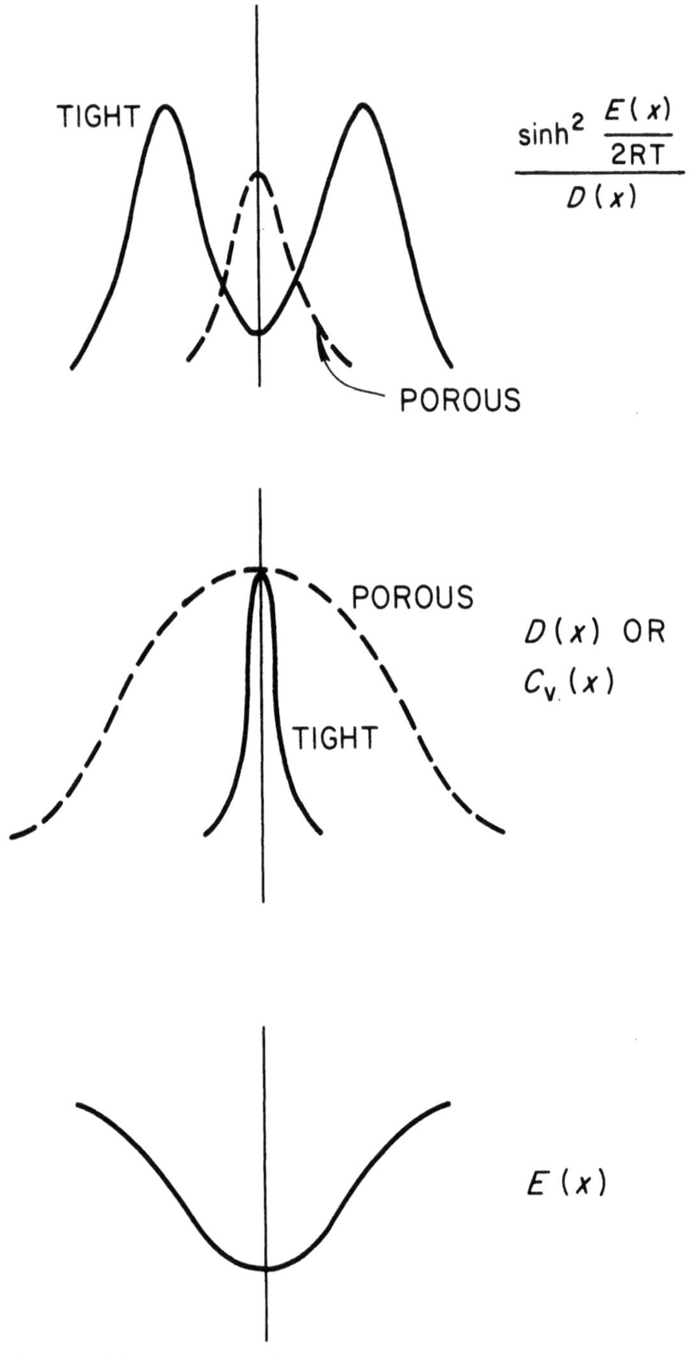

Fig. 11. Schematic interaction energy and diffusivity profiles and the impurity drag function for "tight" and "porous" boundaries.

where Q_L^i is the activation energy for bulk diffusion of the solute atom in the solvent lattice. The second term in Eq. (29) is always larger than the third. Thus the impurity drag theory predicts an activation energy $Q_Y > Q_L^i$ for a "tight" boundary.

On the other hand, a "porous" boundary would be one for which $m' \leq 1$ and the range of enhanced diffusivity is comparable to or exceeds the range of $E(x)$. For this type of boundary it can be shown that

$$Q_Y = Q_{GB}^i + E_o \left[\frac{\cosh \frac{E_o}{RT} - 1}{\sinh \frac{E_o}{RT} - \frac{E_o}{RT}} \right] - 2\,RT\ , \qquad (30)$$

where Q_{GB}^i is the activation energy for grain boundary diffusion of the solute in the solvent.

Recent lattice diffusion measurements by Peterson and Rothman[27] and by Alexander and Slifkin[28] gave for gold diffusing in aluminum Q_L^{Au} equal to 27,790 and 27,000 cal/mole, respectively. Impurity diffusion in aluminum grain boundaries has not, to the author's knowledge, been measured. A reasonable guess based on electromigration studies in aluminum alloys by d'Heurle et al.[29] and grain boundary diffusion measurements in other materials[22] is $Q_{GB} = 0.4$ to $0.5\ Q_L$, i.e., $Q_{GB}^{Au} = Q_{GB}^{Al} = 11{,}000$ to $14{,}000$ cal/mole. Comparison of the measured Q values for solute-controlled boundary migration with these diffusion measurements suggested that "tight" boundaries were migrating during recrystallization, i.e., Eq. (29) gave the most plausible agreement between theory and experiment. "Loose" boundaries, Eq. (30), would require an E_o of about 14,000 cal/mole to explain the data — a rather unlikely possibility. The early stage behavior of the 3.4 ppm gold alloy was a possible exception.

Nevertheless, in spite of this approximate agreement of the experimental Q with the impurity-drag theory (to within a few thousand calories per mole) it is somewhat disturbing that in both the present study and the previous work with copper impurities in aluminum[6] the Q's were less than Eq. (9) would predict by at least 3,000 to 5,000 cal/mole (depending on what is assumed for E_o). It is suggested that this is due to our lack of knowledge about the structure of boundaries migrating through a cold-worked matrix.

Another difficulty arose when an explanation of the early migration rate behavior of the 3.4 ppm alloy was sought. The low activation energy suggested solute-independent migration might have occurred early in recrystallization. Recovery in the cold-worked matrix ahead of the moving boundaries could then have lowered the driving force to cause a transition to solute-dependent behavior with its higher activation energy as recrystallization proceeded.[4] Two experimental facts obviated this consideration. First, the measured migration rates were a factor of 60 lower than expected for solute-independent migration (see Fig. 5). Secondly, the migration rate at 110°C in this alloy was suggestive of "breakaway"[5] or transition region behavior.[10] It would be difficult to explain this on the above mentioned proposal.

These facts do support, however, a conclusion that the early stage migration behavior of the 3.4 ppm gold alloy was in all probability solute-controlled but that the boundaries in this case were "porous" and not "tight". The predicted activation energy for migration (Eq. 30) would be approximately equal to Q_{GB} and thus was in good agreement with observation (see Table III). It must then be postulated that as recrystallization progressed, there was a change in boundary structure from a "porous" or loose structure to a "tight" one brought on probably by a reduction in velocity due to recovery. Evidence of a transition in this alloy is shown in Fig. 12; it was also apparent in the recrystallization kinetics (Fig. 2).

Transformations in grain boundary structure have been postulated theoretically by Hart.[30] Recent measurements of grain boundary energy as a function of temperature in high purity lead by Gleiter[31] showed a discontinuity in energy at a particular temperature. This observation was interpreted as evidence for a transformation in boundary structure. Abrupt changes in the temperature dependence of bicrystal boundary migration rates in lead noted by Aust[32] and of grain growth rates in dilute cadmium and lead-base alloys observed by Simpson et al.[33] were also thought to arise from a boundary structure change.

In the present circumstances the boundary structure transformation, if that indeed was what occurred, appeared to be velocity and composition dependent. No critical temperature was evident; the transition was seen only under isothermal conditions in one alloy. Therefore, criteria(on) other than just thermodynamic considerations may be responsible for this transformation. It is postulated that an enhanced vacancy concentration at the boundary was crucial to the occurrence of the boundary structure change. The remainder of this paper presents an outline and discussion of an elementary theory to explain this effect.

THEORY OF VACANCY-ENHANCED BOUNDARY MIGRATION DURING RECRYSTALLIZATION OF DILUTE ALLOYS

The present approach contains some aspects of an earlier theory proposed by In der Schmitten, Haasen, and Haessner[34] to

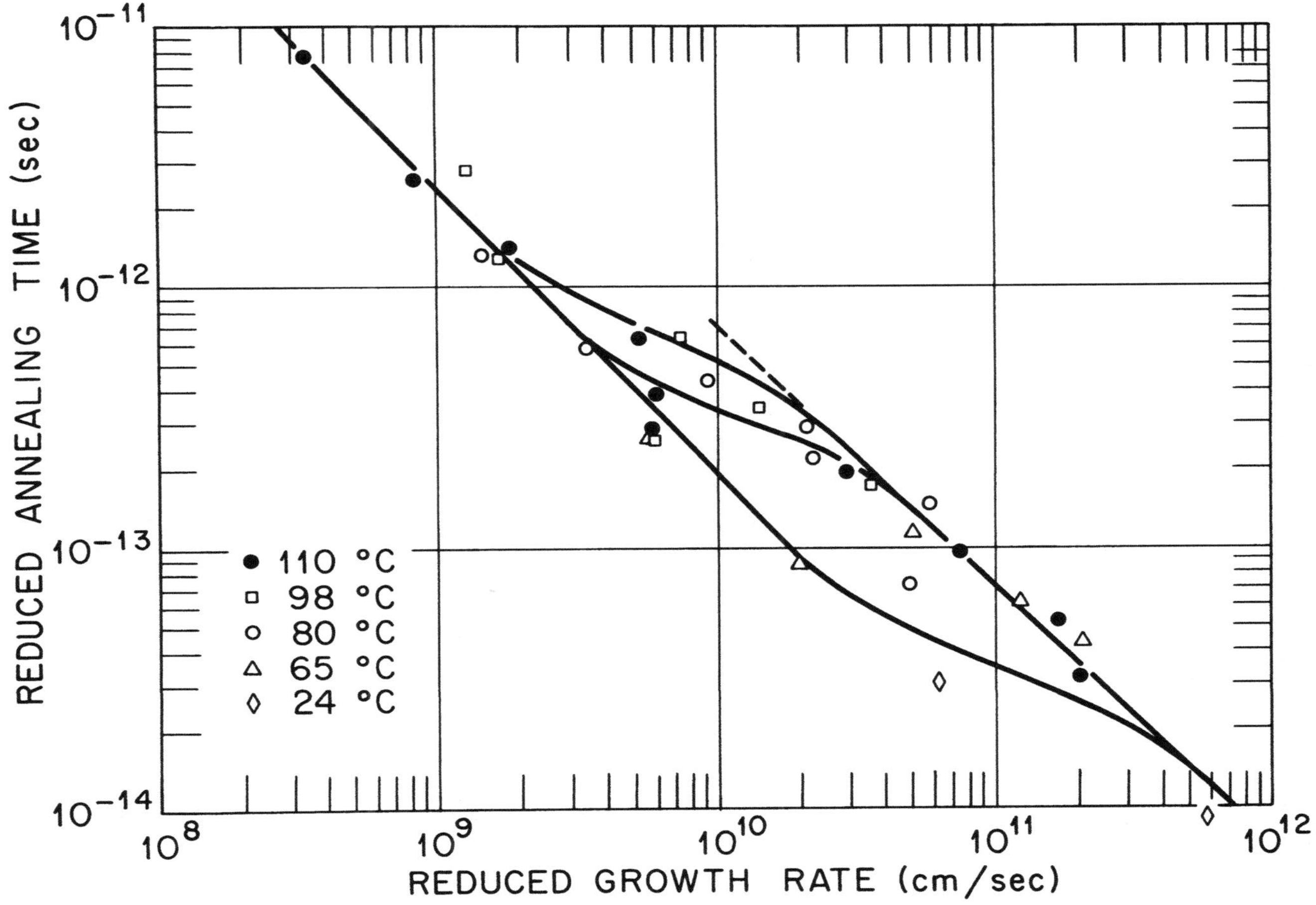

Fig. 12. Growth rate versus time plot for recrystallizing grains in aluminum–3.4 ppm gold suggesting a transition in mechanism of migration.

explain the dependency of boundary migration rate on sample thickness in aluminum bicrystals. These features are incorporated into the equations of the impurity-drag theory. The following important concepts and assumptions are necessary in this development.

1. The elementary diffusion jumps causing the migration of grain boundaries involve the exchange of atoms (solute or solvent) with vacancies. Thus the diffusion coefficient D(x) in Eq. (27) becomes

$$D(x) = a^2 \nu C_v(x) \exp - \left(\frac{\Delta F_m}{RT}\right) , \tag{31}$$

where a is the lattice parameter, ν is the natural atomic vibrational frequency, $C_v(x)$ is the vacancy concentration (atomic fraction) at a distance x from the boundary center and ΔF_m is the free energy of motion of a vacancy.

2. A stationary boundary is a tight boundary, i.e., $m' > 1$. For such a boundary the vacancy concentration at distances $x' > \delta$ and $x' < -\delta$ is C_v^L, the equilibrium vacancy concentration of the lattice, so that

$$C_v(x') = C_v^L = \exp \frac{-\Delta F_F}{RT}$$

where ΔF_F is the vacancy formation free energy per mole. At the boundary center, i.e., in the high porosity region $-\delta < x < \delta$, the normal vacancy concentration is $C_v(x) = C_v^B$ which because of assumption (1) above can be estimated from the equation

$$\frac{C_v^B}{C_v^L} = \frac{D_{GB}}{D_L} . \tag{32}$$

A rough calculation using the diffusion data for silver suggested $C_v^B = 1/8$ or one site out of eight is vacant at the grain boundary center.

3. A grain boundary moving through a cold-worked matrix containing dislocations and leaving behind a relatively dislocation-free, more dense grain must accommodate the release of free volume associated with this change of state. It is assumed the boundary accomplishes this by continuously generating vacancies in excess of the equilibrium number as it moves. If C_1 is the equivalent vacancy concentration per unit volume arising from this state change, then for a dislocation dilation of 1.5 atomic volume per atom plane[35] and a dislocation density of $\sim 10^{10}$ cm^{-2} in the cold-worked matrix, C_1 would be of order 10^{17} to 10^{18} cm^{-3}.

4. These excess vacancies traverse certain diffusion paths in the boundary until they reach a sink and disappear. Grain boundary ledges, steps, or dislocations may be the disposals for this annihilation process.

5. In calculating the concentration of excess vacancies for a given boundary velocity, G, steady state is assumed, i.e., the vacancy production rate equals the annihilation rate. Furthermore if vacancy decay occurs by a first order kinetic process then the excess atom fraction of vacancies, C_v^{ex}, can be estimated by

$$C_v^{ex} \cong \frac{C_1 G}{\rho_s D_v} b^2 , \qquad (33)$$

where ρ_s is the sink density (cm^{-2}), D_v is the vacancy diffusion coefficient and $1/b^2$ is the number of atom sites per unit area of boundary.

6. The total vacancy concentration, C_v, in the boundary's range of influence, i.e., $-m'\delta \le x \le m'\delta$ can never exceed C_v^B. Therefore, $C_v^L < C_v < C_v^B$.

7. Solute atom-vacancy interactions have been neglected.

It should be recognized that α in Eq. (28) and Q_v in Eq. (29) are for boundaries that do not generate vacancies as they move. In accordance with the preceding postulates, the diffusion coefficient in Eq. (28) must be modified to give

$$\alpha = \frac{4\ RT}{V_m} \cdot \frac{m'\delta}{a^2 \nu C_v} \cdot \frac{RT}{E_o} \cdot \exp\frac{\Delta F_m}{RT}\left[8C_v\left\{\sinh\frac{E_o}{RT}\left(\frac{1}{m'} - 1\right) - \frac{E_o}{RT}\left(\frac{1}{m'} - 1\right) + \sinh\frac{E_o}{RT} - \frac{E_o}{RT}\right\} - \left\{\sinh\frac{E_o}{RT}\left(\frac{1}{m'} - 1\right) - \frac{E_o}{RT}\left(\frac{1}{m'} - 1\right)\right\}\right] . \qquad (34)$$

For a boundary moving during recrystallization the vacancy concentration C_v is given by

$$C_v = C_v^L + C_v^{ex} = \exp\frac{-\Delta F_F}{RT} + \frac{C_1 G}{\rho_s D_v} b^2 . \qquad (35)$$

If $\delta C_V \ll 1$ and $m' > 1$, such a moving boundary may still be considered "tight" and

$$\alpha = \frac{4\,RT}{V_m}\,\frac{m'\delta}{a^2\nu}\cdot\frac{RT}{E_o}\cdot\frac{1}{C_V}\exp\frac{\Delta F_m}{RT}\left\{\sinh\frac{E_o}{RT}\left(1-\frac{1}{m'}\right)\right.$$
$$\left.-\frac{E_o}{RT}\left(1-\frac{1}{m'}\right)\right\}. \tag{36}$$

By substituting Eq. (36) into Eq. (24), replacing C_V with Eq. (35) and rearranging, the velocity can be written as

$$G = \frac{\Delta F}{C_s}\,\frac{V_m}{4RT}\,\frac{a^2\nu}{m'\delta}\,\frac{E_o}{RT}\,C_V$$
$$\exp\frac{-\Delta F_m}{RT}\left\{\frac{1}{\sinh\frac{E_o}{RT}(1-\frac{1}{m'}) - \frac{E_o}{RT}(1-\frac{1}{m'})}\right\} \tag{37}$$

or

$$G = \frac{C_V^L G_o}{1 - \frac{C_1 b^2}{\rho_s D_V}\,G_o} \tag{38}$$

where

$$C_V^L G_o = \frac{\Delta F}{C_s}\,\frac{V_m}{4RT}\,\frac{D_L^i}{m'\delta}\cdot\frac{E_o}{RT}\left\{\frac{1}{\sinh\frac{E_o}{RT}(1-\frac{1}{m'}) - \frac{E_o}{RT}(1-\frac{1}{m'})}\right\}.$$

The temperature dependence of G becomes for this case

$$\frac{-d\,\ell n\,G}{d\,\frac{1}{RT}} = Q_Y = Q_L^i$$
$$+\left\{2\,RT - E_o(1-\frac{1}{m'})\left[\frac{\cosh\frac{E_o}{RT}(1-\frac{1}{m'}) - 1}{\sinh\frac{E_o}{RT}(1-\frac{1}{m'}) - \frac{E_o}{RT}(1-\frac{1}{m'})}\right]\right\}\left(\frac{C_V}{C_V^L} - 2\right). \tag{39}$$

Recalling that the second term in the curly brackets in Eq. (39) is larger than the first offers a possibility for $Q_Y < Q_L^i$ if $C_v > 2\, C_v^L$. Thus an enhanced vacancy concentration, i.e., $C_v^V/C_v^L > 2^V$ in the boundary could explain the previously mentioned disturbing experimental observation that Q_Y was less than (see Table III) rather than greater than Q_L^i as theory, Eq. (9), had predicted.

Figure 13 depicts how the velocity varies with the vacancy concentration in the boundary for this proposed theory. But G must remain finite and positive so Eq. (38) is limited to situations where

$$G_o \frac{C_1 b^2}{\rho_s D_v} < 1 \; .$$

Figure 13 implies that as G approaches a critical value G_{crit}, the vacancy concentration becomes very large; C_v approaches infinity. But the vacancy concentration cannot build up inexhaustedly or uncontestedly for the boundary would fracture. Thus postulate (6) was introduced to limit the vacancy concentration in the near boundary region to C_v^B. (An alternate view would be to say that the boundary becomes liquid-like but can't become gaseous-like.) To accommodate the increasing vacancy concentration with increase in migration rate it is envisioned that the high-porosity region at the boundary center widens and spreads out, i.e., δ increases. If the maximum solute atom-boundary interaction distance $m'\delta$ remains constant, then because δ is increasing, m' effectively becomes smaller with increasing vacancy content. When m' reaches 1 a sudden transition in the impurity drag takes place from Case 1 (tight) to Case 2 (porous); see Fig. 11. It is this change that has been referred in this paper as a grain boundary "structure" transformation. The impurity drag parameter α is altered to α' and when $m' = 1$ is given by

$$\alpha' = \frac{4RT}{V_m} \frac{m'\delta}{D_{GB}^i} \frac{RT}{E_o} \left\{ \sinh \frac{E_o}{RT} - \frac{E_o}{RT} \right\} \; . \tag{40}$$

When this transformation in impurity drag occurs, Eqs. (37) and (38) are no longer valid and the boundary migration rate becomes

$$G = \frac{\Delta F}{C_s} \frac{V_m}{4RT} \frac{D_{GB}^i}{m'\delta} \frac{E_o}{RT} \left\{ \frac{1}{\sinh \frac{E_o}{RT} - \frac{E_o}{RT}} \right\} \; . \tag{41}$$

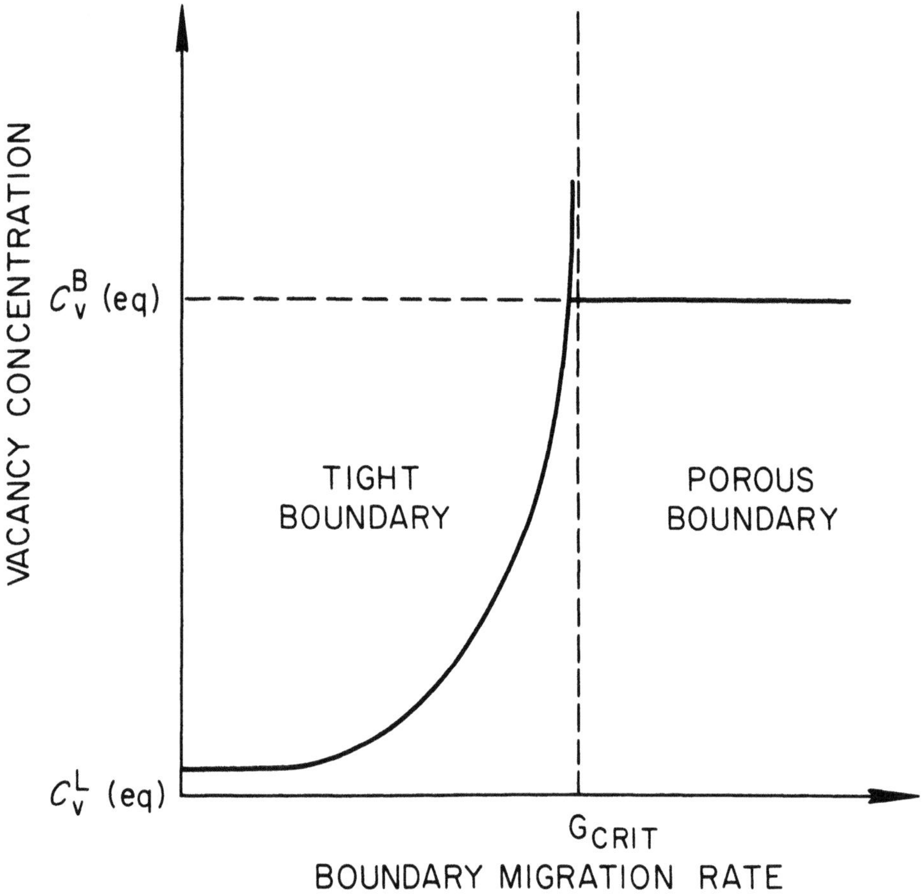

Fig. 13. Grain boundary vacancy concentration versus migration rate for recrystallizing grain boundaries.

For this situation, Eq. (30) describes the activation energy for boundary migration.

A critical velocity for this structural transformation may be defined approximately by the condition that the denominator in Eq. (38) be zero. Thus

$$\frac{C_1 b^2}{\rho_s D_v} \cdot G_o = 1 .$$

Recognizing that G_o is G/C_v and that this change occurs when $C_v = C_v^B \cong 1/8$, then

$$G_{crit} \cong \frac{1}{8} \frac{D_v \rho_s}{C_1 b^2} . \tag{42}$$

Summarizing, when $G < G_{crit}$, Eq. (38) is the expression for G but when $G > G_{crit}$, Eq. (41) describes G.

Values for several parameters entering into these expressions for G are not known with a high degree of certainty ($m'\delta$, E_o, ρ_s, etc.). Therefore, a critical test of this theory is not possible. Nevertheless some assessment of the theory can be made by calculating what the above mentioned parameters would need to be to give satisfactory agreement with experiment. The behavior of the 3.4 at. ppm alloy suggested that G_{crit} was between 10^{-5} and 10^{-6} cm/sec at 100°C for that alloy. Using these values and Eq. (42), a sink density, $9 \times 10^4 < \rho_s < 9 \times 10^5$, may be calculated. In this calculation it was assumed that

$$C_1 = 10^{17} \text{ vacancies/cm}^3,$$

$$D_v = a^2 \nu \exp + \Delta S_m/R \exp -Q_m/RT,$$

$$\nu = 10^{13} \text{ sec}^{-1}$$

$$a = b$$

$$R = 1.987 \text{ cal/mole/°K}$$

$$T = 100°\text{C}; \ 373°\text{K}$$

$$\Delta S_m = 0$$

$$Q_M = Q_{SD}^{Au} - Q_F$$

$$Q_{SD}^{Au} = 27{,}800 \text{ cal/mole}$$

$$Q_F = 0.76 \text{ ev (ref. 36); } 17{,}500 \text{ cal/mole.}$$

This order of magnitude sink density is about 50 times greater than the one estimated by In der Schmitten, Hassen, and Haessner[34] for a specific type of boundary in aluminum. There are, however, significant differences between these two experiments, i.e., polycrystal vs bicrystal, orientation effects, amount of cold work, etc., which could account for this difference. The calculated sink density is close to the dislocation density of annealed metals which may suggest that dislocations trailing and being left behind the moving boundary are the sinks.

Assuming the sink density to be the same for the other alloys, the critical velocity for the 27 ppm gold alloy at 100°C would be

$$G_{crit}(27 \text{ ppm}) = \frac{3.4}{27} G_{crit}(3.4 \text{ ppm}) = 1.3 \times 10^{-7} \text{ cm/sec}$$

assuming the lower limit above for $G_{crit}(3.4 \text{ ppm})$. Since the measured rate of boundary migration (8×10^{-8} cm/sec) at 100°C for the 27 ppm alloy even in the early stage of recrystallization was less than G_{crit}, no boundary structure change should be observed and none was. Furthermore, it can be shown that

$$\frac{C_v}{C_v^L} = \frac{G_{crit}/G}{(G_{crit}/G) - 1} \quad . \tag{43}$$

For the 27 ppm gold alloy this ratio was calculated to be 2.7 using the above mentioned values of G_{crit} and G. Inserting this into Eq. (43) and using the measured $Q_V = 25{,}700$ cal/mole and $Q_L^i = 27{,}000$ cal/mole, a value for E_o was computed and found to be 4500 cal/mole (m' = 4 was assumed).

The reasonableness of the theoretical parameters calculated from the experimental data and the apparent self-consistency of the calculation in going from one alloy to another must be considered to provide some degree of credulence to theory until a more thorough test is made.

CONCLUSIONS

1. The early stage characteristics of recrystallization in five gold-doped, zone-refined aluminum alloys cold-rolled 40% at 0°C conformed to a grain edge-nucleated, growth-controlled phenomenological model[1,2] for recrystallization.

2. The length, ℓ, of the matrix grain boundary edges producing recrystallized colonies (groups of contiguous recrystallized grains strung out along these edges) depended on the penultimate grain size and was longer for larger grain sizes.

3. A modification of the original phenomenological model was proposed which incorporated colony lengthening as well as the usual colony thickening in the growth aspect of recrystallization.

4. Discrepancies between the measured recrystallization parameter, k, and the expected one based on the original model, when $\ell \leqslant 0.030$ cm were semiquantitatively accounted for by the modified model.

5. The presence of $AuAl_2$ precipitates due to aging did not materially alter either the number or the type of nucleation sites for recrystallization but did cause an enormous (10^4) increase in grain boundary migration rate; a result that can be rationalized in terms of the solute-drag theory of boundary migration.

6. The activation energy for grain boundary migration in alloys containing > 3.4 ppm gold was 25,700 cal/mole and did not vary with the extent of recrystallization. This value was slightly less than that predicted for solute-controlled migration of a "tight" boundary.

7. In the 3.4 ppm alloy a transition in activation energy for boundary migration was observed and attributed to a grain boundary "structure" transformation. An excess vacancy concentration in the boundary region was thought to control the transformation by changing the nature of the solute-atom drag from a "porous" to "tight" condition.

8. A theory of vacancy-enhanced grain boundary migration, incorporating the impurity drag concepts in it, was proposed and used to analyze some of the experimental results.

ACKNOWLEDGEMENT

The assistance of E. D. Bolling in the metallographic preparation of the specimens is acknowledged.

REFERENCES

1. R. A. Vandermeer and Paul Gordon, Trans. TMS-AIME, 215, 577–88, (1959).

2. R. A. Vandermeer, Met. Trans., 1, 819–26, (1970).

3. W. C. Leslie, F. J. Plecity, and J. T. Michalak, Trans. TMS-AIME, 221, 691–700, (1961).

4. R. A. Vandermeer and Paul Gordon, in Recovery and Recrystallization of Metals, Interscience Publishers, New York (1963), pp. 211–40.

5. K. Lücke and K. Detert, Acta Met., 5, 628–37, (1957).

6. Paul Gordon and R. A. Vandermeer, Trans. TMS-AIME, 224, 917–28, (1962).

7. C. Frois and O. Dimitrov, Mem. Scient. Rev. Metallurg., 59, 643–48, (1962).

8. John W. Cahn, Acta Met., 10, 789–98, (1962).

9. Kurt Lücke and Hein-Peter Stüwe, in Recovery and Recrystallization of Metals, Interscience Publishers, New York (1963), pp. 171–210.

10. Paul Gordon and R. A. Vandermeer, in Recrystallization, Grain Growth and Textures, Am. Soc. Metals, Metals Park, Ohio (1966), pp. 205–66.

11. E. L. Holmes and W. C. Winegard, Trans. TMS-AIME, 224, 945–49, (1962).

12. Manfred von Heimendahl, Zeit. für Metallkunde, 58, 230–35, (1967).

13. M. von Heimendahl, Acta Met., 15, 1441–52, (1967).

14. Franz Schücker, in Quantitative Microscopy, R. T. DeHoff and F. N. Rhines, Editors, McGraw-Hill Book Co., New York (1968), pp. 201–65.

15. M. Avrami, J. Chem. Phys., 7, 1103, (1939); ibid., 8, 212, (1940); ibid., 9, 177, (1941).

16. John W. Cahn and W. C. Hagel, Acta Met., 11, 561–74, (1963).

17. John W. Cahn, Acta Met., 4, 449-59, (1956).

18. D. Eylon, A. Rosen, and S. Niedzwiedz, Acta Met., 17, 1013-19, (1969).

19. D. C. Larson and U. F. Kocks, Discussion to Ref. 7.

20. S. Weissmann, Trans. ASM, 53, 265-81, (1961).

21. W. C. Leslie, J. T. Michalak, and F. W. Aul, in Iron and Its Dilute Solid Solution, C. W. Spencer and F. E. Werner, Editors, Interscience Publishers, New York (1963), p. 119.

22. C. Zener, see D. McLean: Grain Boundaries in Metals, Oxford University Press, London (1957), pp. 239-40.

23. P. Niessen and W. C. Winegard, J. Inst. Metals, 94, 31-35, (1966).

24. C. J. Simpson, K. T. Aust, and W. C. Winegard, Met. Trans., 2, 993-97, (1971).

25. G. F. Bolling, Trans. TMS-AIME, 239, 193-203, (1967).

26. R. Fromageau and Ph. Albert, Compt. Rend., 260, 895-97, (1965).

27. N. L. Peterson and S. J. Rothman, Phys. Rev. B. Solid State, 1(8), 3264-73, (1970).

28. W. B. Alexander and L. M. Slifkin, Phys. Rev. B Solid State, 1(8), 3274-82, (1970).

29. F. M. d'Heurle, private communication, also this volume.

30. E. W. Hart, Scripta Met., 2, 179-82, (1968); also this volume.

31. Herbert Gleiter, Zeit. für Metallkunde, 61, 282-87, (1970).

32. K. T. Aust, Can. Met. Quart., 8, 173-78, (1969).

33. C. J. Simpson, K. T. Aust, W. C. Winegard, Met. Trans., 2, 987-91, (1971).

34. W. In der Schmitten, P. Haasen, and F. Haessner, Zeit für Metallkunde, 51, 101-8, (1960).

35. L. M. Clarebrough, M. E. Hargreaves, and M. H. Loretto, in Recovery and Recrystallization of Metals, Interscience Publishers, New York (1963), pp. 63-130.

36. T. Federighi, in Lattice Defects in Quenched Metals, R. M. Cotterill et al, eds., Academic Press, New York (1965), pp. 217-68.

MECHANISMS OF ELECTROMIGRATION DAMAGE IN METALLIC THIN FILMS

R. Rosenberg

IBM Watson Research Center
Yorktown Heights, New York 10598

ABSTRACT

Interaction between conducting electrons and thermally diffusing metal atoms causes a net drift of the atoms in the direction of electron flow. In thin films, where joule heating is minimized by substrate cooling and the temperature is relatively low, this "electromigration" has been shown to be confined mainly to grain boundaries, and, in some cases, surfaces. Because of certain irregularities in the boundary network, the flux of atoms is nonuniform, leading to divergencies and localized depletion. Depletion leads to the formation of grain boundary holes by either void nucleation and growth or by accelerated grain boundary grooving. Decrease in the damage rate is achieved by control of atomic mobility in grain boundaries and surfaces. A particularly effective method of controlling mobility is to introduce a solute species which migrates to appropriate defect sites and effectively inhibits motion of host atoms. The solute, however, being in preferred diffusion paths, is susceptible to electromigration and becomes locally depleted. Restriction of solute depletion is critical for reduction of gross damage by loss of solvent.

INTRODUCTION

The special properties and structures characteristic of metallic interfaces have broad implication with respect to the behavior of electronic devices which depend on the integrity of thin film conductors. Ease of atomic transport through grain boundary defects at relatively low temperature can lead to many modes of degradation and ultimate device failure. One such failure mode which was first evidenced in conductor stripes a few years ago was the occurrence of mass depletion and circuit opens related to current-induced migration of atoms within grain boundaries (e.g.: see Fig. 1). Confinement of the electromigration flux to grain boundaries combined with lack of a detailed understanding of mechanisms leading to flux divergences have made analysis of the mass depletion phenomenon a difficult task. Flux measurement techniques usually employed for bulk studies (e.g. marker motion(1), tracer profile(2)) are not applicable in films because of the nonuniformity of the grain boundary network. As a result, lifetime (time to form an open) studies have been used extensively to obtain diffusion information, under the assumption that a direct correlation exists between the magnitude of flux and stripe failure. This, of course, is not necessarily the case as the distribution of divergent sites, which control the latter, can be quite independent of the absolute flux. The purpose of this paper will be to briefly review the mechanisms by which structural irregularities can cause local mass depletion, and to discuss methods of altering the electromigration flux and the contributions of such irregularities. Consideration will be given to mechanistic effects of solute additions as related to diffusion and lifetime measurements, a complex subject that needs clarification.

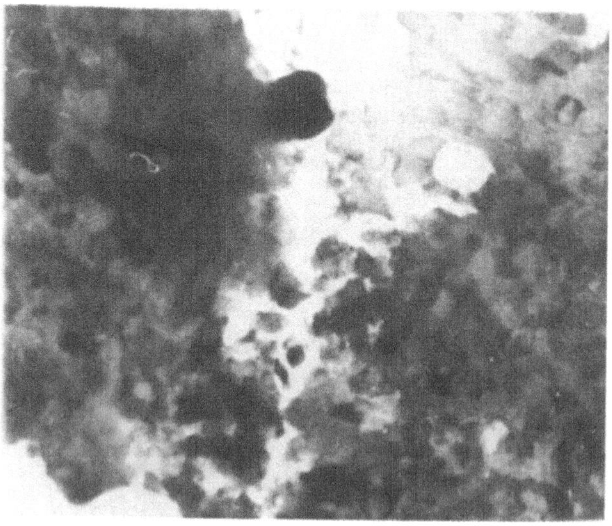

Fig. 1. A transmission electron micrograph of a typical electromigration failure.

MECHANISMS OF DAMAGE

Early work showed much of the damage to be near the cathode end of the conductor, and as a result it was postulated that electromigration through the temperature profile produced by joule heating was responsible.(3) In this case the gradient in diffusivity caused by increasing temperature causes mass depletion. It was then shown(4) that certain structural irregularities accelerated depletion and would lead to earlier failure than typified by the background temperatue profile. Also, poor fabrication procedures can lead to local effects as illustrated by the tests on aluminum stripes deposited on NaCl;(5) local sites of nonadhesion (blistering) caused hot spots and accelerated hole formation.

The most active site in an equiaxed structure is the triple point (three grain boundary junction), where flux in and out must be balanced. The flux divergence can be simply approximated as,

$$J = \frac{Nez^*E}{kT}(D_1 \cos \theta_1 - D_2 \cos \theta_2 - D_3 \cos \theta_3) \quad (1)$$

where N is the density and z^* the effective charge of the diffusing species, eE is the field applied, θ is the angle between the direction of motion of the diffusing atom in the grain boundary and the field direction, D is the boundary diffusivity, and boundary 1 leads into the triple point and boundaries 2 and 3 lead out. Nonuniform boundary diffusivities can lead to either depletion or accumulation,(4) and because of high vacancy concentration and high strain energy, the site is most susceptible to damage modes such as void formation by vacancy condensation or boundary grooving.

In deposited films there is usually a wide distribution of grain sizes which becomes a major source of trouble when the nominal grain size is about one-quarter to one-fifth of the width of the conducting stripe. In this case, there is a high probability that single grains will span the stripe and will lead to depletion at the boundary between the single grain (with no diffusion paths) and the adjacent polycrystalline region.(6) A majority of failures occurs at this type of discontinuity.

Observation of the rate of void growth at grain boundary junctions by transmission(7) or scanning(4) electron microscopy leads to a direct measure of the electromigration flux per boundary. From these measures of flux combined with values of activation energy determined for electromigration in aluminum films,(5,7) it was possible to calculate the multiple $D_o z^*$.[8] This value was constant at about 3 ($\pm$ 0.5) x 10^{-2}cm^2 sec^{-1}. Assuming z^* to be similar to that for bulk aluminum, D_o would be about 10^{-3}, a reasonable value as compared with those reported for silver, for example. Knowledge of $D_o z^*$ allowed calculation of vacancy supersaturations produced at structural defects by various flux divergences.[9] The supersaturations are less than unity for all cases, and where there exists boundary vacancy sinks with less than 10^3 Å average spacing the supersaturation drops below 0.1. The time required to reach the maximum supersaturation is in the minute range under the test conditions where voids are not observed for 10-100 hours. Both the low supersaturation and long incubation time indicate that observable voids come from growth controlled rather than nucleation controlled processes. In the event that heterogeneous nucleating sites exist that would allow vacancy condensation into voids at such low supersaturation levels, this would still occur instantaneously requiring substantial growth prior to observation.

Holes can be produced at any level of supersaturation by an accelerated grain boundary grooving process.[9] Vacancies accumulating at a boundary defect site disperse by diffusing to the surface as atoms flow inward, causing the groove to deepen. Atoms diffuse along the surface into the boundary groove to maintain the equilibrium groove angle. This causes thinning of the stripe (much like sand flowing through a funnel). The incubation time for hole formation is the time required for the groove to reach the substrate. The triple point is especially sensitive to this process because of its high energy and correspondingly deeper grooves. Calculations show this mechanism to be comparable with triple point void growth in terms of the rate of hole formation.

RESTRICTION OF DAMAGE

Preferred Orientation

Although most work has involved change in flux as a means of reducing damage rate it is possible to influence the structural divergent sites for increased lifetime. An example of this is the work on preferred orientation in aluminum stripes.[6] In the case where all grains have a common pole axis the grain boundaries become more uniform in properties and are basically tilt-type. For tilt boundaries, as illustrated at the top of Fig. 2, for example, the high diffusion paths such as ledges or dislocations lie parallel to the tilt axis, or, in the case of test stripes, perpendicular to the applied field. The thermal diffusivity

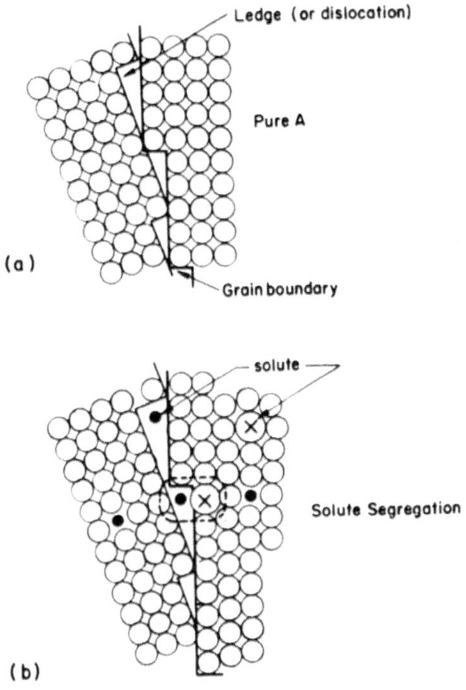

Fig. 2. Schematic representation of grain boundary defects and solute partitioning. a) pure metal, b) segregation of solute into defects and formation of two solute cluster.

thus should become more difficult in the direction of electromigration. Also, because of the uniformity in boundary diffusivities the triple point should not be an effective site for vacancy buildup. These characteristics of preferred orientation were manifested in the aluminum work as an increase in activation energy for failure from an average of 0.6 to about 0.71 eV and a change in the width σ of the lifetime distribution from 0.5 to about 0.2. The significance of σ lies in its dependence on the number of contributing mechanisms, the smaller the number the narrower the distribution. In these stripes about 80 percent of the failures were of the mixed grain size type with no triple point voids in evidence, thus the decrease in σ.

Solute Effects

The electromigration behavior of stripes containing solute additions is quite complex, involving changes in flux and lifetime in different ways. Because of this basic difference, extraction from lifetime data of quantitative information on the mechanisms by which solute may inhibit electromigration is not clear. Essentially, it can be assumed that solute decreases the diffusivity of the host atoms as the major contribution. In surface active media, for example, the damage rate is associated with rapid grain boundary grooving. By effectively limiting surface diffusion by solute adsorption the grooving may be inhibited. This was shown to be the case for silver stripes with a thin nickel layer on the top surface.[10] During electrical powering the uncoated stripe edges damaged quite rapidly while the nickel coated surface remained uneffected. In aluminum, which is less surface active, the increased lifetime observed for copper additions[11] are most likely attributable to grain boundary migration effects. The remainder of this discussion will concentrate on solute: grain boundary interactions which are more generally characteristic of electromigration in films.

The solute contribution is illustrated in the lower half of Fig. 2. In this sketch the solute atoms are shown to segregate to boundary diffusion channels with which they have a binding energy, B_{NS}. If diffusion at the low test temperature ($T < 0.4\ T_m$) is restricted to the channel regions then the diffusing specie "sees" the solute and becomes pinned with energy, B_{VS}. The amount of solute needed for inhibition is small with respect to the number of host atoms available in the channels because of an effective blocking action. At higher temperatures the effectiveness would be limited by motion from the channels into unblocked boundary regions. Figure 3 [10] shows a rapid decrease in the number of unblocked channels as the total boundary solute concentration, S_b^o, becomes of the order of 10^{-2} and $B_{NS} \geq 0.1$ eV. The electromigration flux in the presence of solute can be calculated by considering contributions of free and blocked channels.[12] Results reproduced in Fig. 4 show the flux to be greatly reduced for $B_{NS} = 0.2$ eV, $B_{VS} > 0.05$ eV, and nominal S_b^o of 10^{-2}. The temperature dependence of the flux is shown in Figure 5 to increase from 0.6 eV for pure material and small solute interaction (low B_{NS}) to about 0.8 eV for large interactions ($B_{NS} > 0.2$ eV leads to still larger activation energies). This assumes constant solute concentration

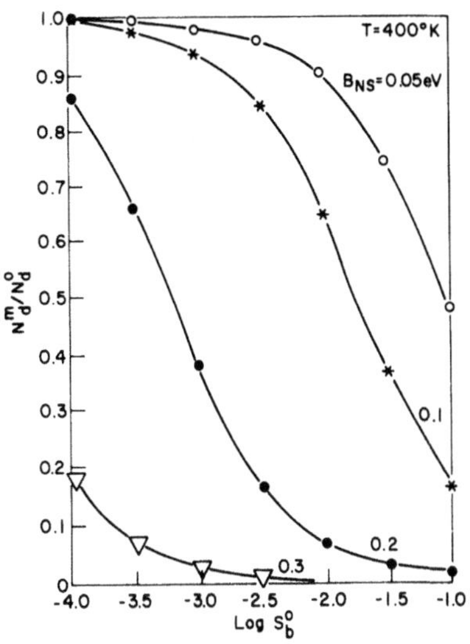

Fig. 3. Reduction in the number of boundary defects, N_d^m, in which the solvent atoms remain mobile as solute is introduced. S_b^o is the total boundary solute concentration.

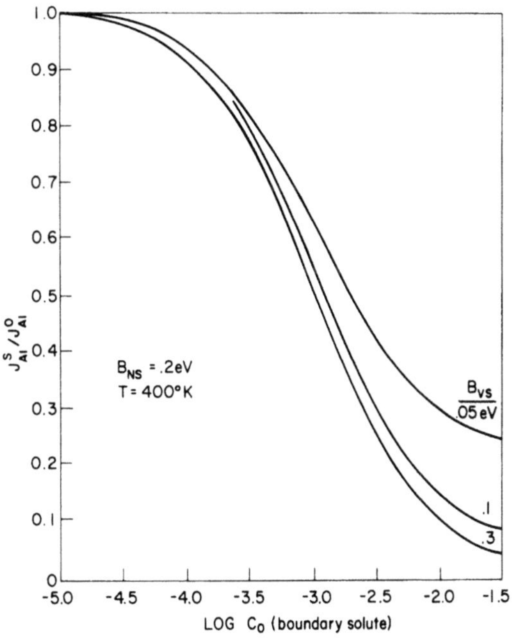

Fig. 4. Reduction in electromigration flux caused by solute as a function of the magnitude of binding between the solute and diffusing species in the grain boundaries.

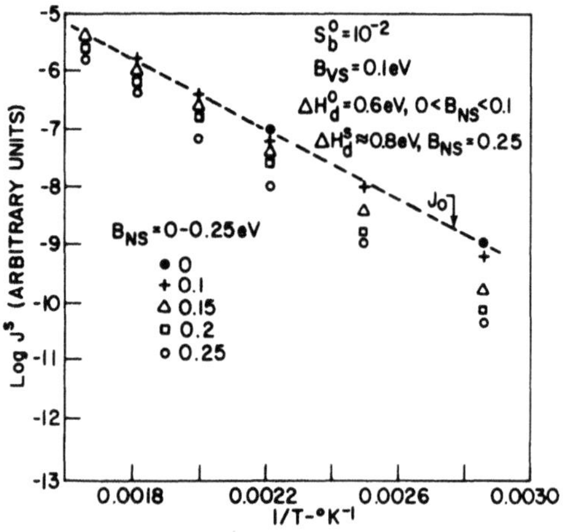

Fig. 5. Temperature dependence of electromigration flux as the solute becomes more effective in pinning defects. The apparent activation energy increases.

with temperature. If the heat of mixing is considered then the value of activation energy will be decreased accordingly. The general inhibition effect was illustrated for the case of magnesium in aluminum[13] by diffusing magnesium atoms into the central region of a stripe to a concentration of about 2 percent and then powering. At the line of intersection between the diffused and undiffused regions mass depletion occurred rapidly when electron flow was directed away from the diffused area. This can be attributed to predictions of Figure 4 where the aluminum flux would be much higher away from the magnesium. Microprobe studies indicated the magnesium concentration at the depletion site to be between 0.1 and 1 percent.

Solute selection should involve maximization of binding energies. As a first approximation, atomic size would seem to be important, where migration to defect sites in the boundary would relieve distortional energies. Either larger or smaller atoms would segregate; e.g. copper and magnesium in aluminum. Electronic contributions, e.g. palladium in silver,[14] could also be important. Values of binding energies of solute to lattice vacancies are not good indicators for boundary contributions; copper has almost zero binding energy in aluminum[15] while having a large effect on electromigration. A better indicator would seem to be the effects of the solute on low temperature grain boundary migration or recrystallization rates. In recrystallization tests copper has proven to be effective.[16] If this assumption is valid then transition metals should be very effective inhibitors to electromigration, as small percentages do the job, e.g. iron in aluminum[17] and gold.[18]

The lifetimes expected of stripes containing an effective solute should be prolonged by an amount equivalent to the increased energy. The basic problem, however, is the instability of the solute with respect to the electron flow. Being in a somewhat interstitial position, solute electromigrates rapidly[19,20] and becomes depleted in local structural irregularities much the same way as does pure metal. This depletion creates a situation quite similar to the magnesium experiment described above in which material in the depleted region rapidly moves away from material which is undepleted. The resultant stripe failure takes place in a relatively pure region, and the lifetime becomes associated mainly with the time required for solute depletion. It should be emphasized that the lifetime for this failure mode is independent of the manner in which the solute is restricting electromigration. Therefore, the mechanism cannot be defined by lifetime studies.

Future experimentation should be designed to limit solute migration such that full effectiveness can be achieved. Solution of boundary precipitate particles to replace solute does not appear to be sufficient although depletion is prolonged. No correlation is found between precipitate distribution and lifetime. A possible approach is illustrated in Figure 2 where a two-solute complex is formed in the channel region. In this way the host atom is blocked as before but now the solute is pinned. Choice of solute requires a high binding energy between the two and a tendency for both to migrate to boundary defects. One larger and one smaller than the host atom, for example, may be of interest.

REFERENCES

1. H. B. Huntington and A. R. Grone, J. Phys. Chem. Sol. 20, 76 (1961).

2. H. M. Gilder and D. Lazarus, Phys. Rev. 145, 507 (1966).

3. D. Chaabra, N. Ainslie, and D. Jepson, The Electrochem. Soc. Meeting, Dallas, Texas, May, 1967.

4. L. Berenbaum and R. Rosenberg, Thin Sol. Films 4, 187 (1969).

5. R. Rosenberg and L. Berenbaum, Appl. Phys. Letters 12, 201 (1968).

6. M. J. Attardo and R. Rosenberg, J. Appl. Phys. 41, 2381 (1970).

7. I. A. Blech and E. S. Meieran, J. Appl. Phys. 40, 485 (1969).

8. R. Rosenberg, Appl. Phys. Letters 16, 27 (1970).

9. R. Rosenberg and M. Ohring, J. Appl. Phys. 42, December, 1971.

10. R. Rosenberg and L. Berenbaum, in Proc. Europhys Conf. on Atomic Transport (published z. Naturforsch, Germany) Marstrand, Sweden, June, 1970.

11. I. Ames, F. M. d'Heurle, and R. E. Horstmann, IBM J. Res. and Dev. 14, 461 (1970).

12. R. Rosenberg, J. Vac. Sci. Tech. (to be published).

13. R. Rosenberg, A. F. Mayadas, and D. Gupta, in Proc. Int. Conf. on Grain Boundaries, IBM Watson Research Center, Yorktown Heights, New York, August, 1971.

14. A. D. LeClaire, Phil. Mag. 7, 141 (1962).

15. T. R. Anthony, Phys. Rev. B2, 264 (1970).

16. P. Gordon, in Energetics in Metal Phenomena, ed. W. M. Mueller, Gordon and Breach, New York, p 205 (1965).

17. P. Gordon and R. A. Vandermeer, Trans. TMS-AIME 224, 917 (1962).

18. W. Grünwald and F. Haessner, Acta Met. 18, 217 (1970).

19. F. M. d'Heurle, Metal. Trans. 2, 683 (1971).

20. L. Berenbaum and R. Rosenberg, in Proc. of IEEE Reliability Phys. Symposium, Las Vegas, Nevada, April, 1971.

SOLUTE EFFECTS ON GRAIN BOUNDARY ELECTROMIGRATION AND DIFFUSION

F. M. d'Heurle and A. Gangulee

IBM Thomas J. Watson Research Center

ABSTRACT

The grain boundary transport of aluminum chromium, and copper resulting from electromigration at 175°C in aluminum-chromium and aluminum-copper thin film conductors has been measured. The forces acting on the various types of atoms are estimated from other experiments or from theoretical derivations. It is then possible to calculate the diffusion constants of the different atomic species. The values obtained for aluminum grain boundary diffusion in aluminum-chromium correspond well to values which have been found through different means in the grain boundary of pure silver. It is concluded that chromium does not reduce the rate of grain boundary diffusion of aluminum, but copper reduces this rate by a factor of about 80. The results of a series of electromigration experiments in aluminum-copper thin films are interpreted in terms of the information they yield on the diffusion of adsorbed copper atoms in aluminum-copper grain boundaries. The conclusions reached are compared to the previously reported results on the effect of alloy additions on grain boundary diffusion in a variety of different elements.

INTRODUCTION

Electromigration, the transport of matter as a result of the passage of direct current has been the object of many publications and reviews [1-7]. One needs only mention here that this effect is also observed in liquid metals (see for example references 8-10). In crystalline metals the object of past studies has generally been electromigration in bulk (as opposed to thin film) samples - at relatively low current densities (10^3 to 10^4 Amp/cm^2) - and at high temperatures ($T > T_m/2$) such that material transport occurs via a lattice diffusion mechanism. During the last few years failures occurring in thin film conductors used in electronic devices led to a great deal of interest in electromigration phenomena in such thin films where at high current densities (10^5 to 10^6 Amp/cm^2) and at low temperatures ($T < T_m/2$) transport occurs via grain boundary diffusion mechanisms. A number of observations mostly in aluminum films, but also in gold and silver films have been reviewed.[5]. More recent publications have considered electromigration phenomena in aluminum [12-16], gold [13-17], copper and silver films [17], theoretical aspects of electrotransport at surfaces [18], nucleation of voids [19-21], and the reduction of failure rates [5,22]. In experiments conducted on tin-indium[23] films (with a low melting temperature) the activation energy [24] values indicate that electromigration proceeds by lattice diffusion, confirming initial results obtained with pure indium [25] films.

The addition of copper[26] in the amount of about 4 wt% to aluminum film conductors decreases the rate of failure by electrotransport and crack formation by a factor of about 70. This beneficial effect of copper alloying in aluminum films has been the object of a number of studies [27-34]. Silver and gold [27] additions in aluminum films were found to be without effect in electromigration tests, but chromium, nickel and magnesium were found to increase the lifetimes of aluminum thin film conductors [35]. In the case of chromium additions amounting to about 2 wt% a lifetime improvement over pure aluminum by a factor of 300 was obtained.

Because of the heterogeneous character of electrotransport along the grain boundaries of aluminum thin films quantitative determinations of transport parameters are difficult to make. Accelerated electromigration tests for reliability purposes are usually carried at high current densities which result in early failure. The time for crack formation varies as J^{-n} (J is the current density and n an exponent with values from about 2 to 6) [20,27], however, transport [36,37] varies only as J. Thus, high current densities which allow to shorten lifetime tests tend to cause thin film conductors to become discontinuous before measurable mass movement can take place. In this study quantitative measurements of electrotransport in aluminum-chromium films, and aluminum-copper will be presented. The results will be analysed in the light of a variety of experiments conducted with aluminum-copper films. Conclusions about the effect of impurities on diffusion in grain boundaries shall be compared to results already reported in the literature.

EXPERIMENTAL PROCEDURES AND RESULTS

Chromium Additions

Details of the film deposition technique have been previously described elsewhere [38]. Aluminum was first deposited to a thickness of 3000 Å, in a vacuum of about 2×10^{-7} torr from an rf heated BN-TiB_2 crucible. Chromium was deposited next from a molybdenum crucible, also heated from the same rf power supply and coil. A rotating crucible support endowed with vertical motion allowed the sequential positioning of up to four different crucibles in the same rf coil without opening the vacuum system [39]. Another 3000 Å layer of aluminum was deposited after the chromium layer, for a total film thickness of about 6000 Å. During the initial stages of each deposition the substrates were protected from evaporating impurities by a water cooled shutter which permitted thorough outgassing of the sources prior to actual deposition. During and after the chromium deposition the pressure level decreased to the high 10^{-8} torr scale on account of the gettering properties of this metal. The chromium concentration amounted to about 2 wt% of the total film. After outgassing at 250°C, the oxidized silicon substrates (178 microns thick) were maintained at 200°C during film deposition. Upon removal from the vacuum system the films were annealed at 560°C in a nitrogen atmosphere for 20 minutes for the purpose of homogenization.

The samples for electromigration testing were prepared by usual photoetching procedures. The silicon substrates were diced into 1.78 mm square chips, and each chip was mounted with conductive silver epoxy unto a TO-5 transistor header in order to maximize the dissipation of Joule heat. The sample configuration is shown in Fig. 1. The conductor was about 7.5 microns wide and 250 microns long. The terminal pads were 1270 microns in their largest dimension, and their width varied from 390 to 500 microns. At each terminal four ultrasonically bonded wires (visible in Fig.1) were used for current leads, while two others could be used for voltage monitoring. For electromigration testing two sets of eleven samples were placed in silicone oil baths at temperatures of 166°C and 218°C, respectively. The samples were used as their own resistance thermometers. During testing at a current density of 2×10^6 Amp/cm^2 the Joule effect caused the temperature of the stripes to increase by 6°C. Thus the average stripe temperatures for the two electromigration tests were 172 and 224°C. Each sample was tested under constant current until such a time as crack propagation across its width caused it to become electrically discontinuous, and the time

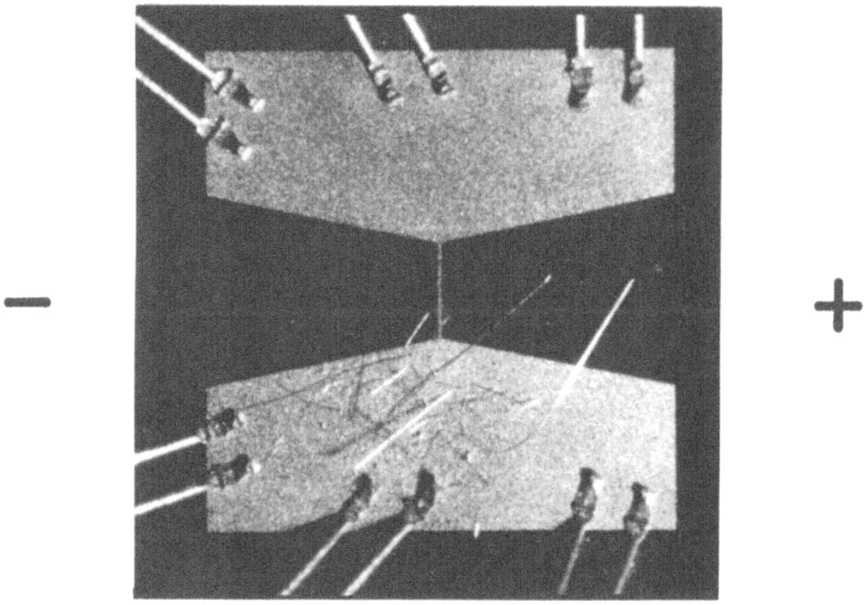

Figure 1. Photograph of an aluminum-chromium thin film conductor after testing at 175°C and $2x10^6$ Amp/cm^2 for 8300 hours. The conductor is 250 microns long and 7.5 microns wide.

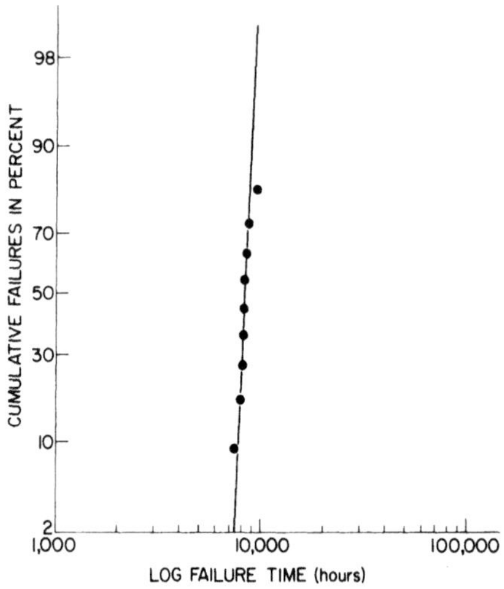

Figure 2. Cumulative failure percentage data as a function of the logarithm of the failure time for aluminum-chromium thin film conductors tested at 175°C and $2x10^6$ Amp/cm^2.

of failure was recorded. Prior to testing, but after the 560°C anneal, the resistivity of the samples was about 6 x 10^{-6}ohm-cm, and the resistivity ratio(ρ at room temperature/ρ at liquid helium temperature) was 3.0. For pure aluminum films comparable values would be about 3 x 10^{-6}ohm-cm, and 30.

Results of the test carried at 172°C are shown on Fig.2 where the order of each failure is plotted on a probability scale versus the logarithm of the failure time. There was one early failure at 58 hours, which has been ignored in plotting Fig.2. The test was terminated before failure of the tenth remaining sample. The median failure time was found to be about 8300 hours. For similar test conditions the median lifetime of pure aluminum conductors would be about 30 hours. For the test carried at 224°C the data points are more scattered because several of the conductors failed not by crack formation, but by failure of the leads at the contact terminals. However, the median failure time could be determined as being about 1500 hours.

Examination of the samples tested at 172°C revealed that the positive terminals were all covered with whiskers as seen in Fig.1. Moreover, the whiskers seemed to be relatively uniformly distributed over the whole positive terminal area rather than preferentially located at the end of the stripe where the current density was maximum. One whisker with a diameter of about 2.5 micron and a length of 0.5 mm was found about at the center of the positive terminal. Some whiskers were found along the stripes themselves. The same samples were also carefully examined with the scanning electron microscope and the electron microprobe. Even with the scanning electron microscope it was not possible to locate precisely the position of the crack discontinuity across the stripes. The cracks are undoubtedly very fine and those which caused the electrical failure of the samples could not be distinguished from other incipient cracks randomly distributed along the length of the stripes. Fig.3 shows a possible crack across one of the samples. However, the scanning electron microscope revealed extensive damage over most of the negative terminal area. As shown in Fig.4a, this damage took the form of a grooving along the grain boundaries which clearly displays the grain size of the films, 0.5 micron, about equal to the film thickness. In contrast the positive terminal areas showed no grooving, but a somewhat irregular surface marked with hillocks as represented in Fig.4b.

The electron microprobe was not more helpful than the scanning electron microscope in locating the crack discontinuities across the conducting stripes. Sweeping the electron beam perpendicularly to the length of the stripe at one micron intervals and monitoring both the chromium and the aluminum x-ray fluorescences was one mode of analysis which was used. The results are shown for one half of a stripe (which had been determined to be electrically discontinuous) in Fig.5. The electrons were accelerated to a voltage of 15 kV;

Figure 3. Scanning electron microscope image of a possible failure site in an aluminum-chromium thin film conductor tested at 175°C and 2×10^6 Amp/cm^2. The width of the conductor is 7.5 microns, and the angle of viewing is 45°.

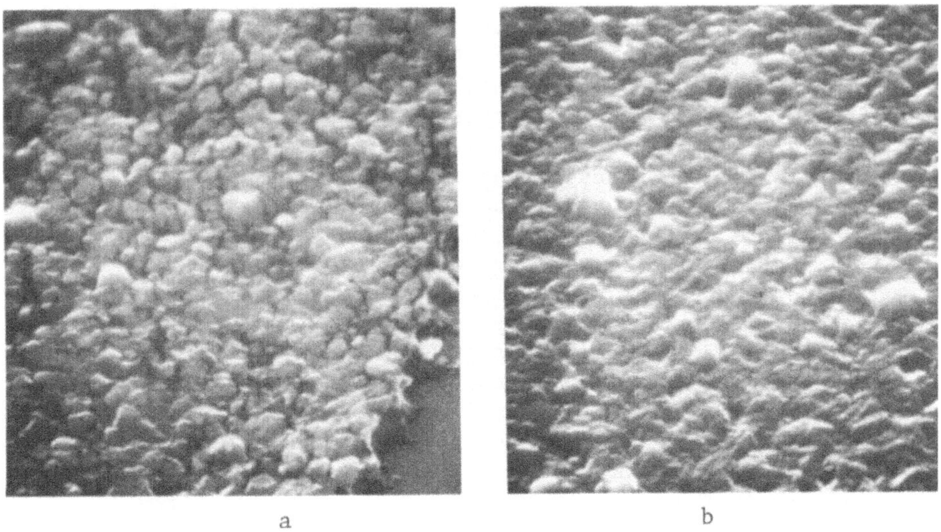

Figure 4. Scanning electron microscope images of the terminal areas of an aluminum-chromium thin film conductor tested at 175°C and 2×10^6 Amp/cm^2 for 8300 hours. (a) Negative terminal, (b) positive terminal. Magnification 9000 X, angle of viewing 45°.

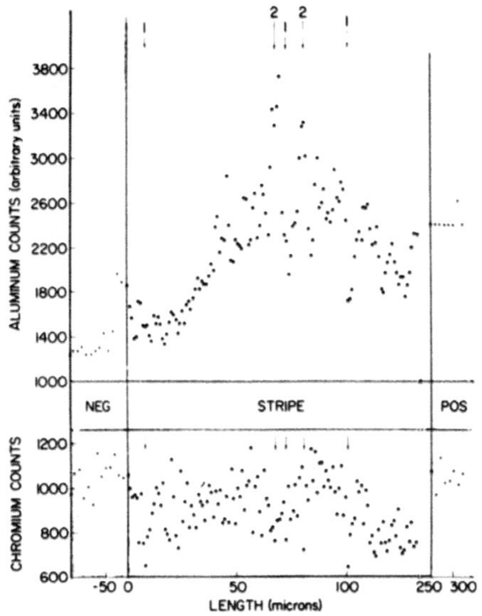

Figure 5. Distribution of aluminum and chromium counts along the length of an aluminum-chromium thin film conductor after testing at 175°C and 2×10^6 Amp/cm^2 for 8300 hours.

the chromium Kα and the aluminum Kα lines were monitored. The chromium counts were corrected for background noise. At each end in Fig.5, crosses represent counts obtained at different positions in the terminal areas. The arrows marked 1 correspond to points of low chromium counts and low aluminum counts, while the arrows marked 2 correspond to points of low chromium counts also, but high aluminum counts. The chromium counts along the length of the stripe (dots) are somewhat lower than the counts in the terminals (crosses), however, the difference about 15% can be accounted for by the fact that in sweeping the electron beam across the stripe care was taken to sweep somewhat beyond the stripe width, about 1 micron on each side. There was a sharp difference in the aluminum counts which indicated depletion in the negative terminal area and accumulation in the positive terminal area. As for the location of whiskers, it is remarkable that both depletion and accumulation seemed to have spread in the terminals almost independently of current density variations. Accumulation and depletion extended to the very vicinity of the contact wires. A systematic probing of the terminal area of another sample with 20 kV electrons gave aluminum counts varying from 7900 to 12500 for the negative terminal, and from 11400 to 24500 for the positive terminal. Averaged over the whole terminal areas aluminum counts varied as 29 to 39 for the negative and positive ends, not counting the whiskers. When these are taken into account, the total aluminum transport (assuming that mass is proportional to counts) amounts to 4/17 of the total terminal volume. Chromium transport was not as extensive as aluminum transport, and because of the smaller counts could not be determined as accurately as the transport of aluminum. It is estimated that the total amount of chromium in the negative terminal decreased by about 2%.

The electron microprobe was used for two other studies. Analysis of the whiskers showed them to be chromium free; within the limits of accuracy obtained, the concentration of chromium in the whiskers was less than 10% of the concentration in the film. (However, this could well be above the solubility limit at 175°C.) The probe was also used to investigate the distribution of chromium in the films, after the 560°C anneal, but prior to electromigration testing. A series of counts were taken with the electron beam reduced to a size of 0.6 micron. Over twelve points at 1 micron interval aluminum counts varied from 149000 to 159000 while chromium counts varied from a low value of 3600 to a high value of 7400, which indicates that considerable chromium diffusion took place during the annealing treatment. In order to determine the chromium distribution through the thickness of the film the beam was swept over an area 8 by 10 microns, and a series of counts were taken for different excitation voltages. The results indicate that although complete homogenization had not occured during the annealing treatment the upper 1000 A of the film contained more than 1 wt% chromium. The average chromium concentration was also greater than 1 wt%, which is in keeping with an intended concentration of 2 wt%.

Copper Additions

In a first series of experiments the activation energy for electromigration in aluminum-copper films was determined by monitoring the effect of sudden changes in temperature on the rate of resistance increase of samples undergoing electromigration tests [32]. In all cases the measurements were made during the initial period of the tests, before the formation of extensive macroscopic damage such as cracks, or hillocks. Results indicate that the rate of resistance changes under similar test conditions (4 x 10^6 Amp/cm^2

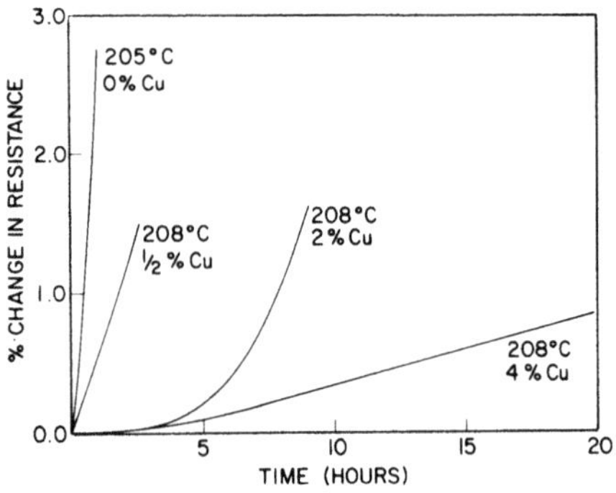

Figure 6. The effect of copper concentration on the rate of change of electrical resistance in aluminum-copper thin film conductors undergoing electromigration tests at about 208°C and 4x10^6 Amp/cm^2 (after ref. 32).

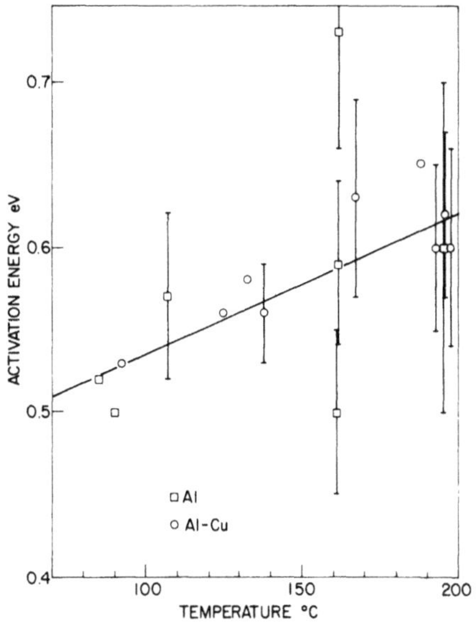

Figure 7. The activation energy for electromigration determined as a function of temperature by the electrical resistance technique for pure aluminum and aluminum-copper thin film conductors (after ref. 32).

and 210°C) decreases markedly in samples with increasing copper concentrations, at least up to 4 wt%, as shown in Fig.6. The activation energy values which were derived were found to be independent of the copper concentration, from 0 to 8 wt%, but to increase slightly with increasing temperature, as shown in Fig.7. The temperature range was from about 100°C to 200°C, and the mean activation energy was about 0.55 eV.

In another series of electromigration experiments the activation energy for failure, by crack propagation and open circuit formation, was determined as a function of copper concentration [34]. The results, shown in Fig. 8, indicate that in samples containing about 2 to 4 wt% copper the activation energy for failure is about 0.2 to 0.3 eV greater than in pure aluminum samples.

An ingenious experiment [30], performed in a different laboratory, demonstrated that the addition of copper to aluminum films decreases the rate of aluminum transport during electromigration. A dot of copper was diffused into a thin film aluminum conductor prior to electromigration testing. As shown schematically in Fig.9, when such a conductor is subjected to the passage of current, aluminum atoms transported from the negative terminal are effectively

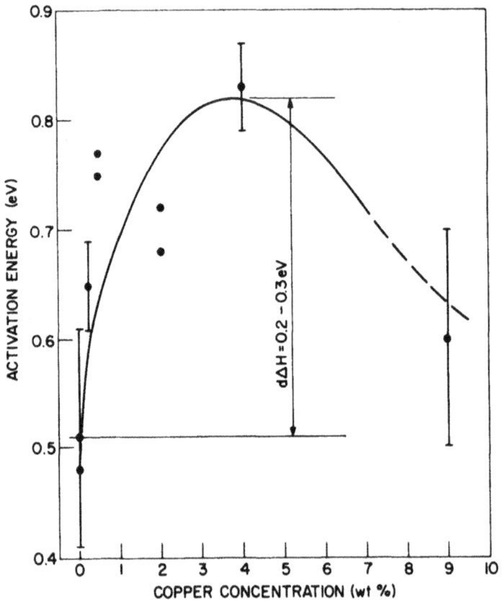

Figure 8. The activation energy for electromigration failure of aluminum-copper thin film conductors plotted as a function of copper concentration (after ref. 34).

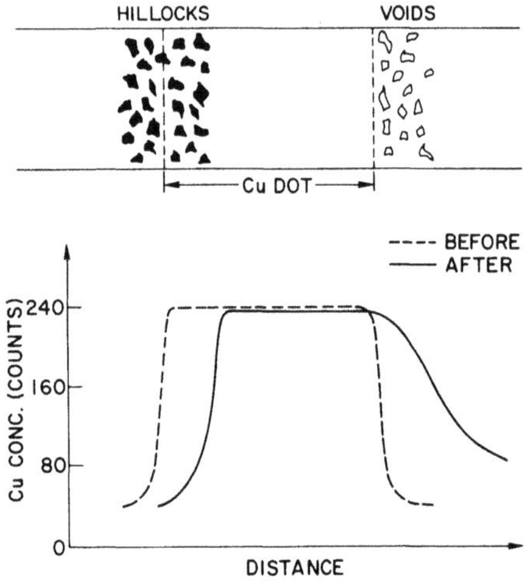

Figure 9. Schematic diagram showing the retarding effect of copper on the rate of grain boundary diffusion of aluminum. The motion of electrons was from left to right (after ref. 30).

stopped at the copper rich dot, where they form hillocks. Further down the electron stream, at the junction between the copper rich dot and pure aluminum, the transition from a zone of slow aluminum transport to one of fast aluminum transport results in void formation. Furthermore, the displacement of the copper rich dot in the course of electromigration test, provides evidence for the relatively high mobility of the copper atoms.

Measurements of transport quantities, similar to those reported above for aluminum-chromium, were made on a sample with the same configuration as in Fig.1, but with a copper concentration of 2 wt%. The sample had undergone test at 175°C, at a current density of 1×10^6 Amp/cm^2, for 25,000 hours. The grain size of the sample as tested was revealed by etching to be about 2.5 microns. Transport of aluminum from the negative to the positive terminal amounted to about 0.5% of the terminal volume, whereas the transport of copper caused a decrease in the total copper content of the negative terminal of about 0.6%. No whiskers were detected.

Because some alloying elements are known to interfere with vacancy motion either in quenched aluminum [40], or in quenched aluminum-copper [41], some experiments were carried out with the aim of investigating whether such additions affect the lifetimes of aluminum-copper films. A series of five aluminum-copper films with ternary alloy additions were prepared and tested for this purpose. All of the five alloy films contained 5 wt% copper, the first one did not have any other additions, the second one contained 0.06 wt% tin, the third contained 0.12 wt% tin, the fourth and fifth contained the same respective concentrations of indium. The films were annealed at 530°C for twenty minutes in a nitrogen atmosphere. Testing of the stripes made from the different films was carried out at a temperature of 175°C, and a current density of 4×10^6 Amp/cm^2. The respective median lifetimes were: 740, 140, 130, 400, and 150 hours.

DISCUSSION

Diffusion of Aluminum in Aluminum-Chromium

The amount of aluminum A transported during a time t along a conductor is given by the relation [36, 37]:

$$A = n.\ell.\delta.D_o.\exp(-\Delta H/kT).(1/kT).Z^*.e.J.\rho.t \qquad (1)$$

where the quantities n, ℓ, and δ are the number, the length and the width of the grain boundaries in the stripe cross section, D_o and ΔH are the preexponential factor and the activation energy for grain boundary diffusion, Z* is the effective charge on the moving aluminum atoms, e the charge on the electron, J the current density, and ρ is the resistivity of the sample. From the size of the terminals, and the experimentally determined transport of aluminum reported above (4/17), A is about equal to 9×10^{-8} cm^3. From the grain size n would be about 16, and ℓ can be taken as equal to the thickness of the film 6×10^{-5} cm. The current density was 2×10^6 Amp/cm^2, and at the test temperature (T = 445°K), the product J.ρ was 18 volt/cm. The value of Z* can be extrapolated from measurements [42] on lattice electromigration in pure aluminum (at higher temperatures) to be about 60 at the test temperature. The time t was 3×10^7 seconds. One can therefore calculate the quantity $\delta.D_o.(\exp{-\Delta H/kT})$, which is a measure of the grain boundary diffusion coefficient, to be equal to 1.1×10^{-16} cm^3/sec. The median lifetimes at 172 and 224°C are compatible with a value of ΔH of about 0.55 eV, as found for pure aluminum [32, 43-45]. Thus one obtains a value for $\delta.D_o$ of 1.8×10^{-10} cm^3/sec. If δ is taken to be 5×10^{-8} cm, D_o would become 3.6×10^{-3} cm^2/sec. Actually these values represent minima. Because the resistivity of the samples was about twice that of pure aluminum the value of Z* should have been decreased proportionally according to the relation [36]:

$$Z^* = -1/2.Z\ (N/N_d \cdot \rho_d/\rho.S - 2) \qquad (2)$$

where Z is the valency of the host atoms (aluminum in this case), ρ is the resistivity of the conductor, S is a factor with a value of about 0.6, and the quantity $N/N_d \cdot \rho_d$ is the excess resistivity contributed by the moving atoms. Thus, if Z* is taken to be 30, $\delta.D_o$ would become equal to 3.6×10^{-10} cm^3/sec, and D_o would become equal to 7.2×10^{-3} cm^2/sec. These values appear to be in good agreement with values of 1.2×10^{-10} cm^3/sec for $\delta.D_o$ [46], and 3×10^{-3} cm^2/sec for D_o [47] which have been reported for silver. More significantly, the value of D_o for aluminum which has been derived in the case of aluminum containing chromium appears to be somewhat larger, not smaller, than the value of 5×10^{-4} cm^2/sec which can be derived from previous work on pure aluminum [48]. It must be concluded, therefore, that the enhanced lifetime (8300 hours for aluminum-chromium conductors, 30 hours for pure aluminum conductors tested in the same conditions: 2×10^6 Amp/cm^2, and 175°C) in aluminum-chromium conductors results from the effect of chromium on the formation of voids and the propagation of cracks, rather than from a reduction in the rate of diffusion.

Although the geometry of whiskers is quite well defined, it is difficult to calculate diffusion parameters from whisker growth because the current density at various points of the positive terminals is improperly known. In order to obtain values in agreement with the diffusion parameters obtained above, it was found necessary to assume that individual whiskers grew from simultaneous transport along several grain boundaries. Even then the derived values of D_o tended to be high. It is possible that whisker growth requires not only singular grain boundary configurations but current crowding as well, or possibly the cooperation of some mechanical stress mechanism.

Ideally, in a conductor without temperature gradients and gross material inhomogeneities, electromigration should result in the slow displacement of the conductor in the direction of the charge [carrier] flow. However, in aluminum thin film conductors of the type depicted in Fig. 1, electromigration usually results in the early formation of a continuous crack somewhere across the narrow portion of the conductor, where the current density is the highest. In aluminum-chromium on the contrary, damage extends over the terminal areas, as well as over the stripe, and failure is greatly retarded. Apparently, the increase in lifetime due to chromium does not result from a decrease in the rate of transport of aluminum; rather it seems that chromium has caused the aluminum film to react as if it were more ideal, less inhomogeneous. The divergence in grain boundary diffusion seems to have been greatly reduced. Tentatively this may be explained if it is assumed that the boundaries of aluminum films are made of "normal" and "random" boundaries [49, 50] and that a predominant cause of diffusion divergence in aluminum films is due to the difference of transport characteristics of the two types of boundaries. Chromium atoms would affect primarily the random boundaries on which they would be preferentially adsorbed than on the special boundaries. Thus, by reducing the rate of diffusion in the normal grain boundaries, and leaving the others essentially unaffected the atomic flux divergence would be reduced and the rate of failure would be proportionally decreased. If the atomic flux divergence is considered to be the difference between two large quantities, this divergence could be considerably reduced without significantly affecting either of the two quantities. Hence, although chromium additions could, from such an effect, reduce the rate of failure, the difference in the average rate of grain boundary diffusion between aluminum, and aluminum-chromium can be negligible.

Since the effective rate of migration of chromium was smaller than the effective rate of migration of aluminum, the chromium concentration at the negative terminal increased, while the concentration at the positive terminal decreased, as a result of electromigration. Analysis of the diffusion of chromium during electromigration shall be considered later, in comparison to data obtained on the migration of copper in aluminum-copper films. Suffice it to note at this point that in considering the grain boundary migration of alloy additions one must take into account the solubility limit and the adsorption of the additive atoms on grain boundaries. Thus equation 1 should be modified as follows:

$$A_i = n.\ell.\delta.D_{oi}.\exp(-\Delta H_i/kT).(1/kT).z^*_i.J.\rho.t.$$

$$c_{oi}.\exp(-H_i/kT).\alpha_i.\exp(Q_i/kT) \qquad (3)$$

where the subscript i refers to the species of the migrating atoms, $c_{oi}.(\exp{-H_i}/kT)$ is the solubility limit c_i at the temperature T and H_i the heat of solution of atoms of species i in the aluminum matrix, α_i, and Q_i are a preexponential factor and the heat of adsorption of atoms of species i on the grain boundaries [51]. A more precise formulation for the adsorption of atoms is given in ref. 51, but for the purpose at hand the formulation used here is accurate enough.

Diffusion of Aluminum in Aluminum-Copper

In order to facilitate the comparison between the transport data obtained with aluminum-copper and with aluminum-chromium the conditions of the aluminum-chromium test will be considered as standard. Then, according to equation 1, the transport amounts obtained for aluminum-copper will be multiplied by 4, to account for the grain size difference, and by 2, to account for the current density difference, divided by 3, to account for the time difference. No corrections need be made for Z* and ρ since as a first approximation the product $Z^*.\rho$ can be considered to be constant. Thus under identical sets of conditions (8300 hours, 2 x 10^6 Amp/cm^2) in aluminum-copper aluminum transport amounts to about 0.3% of the terminal volume, as compared to about 23% in the case of aluminum-chromium; the copper transport which amounts to about 1.6% of the initial copper content, is quite close to the transport of chromium, about 2% of the initial chromium content.

It has been shown above that the transport of aluminum in aluminum-chromium is about equal to the transport which would be anticipated in pure aluminum films. In aluminum-copper the transport of aluminum is about 80 times smaller. This is quite compatible with electromigration studies which have shown that under identical test conditions the lifetime of samples containing about 4 wt% copper is about 70 times longer than the lifetime of pure aluminum samples [26]. In conjunction with the results displayed in Fig. 9, these results conclusively demonstrate that copper additions to aluminum films reduce the rate of grain boundary diffusion of aluminum. Thus, one may write, from equation 1:

$$\delta.D_o{}^{Al}_{Al-Cu}.\exp(-\Delta H^{Al}_{Al-Cu}/kT) = 1/80.\delta.D_o{}^{Al}_{Al}.\exp(-\Delta H^{Al}_{Al}/kT) \quad (4)$$

where the superscripts refer to the diffusing species, and the subscripts to the material in which diffusion takes place. The problem of defining the mechanism of the effect of copper on aluminum diffusion is to define which of the terms in equation 3 are affected by copper. Precise determination of activation energies would be required since a change of ΔH by 0.13 eV would account for the difference in the two sides of equation 3. The results of the resistance change measurements displayed in Fig. 7 have not been unequivocally interpreted, but they do not exclude the following relations:

$$\Delta H^{Cu}_{Al-Cu} = \Delta H^{Al}_{Al-Cu} = \Delta H^{Al}_{Al} \quad (5)$$

Some information can also be derived from the results of lifetime studies [34] displayed in Fig. 8. A number of experimental observations on aluminum-copper thin film conductors indicate that failure formation, at least in samples with relatively low copper concentration (2-4 wt%), is preceded by copper depletion. Thus the rate of failure is proportional to the flux of copper atoms, and the activation energy for failure is given by summing the exponential terms in equation 3. The increment of activation energy for failure in aluminum-copper dΔH over the value for pure aluminum is given by:

$$d\Delta H = (\Delta H^{Cu}_{Al-Cu} - \Delta H^{Al}_{Al}) + (H^{Cu}_{Al} - Q^{Cu}_{Al}) \quad (6)$$

where H and Q, are the heat of solution of copper in aluminum, and the heat of adsorption of copper on aluminum grain boundaries. The current estimates of the terms in equation 6 are too imprecise for making positive statements, although the equality between the activation energies for grain boundary diffusion of copper in aluminum-copper, and for aluminum in aluminum, as written in equation 5, does not appear to be excluded. This in turn would imply equality between the activation energies for grain boundary diffusion of aluminum in aluminum-copper, and of aluminum in aluminum. Thus, returning to equation 4, one may tentatively ascribe the effect of copper in reducing the rate of diffusion of aluminum to a decrease in the product $\delta.D_o$.

Interpretation of Electromigration Experiments in Aluminum-Copper

The new data about the effect of copper on the rate of diffusion of aluminum, and on the relative rate of diffusion of copper during electromigration allow one to reexamine some of the previously obtained results [27, 32-34]. Analysis of the effect of copper on the activation energy for failure [34], leading to the derivation of equation 6, was based on the observation that copper depletion occurs faster than aluminum depletion [27]. It has now been found that the ratio between the two effective rates of depletion is about 6 for the concentration considered (2 wt%), which is in agreement with the observation that after electromigration testing of a conductor, areas could be found where the initial copper concentration had increased by a factor of 4 to 5 (see Fig. 9 in reference 27). Moreover, the fact that the presence of copper reduces the rate of aluminum diffusion by a factor of about 80 provides a coarse correlation with other experimental observations. The decrease in activation energy for failure at high copper concentration in Fig. 6 can be explained if at such concentrations failure occurs by aluminum diffusion prior to total copper depletion from the crack areas whereas at low concentrations the contrary would be true. Since lifetimes are increased by a factor of about 70 for a copper concentration of 4 wt% [26], and copper reduces the aluminum diffusion rate by a factor of 80, it would seem that for copper concentrations above 4 wt% the activation energy for failure indeed should show a decline.

The relationship between activation energy for failure and concentration of copper will vary with temperature range as shown schematically in Fig. 10 where the logarithm of lifetime, at a given current density, is plotted against the reciprocal of the absolute temperature. The lower line pertains to pure aluminum;

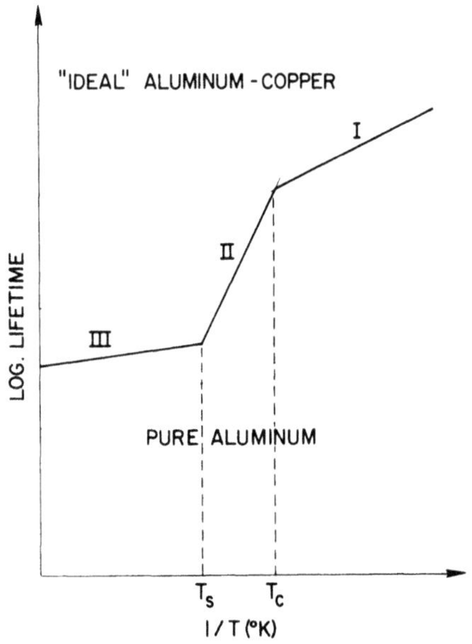

Figure 10. Schematic representation of the anticipated effect of different temperature ranges on the activation energy for electromigration failure in aluminum-copper thin films.

its slope is a measure of ΔH_{Al}^{Al}. The upper line corresponds to an ideal aluminum-copper, where an external source of copper would insure that failure would never occur through copper depletion; its slope is proportional to ΔH_{Al-Cu}^{Al}. (These two slopes are deliberately shown as being unequal in Fig. 10 to maintain the generality of the present discussion, although for copper additions it is believed that they are approximately equal). For samples with a given copper concentration the lifetime curve will vary between these two extremes, and assume different slopes at different temperatures. At low temperatures failure will occur prior to copper depletion, and the activation energy will be equal to ΔH_{Al-Cu}^{Al} (Zone I). At intermediate temperatures (above a critical temperature T_c which increases with concentration but below the temperature T_s at which complete solubility is achieved) failure will occur via local copper depletion, including the dissolution of the Al_2Cu precipitates; the activation energy will be equal to $\Delta H_{Al-Cu}^{Al} + H_{Al}^{Cu} - Q_{Al}^{Cu}$ (Zone II). At high temperatures, above the solubility limit, failure would still happen through copper depletion, but the activation energy would now assume the value $\Delta H_{Al-Cu}^{Cu} - Q_{Al}^{Cu}$ (Zone III). Obviously the part played by lattice diffusion at higher temperatures has been consciously eliminated from this scheme. Note that for low copper concentrations the temperature range with the highest slope (Zone II) should be moved towards low temperatures, i.e., to the right in Fig. 10.

It had been noted that although lifetimes seem to increase somewhat linearly with copper concentration up to about 4 wt%, above this value they seem to be less dependent on concentration [27]. It has even been suggested that there is no relationship between concentration and lifetime in the concentration range between 4 and 20 wt% [33]. These observations also are quite in agreement with the fact that copper reduces the rate of migration of aluminum by a factor of 80, and that most of the benefit of copper additions on lifetime is obtained with a copper concentration of about 2-4 wt%. When the concentration is increased beyond this level failure occurs by aluminum depletion prior to copper depletion, rather than the other way around. From a practical point of view, it appears tempting to increase the lifetime of aluminum-copper thin film conductors by the addition of ternary alloying elements which would reduce the rate of copper diffusion in the grain boundaries. However, it should not be anticipated that the lifetimes would be increased by a factor greater than about 80 (as compared to pure aluminum) unless the ternary additions themselves play a significant role in decreasing the rate of grain boundary diffusion of aluminum.

It has been shown that whereas the effective rates of copper and chromium depletion in the samples which have been analyzed are essentially equal, differences in the rates of diffusion of aluminum cause different effects in the two samples. In the copper alloyed sample electromigration caused the concentration of copper to decrease in the areas from which copper was being removed (negative terminal), whereas in the chromium alloyed sample the concentration of chromium increased in the same area. Thus the fact that in the aluminum-chromium samples the failure times at 172 and 224°C appear to correspond to a relatively low value of activation energy (about 0.55 eV, as for pure aluminum), gives support to the analysis of failure in aluminum-copper which has been presented above.

Interpretation of the results of the resistance change measurements [32], shown in Fig. 6, requires a satisfactory explanation for the apparent equality of the activation energies found for pure aluminum and aluminum-copper. If in aluminum-copper the resistance changes resulting from electromigration are due to the migration of copper atoms, the activation energy should be as high as about 0.8 eV, as found for failures. Since the values experimentally determined are the same as for pure aluminum, one must, therefore, conclude that the resistance changes in aluminum-copper films result from aluminum migration. The dependence of the rate of resistance change on copper concentration could be explained on the basis of structural heterogeneities. For example, aluminum has been shown to migrate more slowly in the presence of copper; one may postulate that the measured resistance changes result from aluminum migration in local zones which have been rapidly depleted of copper. The frequency of occurrence of such zones would obviously be inversely proportional to the copper concentration.

Since additions of tin or indium have been reported to reduce the rate of precipitation in quenched aluminum samples [40, 41], such additions might be anticipated to have similar effects on the rate of electromigration in aluminum films. It was found that, if anything, such additions reduce the lifetime of aluminum-copper films subjected to current stressing, and it is likely that this reduction in lifetime is due to an increase in the rate of electromigration. It has already been shown that the interpretation of results obtained with quenched samples, with their supersaturation of vacancies and solute atoms is a complex matter, so that in those cases which have been tested the reports of high binding energies between atoms and vacancies [41] in such samples were not validated by experiments conducted under conditions of thermal equilibrium [52, 53]. Thus, the effect of additions of tin or indium on the lattice electromigration in aluminum would be quite unpredictable.

In the films under consideration, however, the matter is further complicated since one is concerned with grain boundary diffusion. Therefore, there probably is no direct relationship between the effects of indium and tin additions to aluminum thin film conductors, and the reported effects of similar additions to quenched bulk specimens.

The Electromigration of Alloy Elements: General Considerations

In order to analyse the transport of alloy additions equation 1 must be modified to take into account the concentration of alloy atoms in solution, and since the diffusion phenomenon of interest is grain boundary diffusion one should also consider the adsorption of such atoms on the grain boundaries. The amount of solute transport is then given by equation 3, where the preexponential term α_i is a measure of the entropy change which results from the adsorption of an atom, and of geometrically determined frequency of available adsorption sites on the grain boundaries. As written, the above equation is based on the assumption that equilibrium was maintained throughout the course of the experiments. The concentration of solute atoms should be taken as the solubility limit c_i at the test temperature, since only atoms in solution should be considered mobile, and in the cases under consideration the actual concentration is well above the solubility limit. The solubility limits [54] for copper and chromium in aluminum is given as a function of temperature in Fig. 11. At 175°C the solubility limit for copper is about 2 x 10^{-4}, and that for chromium is about 30 times less.

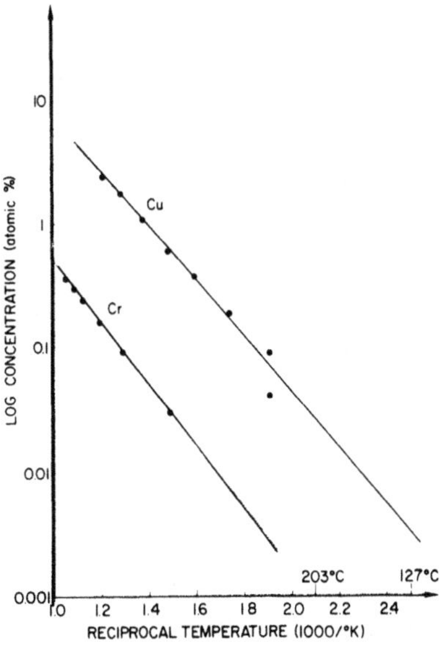

Figure 11. Solid solubility of chromium and copper in aluminum plotted as a function of the inverse absolute temperature.

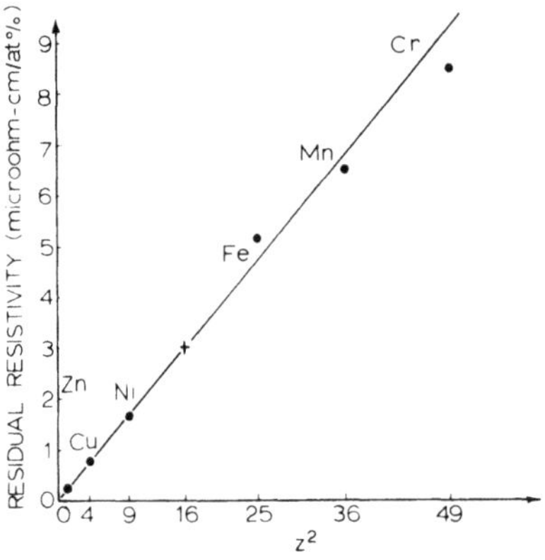

Figure 12. The excess electrical resistivities of alloying elements in aluminum plotted as a function of the square of their relative valencies z.

In calculating the diffusion constant for aluminum, use was made of the value of Z* obtained in experiments conducted with bulk samples [42]. However, the values of z* for copper or chromium in aluminum have not been experimentally determined. Such values, therefore, will have to be estimated from a theoretical relation [37] similar to the one given above for Z* (equation 2):

$$z^* = -1/2.Z.1/\rho.[\lambda z^2 + \lambda(z + Z)^2 - 2\rho/100].100 \qquad (7)$$

where z + Z is the valency of the impurity under consideration and λz^2 is equal to the residual resistivity (per at %) of such an impurity in the host material of valency Z. Experiments have shown that this relation is formally correct for cadmium, indium, tin and antimony in silver [55]. Although it appears correct to attribute a value of -2 for z in the case of copper, there is much more uncertainty in the case of chromium. A clue to the value of z for impurities in aluminum is provided by Fig. 12. The residual resistivities of materials in the series Zn, Cu, Ni, (Co), Mn, and Cr have been plotted against z^2, where z has been given the values -1, -2, -3, etc. Thus the value of z for copper is confirmed to be -2 and that for chromium is shown to be about -7. This may be justified [56] by considering that transition elements have unfilled d shells which may accept up to 10 electrons. Thus for copper z + Z is equal to 1, the values for λz^2 and $\lambda(z + Z)^2$ can be obtained from Fig. 12 and the value of z* is found to be about 30. For chromium z + Z is equal to -4 and the corresponding value of $\lambda(z + Z)^2$ (given by the cross on Fig. 12) is about 3×10^{-6} ohm-cm. With λz^2 equal to 8.5×10^{-6} ohm-cm one obtains, after taking into account the increase in ρ, a value of z* for chromium in aluminum of about 225.

The Diffusion of Copper

Since measured transport quantities in aluminum-copper have already been normalized to the conditions of the aluminum-chromium test (0.6 micron grain size, 8300 hours, 2×10^6 Amp/cm^2) in analyzing copper transport the values for most of the terms in the transport equation should be the same as used for aluminum transport in aluminum-chromium. However, the product $\rho.J$ should be only 10.4 volt/cm (rather than 18 volt/cm) because the resistivity of aluminum-copper is less than that of aluminum-chromium. The volume value of A is 6×10^{-11} cm^3. The values for the solubility limit and z* have already been given as 2×10^{-4} and 30, respectively. It follows therefrom that the product $\delta.D_o.\exp(-\Delta H/kT).\alpha.\exp(Q/kT)$ would be equal to 1.2×10^{-15} cm^3/sec. In keeping with previous discussion, ΔH and Q may be considered equal to 0.55 ev and 0.2 ev respectively, then $\delta.D_o.\alpha$ would be equal to 1×10^{-11} cm^3/sec. The value of α may be chosen as about 1/10 since it is the product of a geometrical factor estimated [51] to be about 1/3 and a factor corresponding to the decrease of entropy when an atom is removed from solution in the lattice and adsorbed on a boundary (a value of 1/4 has been given to this term) [57]. Then for copper in aluminum-copper one derives a value of about 1×10^{-10} cm^3/sec for $\delta.D_o$. This is more than one order of magnitude larger than the corresponding value for aluminum in aluminum-copper, and only slightly smaller than the corresponding value for aluminum in aluminum-chromium.

The Diffusion of Chromium

With respect to aluminum a number of known features make chromium strongly different from copper, e.g. solubility limit, diffusion, and specific resistivity. As shown in Fig. 11 the solubility limit of chromium is about 30 times smaller than that of copper at 175°C. The lattice diffusion of copper in aluminum is "normal", with an activation energy varying but little from the value of self diffusion; however, chromium diffuses extremely slowly in the aluminum lattice [58], and it probably belongs with other transition elements to a family of metals characterized by an anomalously high activation energy (Fe:2eV [59], 2.7eV [60-62]; Co:1.8eV [58]; Mn:2.2eV [63]) for lattice diffusion in aluminum.

In utilizing equation 3 to determine diffusion characteristics of chromium the value of A_i is 8×10^{-11} cm^3. The solubility limit is 7×10^{-6} at 175°C. The value for z* has been estimated to be about 225. As a first approximation the adsorption terms for chromium atoms on aluminum grain boundaries may be assumed to be equal to those for copper (because of their relative atomic sizes, 2.556Å for copper, 2.498Å for chromium, as compared to 2.862Å for aluminum), so that $\alpha.\exp(Q/kT)$ is about equal to 20. Thus one obtains for $\delta.D_o.\exp(-\Delta H/kT)$ a value of 2×10^{-6} cm^3/sec, which is about equal to the corresponding value for aluminum in aluminum-chromium, and about 3 times larger than the corresponding value for copper in aluminum-copper. Therefore, it would appear that the grain boundary diffusion of chromium in aluminum is "normal", unlike the lattice diffusion of chromium in aluminum.

ALLOY ADDITIONS AND GRAIN BOUNDARY DIFFUSION

In the study of electromigration phenomena in alloyed thin film conductors it is necessary to consider the diffusion of the host element, aluminum, and how it is affected by alloy additions, as well as the diffusion of alloy additions themselves. Thus one finds that copper reduces the rate of diffusion of aluminum, but chromium does not. It is perhaps to be expected that chromium with a solubility limit some 30 times smaller than that of copper, should have little effect on aluminum diffusion. That copper with a solubility limit of only 2×10^{-4} should be found to reduce the rate of aluminum grain boundary diffusion by a factor of about 80 is surprising. However, since the effect is presumed to result from the adsorption of copper atoms on the grain boundaries, the solubility limit to be considered should be larger than the lattice value. At the temperature of the experiments considered (175°C), if the preexponential term α and the energy of adsorption Q are taken to be 1/10 and 0.2 eV the solubility limit of copper on aluminum grain boundaries would be of the order of 4×10^{-3}. The preexponential factor in grain boundary diffusion, of the order of 3×10^{-3} cm^2/sec (this paper and ref. 47) may be taken as a measure of the excess vacancy concentration in the grain boundaries, at 175°C. Thus, there would be about one copper atom per vacancy in the grain boundaries. The effect of copper on aluminum diffusion would then require that almost all copper atoms be associated with grain boundary vacancies, which is perhaps compatible with an over-simplified view of the elastic interaction of grain boundaries with adsorbed atoms (quite different from the equilibrium atom-vacancy interaction in the lattice). One may still ask why such an effect would not be strongly temperature dependent so that the activation energy for diffusion of aluminum in aluminum-copper should be about the same as that of aluminum in aluminum. It is not inconceivable that at lower temperatures, (where the concentration of adsorbed copper atoms should be smaller than at 175°C), copper atoms in some state of precipitation would be just as effective in slowing aluminum diffusion as atoms in thermal equilibrium. With respect to the flux of copper atoms, as in the determination of the activation energy for failure in aluminum-copper, the situation is different, since one is now exclusively considering mobile copper atoms which must be in a condition approaching thermodynamic equilibrium. Hence while the diffusion of aluminum in aluminum-copper may not be a direct function of the equilibrium concentration of copper atoms in the grain boundaries, the flux of copper atoms on the contrary would be directly proportional to the concentration.

The above discussion is a tentative analysis of the effect of copper and chromium on the grain boundary diffusion of aluminum. It is, of course, of a very preliminary nature and should be considered together with the few other data which have been obtained about the effect of alloy additions on grain boundary diffusion. Such data have been summarized in Table I, where the nature of the host material, the nature of the diffusing species, and the nature of the alloy additions have been identified. The effect of the alloy addition on diffusion has been given a minus sign if the alloy addition has been reported to retard the migration of the diffusing atoms, a plus sign if the contrary is true, and zero if no effect was detected. In quite a few cases impurity additions have been found to enhance (as perhaps indium and tin in the present work) grain boundary diffusion rather than retard it. It would seem that in order to explain such an effect a change in grain boundary structure would have to be invoked. Nothing then would preclude that the retarding effects of other alloy additions also result from a change in grain boundary structure.

Analysis of the failure mode of aluminum-copper conductors subjected to electromigration required considering the diffusion of copper in aluminum-copper grain boundaries. In keeping with some experimental observations it was assumed as a first approximation that the activation energy for such diffusion is the same as the activation energy for grain boundary diffusion of aluminum in aluminum. On such a basis it was found that the product $\delta.D_o$ for copper in aluminum-copper is about equal to the corresponding value for aluminum in aluminum. The literature apparently does not contain any information of a quantitative nature on grain boundary diffusion of the type copper in aluminum-copper; however, some systematic investigations of diffusion of the type copper in aluminum have been reported. In Fig. 13, where the activation energies for grain boundary diffusion are plotted against the activation energies for lattice diffusion, the results of a series of experiments on the diffusion of silver [46], cadmium [80], indium [81], tin [82], and antimony [83] in silver, have been summarized. A strong positive correlation between the activation energies for lattice and grain boundary diffusion is clearly displayed, as well as a strong negative correlation between atomic number and activation energy. The reported values for the preexponential product $\delta.D_o$ do not seem to vary in any systematic way; the lowest value is that for silver in silver.

TABLE I. Data on Grain Boundary Diffusion

Host Metal	Diffusing Species	Alloy Addition	Effect[†]	Reference
Feγ	Ni,Pd,Cu	Ti,Nb,B,Mo	-	64
Feγ	Ag	Pd	+	64
Feγ	Ag	Cu	-	64
Feγ	Fe	x	+	65
Feα	Fe	x	+	66
Ni	Fe	x	-	65
Ni	Sn	B	-	67
Ni	Sn	Ce,Mo,W	0	67
Cu	Au	Au	-	68
Cu	Ag	Be,Sb	+	69,70
Cu	Ag	Zr,Ti	-	71
Cu	Ag	Cr,Te	±*	71
Ag	Ag	Tl,Cd	+	72,73
Ag	Ag	x	-	74
W	W	Ni	+	75
Feα,γ	Fe	B	-	76
Fe-15%Cr	Fe,Cr	C	-**	77
Feα	S	S,O	+	78,79
Feα	S	C	-**	78,79
Al	Al	Cu	-	This
Al	Al	Cr	0	work
Al	Al	Ag,Au	0	27

† A plus sign indicates that the impurity increases grain boundary diffusion; the contrary holds for a minus sign.

* The effect is a decrease in the rate of diffusion only at low temperatures.

** Carbon seems to be most effective only above the solubility limit.

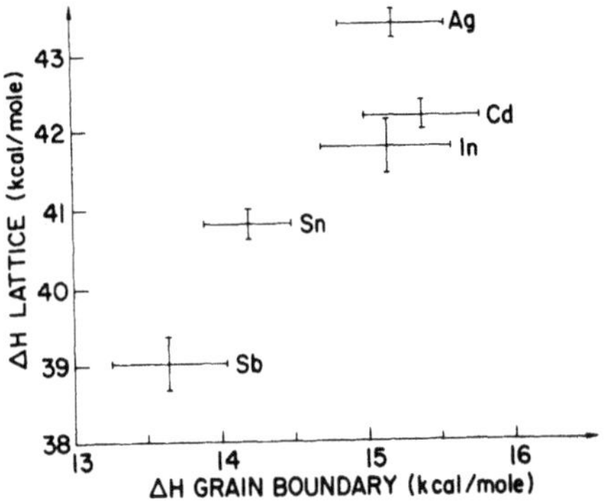

Figure 13. The correlation between the activation energies for grain boundary diffusion and lattice diffusion of several elements in silver.

For the purpose at hand, the correlation between the activation energies for lattice and grain boundary diffusion may be considered valid for the case of copper in aluminum. Because the activation energies for lattice diffusion of copper $\Delta H^{Cu}_{lat.Al}$ (1.33-1.40 eV) [84, 58], and aluminum $\Delta H^{Al}_{lat.Al}$ (1.26-1.46 eV) [58, 85, 86] in aluminum are about equal, the same is likely to be true for grain boundary diffusion. By extension one may assume that the activation energies for grain boundary diffusion of aluminum in aluminum ΔH^{Al}_{Al} and copper in aluminum-copper ΔH^{Cu}_{Al-Cu} are also about equal. If the high value of $\Delta H^{Cu}_{lat.Al}$ and the low value of $\Delta H^{Al}_{lat.Al}$ are to be preferred [58] it would follow that ΔH^{Cu}_{Al-Cu} may be as high as 0.65 eV, rather than 0.55 eV as selected previously. In this case the value of $\delta.D_o$ for the grain boundary diffusion of copper in aluminum-copper would be about 14 x 10^{-10} cm^3/sec, which is about 4 times greater than the corresponding value for aluminum in aluminum. This later result would be in good agreement with the analysis of impurity diffusion in silver [46, 78-81].

SUMMARY

A series of experiments on electromigration in pure aluminum and alloyed aluminum films, including transport measurements in aluminum-chromium and aluminum-copper films have been analysed. The effective charge Z* on aluminum atoms was estimated from lattice experiments to be about 60 in pure aluminum at 175°C. For copper and chromium values of z* of 30 and 225 respectively were estimated from resistivity data. The results are consistent with the following conclusions:

a) Aluminum, copper and chromium atoms migrate from the negative to the positive terminal.
b) The grain boundary diffusion of aluminum in aluminum-chromium is about the same as in pure aluminum.
c) The activation energy for grain boundary diffusion of aluminum in aluminum-chromium is probably about the same as that for aluminum in aluminum, i.e. approximately 0.55 eV.
d) The value of $\delta.D_o$ for the grain boundary diffusion of aluminum in aluminum-chromium was found to be 3.6×10^{-10} cm^3/sec, which is about the same as found for the grain boundary diffusion of silver.
e) The grain boundary diffusion of chromium in aluminum-chromium does not appear to be significantly different from the grain boundary diffusion of copper in aluminum-copper. This is in sharp contrast with the results of experiments on lattice diffusion in aluminum.
f) If it is assumed that the adsorption of chromium on aluminum grain boundaries is quantitatively equivalent to that of copper, the value of the diffusion coefficient of chromium in aluminum-chromium is about equal to the value of the diffusion coefficient of aluminum in aluminum-chromium.
g) The grain-boundary diffusion of aluminum in saturated aluminum-copper is about 80 times slower than in pure aluminum. It seems likely that this is due to a reduction of the preexponential term $\delta.D_o$ rather than to an increase in activation energy. This explanation, however, requires that at low temperatures copper atoms in some form of precipitation should be as effective as atoms in equilibrium in slowing the motion of aluminum atoms.
h) Results of measurements of copper atom diffusion indicate that copper atoms are adsorbed on aluminum grain boundary. The partition between atoms adsorbed on the boundaries and dissolved in the lattice is given by the relation (0.1)exp(0.2/kT).
i) Failure data and transport data indicate that the activation energy for copper diffusion in aluminum-copper is between 0.55 and 0.65 eV, and the corresponding values of $\delta.D_o$ would be between 1×10^{-10} and 14×10^{-10} cm^3/sec. In either case copper atoms in aluminum-copper would affect the grain boundary diffusion of copper markedly less than they affect the diffusion of aluminum.

Two observations could not be explained on the basis of thermal equilibrium only: the dependence of the rate of resistance change on the copper concentration, and the temperature independence of the effect of copper on $\delta.D_o$ for grain boundary diffusion of aluminum. Apparently the values of effective charges in the grain boundaries are not very different from the values in the lattice. Results were compared with the reports contained in the literature on the effects of alloy additions on grain boundary diffusion. Further study of such effects should yield precious information about the interaction of grain boundaries and alloy atoms, as well as significant clues about the structure of grain boundaries.

ACKNOWLEDGEMENTS

The authors gratefully acknowledge their indebtedness to many co-workers: Mrs. A. Ginzberg, Messrs. H. Luhn, V. Ranieri and W. Schug for preparing and testing samples; Messrs. C. Aliotta, C. Bremer and J. Kuptsis for their work on the scanning electron microscope and the electron microprobe; Messrs. J. K. Howard, R. Ross and M. C. Shine for permission to reproduce diagrams; and to other colleagues for helpful discussions.

REFERENCES

1. W. Jost, 'Diffusion in Solids, Liquids and Gases', Academic Press, New York, (1952), p. 324.

2. J. Verhoeven, Metal, Reviews, 8, 311-368 (1963).

3. H. B. Huntington, TMS-AIME, 245, 2571-2579 (1969).

4. Y. Adda and J. Philibert, 'La Diffusion dans les Solides', Presses Universitaires, Paris (1966), p. 893.

5. F. M. d'Heurle, Proc. IEEE, 59, 1409-1418 (1971).

6. M. A. Dayananda, Met. Trans., 2, 334-335 (1971).

7. M. Gerl, Z. Naturforsch, 26a, 1-9 (1971).

8. S. G. Epstein, Adv. Phys., 16, 325-332 (1960).

9. J. R. Wilson, R.G.R. Sellors and J.N. Pratt, Adv. Phys., 16, 357 (1960).

10. D. O. Olson, J. L. Blough and D. A. Rigney, Conference Abstracts, Joint AIME and ASM Fall Meeting, Detroit, (Oct., 1971), p. 1571.

11. J. C. Blair, P. B. Ghate and T. C. Haywood, Appl. Phys. Letters, 17, 281-283 (1970).

12. R. E. Hummel and H. M. Breitling, Z. Naturforsch, 26a, 36-39 (1971).

13. P. S. Ho and L. D. Glowinski, Z. Naturforsch, 26a, 32-35 (1971).

14. L. Berenbaum, J. Appl. Phys., 42, 880-882 (1971).

15. J. K. Howard and R. Ross, J. Appl. Phys., 42, 2996-2998 (1971).

16. A. Gonzales and E. M. Philofsky, Proc. IEEE, 59, 1429-1433 (1971).

17. R. E. Hummel and H. M. Breitling, Appl. Phys. Letters, 18, 373-375 (1971).

18. P. Adam, Z. Naturforsch, 26a, 40-44 (1971).

19. M. Ohring, J. Appl. Phys., 42, 2653-2661 (1971).

20. M. Ohring, Mat. Sci. and Eng., 7, 158-167 (1971).

21. R. Rosenberg and M. Ohring, J. Appl. Phys., to be published.

22. R. Rosenberg, J. Vac. Sci. Technology, 9, Jan/Feb (1972).

23. M. Ohring and P. H. Sun, J. Vac. Sci. Technology, 9, Jan/Feb (1972).

24. M. Ohring and P. H. Sun, Thin Solid Films, to be published.

25. D. L. Kennedy, J. Appl. Phys., 39, 6102-6104 (1968).

26. I. Ames, F. M. d'Heurle and R. E. Horstmann, IBM J. Res. Develop., 14, 461-463 (1970).

27. F. M. d'Heurle, Met. Trans., 2, 681-689 (1971).

28. L. Berenbaum and B. Patnaik, Appl. Phys. Letters, 18, 284-286 (1971).

29. L. Berenbaum and R. Rosenberg, IEEE Proc. 9th Reliability Physics Symposium, Las Vegas (1971), pp. 136-141.

30. J. K. Howard and R. Ross, J. Appl. Phys., 42, 2996-2998 (1971).

31. E. Hall, E. Philofsky and A. Gonzales, J. Electronic Mat., to be published.

32. M. C. Shine and F. M. d'Heurle, IBM J. Res. Dev., 15, 378-383 (1971).

33. B. N. Agarwala, B. Patnaik and R. Schwitzel, J. Vac. Sci Technology, 9, Jan/Feb (1972).

34. F. M. d'Heurle, N. G. Ainslie, A. Gangulee and M. C. Shine, J. Vac. Sci. Technology, 9, Jan/Feb (1972).

35. A. Gangulee and F. M. d'Heurle, Appl. Phys. Letters, 19, 76-77 (1971).

36. H. B. Huntington and A. R. Grone, J. Phys. Chem. Solids, 23, 76-87 (1961).

37. C. Bosvieux and J. Friedel, J. Phys. Chem. Solids, 23, 123-126 (1962).

38. F. M. d'Heurle, L. Berenbaum and R. Rosenberg, TMS-AIME, 242, 502-511 (1968).

39. I. Ames, J. Hoekstra and G. Folchi, Rev. Scient. Instrum., 42, 1049-1051 (1971).

40. T. Federighi, 'Lattice Defects in Quenched Metals', R. Cotterill, M. Doyama, J. Jackson and M. Meshii, Eds, Academic Press, New York (1965), pp. 217-268.

41. H. K. Hardy, J. Inst. Met., 80, 483-492 (1951-52).

42. R. V. Penny, J. Phys. Chem. Solids, 25, 335-345 (1964).

43. R. Rosenberg and L. Berenbaum, Appl. Phys. Letters, 12, 201-204 (1968).

44. I. A. Blech and E. S. Meieran, J. Appl. Phys., 40, 485-491 (1969).

45. M. J. Attardo and R. Rosenberg, J. Appl. Phys., 41, 2381-2386 (1970).

46. V. N. Kaygorogdov, S. M. Klotsman, A. N. Timofeyev and I. S. Trakhtenberg, Fiz. Met. Metalloved 25, 910-923 (1968).

47. D. Turnbull and R. E. Hoffman, Acta Met., 2, 419-426 (1954).

48. R. Rosenberg, Appl. Phys. Letters, 16, 27-29 (1970).

49. K. T. Aust and J. W. Rutter, TMS-AIME, 215, 119-127 (1959).

50. K. T. Aust and J. W. Rutter, TMS-AIME, 215, 820-830 (1959).

51. D. McLean, '<u>Grain Boundaries in Metals</u>',Oxford U. Press, London (1957), p.125.

52. D. R. Beaman, R. W. Baluffi and R.O. Simons, Phys. Rev., 134A, 532-542 (1964).

53. D. R. Beaman, R. W. Baluffi and R. O. Simons, Phys. Rev., 137A, 917-924 (1966).

54. M. Hansen, 'Constitution of Binary Alloys', McGraw Hill Co., New York (1958), pp. 81-84.

55. Nguyen Van Doan, J. Phys. Chem. Solids, 31, 2079-2085 (1970).

56. J. Friedel, Can. J. Phys., 34, 1190-1211 (1956).

57. P. Gordon and R. A. Vandermeer, TMS-AIME, 224, 917-928 (1962).

58. N. L. Peterson and S. J. Rothman, Phys. Rev.,B1,(1970).

59. W. Alexander and L. Silfkin, Phys. Rev., B1, 3275-3282 (1970).

60. J. E. Hilliard, B. L. Averbach and M. Cohen, Acta Met., 7, 86-92 (1959).

61. G. M. Hood, Phil. Mag., 21, 305-328 (1970).

62. G. P. Tiwari and B. S. Sharma, Phil. Mag., 24, 739-743 (1971).

63. G. M. Hood and R. J. Schultz, Phil. Mag., 23, (1971).

64. V. I. Arkharov, K. A. Efremova, S.I. Ivanovskaya, A.K. Sholts and B. A. Yulinov, Doklady Akad. Nauk. SSSR, 89, 269-270 (1953).

65. P. Guivaldenq and P. Lacombe, 'Proprietes des Joints de Grains', 4° Colloque de Metallurgie, Saclay (1960), Presses Universitaires de France, pp. 105-114.

66. C. Leymonie and P. Lacombe, Mem. Sci. Rev. Met., 57, 285-290 (1960).

67. S. Z. Bokshtein, 'Diffusion Processes, Structure and Properties of Metals', S. Z. Bokshtein Ed., Consultants Bureau, New York (1965) pp. 2-8.

68. A. Austin and N. Richards, J. Appl. Phys., 33, 3569-3574 (1962).

69. V. I. Arkharov, S. M. Klotsman and A. N. Timofeyev, Fiz. Met. Metalloved, 6, 255-260 (1958).

70. V. I. Arkharov and A. A. Pentina, Fiz. Met. Metalloved, 5, 52-56 (1957).

71. G. Barreau, G. Brunel, G. Cizeron and P. Lacombe, Mem. Sci. Rev. Met., 68, 358-366 (1971).

72. V. I. Arkharov, S. M. Klotsman and A. N. Timofeyev, Fiz. Met. Metalloved, 8, 709-713 (1959).

73. S. M. Klotsman, A. N. Timofeyev and I. S. Trakhtenberg, Fiz. Met. Metalloved, 20, 78-83 (1965).

74. R. E. Hoffman and D. Turnbull, J. Appl. Phys., 22, 634-639 (1951).

75. W. Schintlmeister and K. Richter, Planseeberichte für Pulvermetallurgie, 18, 3-6 (1970).

76. V. T. Borisov, V. M. Golikov and G. V. Scherbedinsky, Fiz. Met. Metalloved, 17, 881-885 (1964).

77. A. M. Huntz, P. Guivaldenq, M. Aucouturier and P. Lacombe, Mem. Sci. Rev. Met., 66, 85-104 (1969).

78. T. Rosso, M. Aucouturier and P. Lacombe, Scripta Met., 2, 393-398 (1968).

79. M. Aucouturier, T. Araki. T. Rosso and P. Lacombe, Mem. Sci. Rev. Met., 65, 255-265 (1968).

80. V. N. Kaygorodov, S. M. Klotsman, A. N. Timofeyev and I. S. Trakhtenberg, Fiz. Met. Metalloved, 27, 1048-1053 (1969).

81. V. N. Kaygorodov, Y. A. Rabovsky and V. K. Talinsky, Fiz. Met. Metalloved, 24, 117-124 (1967).

82. V. N. Kaygorodov, S. M. Klotsman, A. N. Timofeyev and I. S. Trakhtenberg, Fiz. Met. Metalloved, 28, 120-128 (1969).

83. V. N. Kaygorodov, Y. A. Rabovsky and V. K. Talinsky, Fiz. Met. Metalloved, 24, 661-668 (1967).

84. J. B. Murphy, Acta Met., 9, 563-569 (1961).

85. T. S. Lundy and J. F. Murdock, J. Appl. Phys., 33, 1671-1673 (1958).

86. M. Beyeler and D. Lazarus, Mem. Sci. Rev. Met., 67, 395-400 (1970).

Note

It has come to our attention that work on the effect of interstitial impurities on grain boundary diffusion in iron is being pursued. A recent report [87] indicates that in the presence of carbon the grain boundary diffusion of phosphorus in α Fe is greatly reduced. As in the case of the effect of carbon on the diffusion of sulfur [78, 79], it seems that the decrease in the diffusion coefficient is mostly due to precipitated carbon. The effect appears to vary linearly with the amount of precipitated carbon only at high temperatures, for small amounts of precipitates.

87. T. Rosso and C. Sabatini, Scripta Met. 6, 51-54 (1972).

GROWTH SELECTION IN HIGH PURITY CADMIUM

E. A. Grey and G. T. Higgins

Department of Metallurgical Engineering, Illinois Institute of Technology, Chicago; formerly of the University of Liverpool U. K. where the work was carried out

ABSTRACT

Growth selection studies have been performed with 99.9999% cadmium and a cadmium-5 atom p.p.m. lead alloy. After severe local deformation single crystals of a standardized orientation were annealed resulting in the recrystallization of the heavily deformed region, and the competitive growth of these recrystallized grains down the strain gradient into the single crystal matrix. A three dimensional orientation space has been employed to describe the distribution of selected orientations and six preferred orientations have been identified. The relative preference for these preferred relationships is shown to be dependent on the single crystal orientation, the solute concentration and the annealing temperature.

INTRODUCTION

The object of growth selection experiments is to identify grain boundary orientations of high mobility. It is assumed that during growth selection the competitive growth of grains of widely differing orientations into a single crystal matrix results in the eventual dominance of the fastest growing grain. The driving force for grain boundary migration may be a deformation or solidification substructure within the single crystal matrix. Previous studies in hexagonal metals have demonstrated the existence of more than one preferred orientation and that the relative preference for these preferred orientations is a function of specimen rod axis (1), annealing temperature and solute concentration (2). For this latter investigation the specimen rod axis was not standardized. In both studies two preferred relationships were identified, these being described to a first approximation by rotations of 30° [0001] and 90° [10$\bar{1}$0].

In the current work the object has been to identify specifically the preferred orientation relationships developed during growth selection in high purity cadmium, and to study the effects of annealing temperature and solute concentration on the relative preference of the observed relationships.

MATERIALS

Zone refined cadmium supplied by Vieille-Montagne Laboratories and specified at 99.9999 atom % purity (0.7 ppm Pb, 0.1 ppm Cu and 0.1 ppm Ag) was used as a base material. This material is referred to as 6NB and an alloy which contained 5 ppm lead is referred to as 5N5.

EXPERIMENTAL

Oriented single crystal seeds of 6NB and 5N5 were grown from the melt under static argon in an open graphite boat at a rate of 0.1 mm/s. The rod axis of the standardized orientation is illustrated in Fig. 1. All specimens of both 6NB and 5N5 had this orientation except for a small batch of high purity cadmium specimens. For these the c-axis was again normal to the growth direction but the rod axis was not standardized. This sample will be referred to as RA.

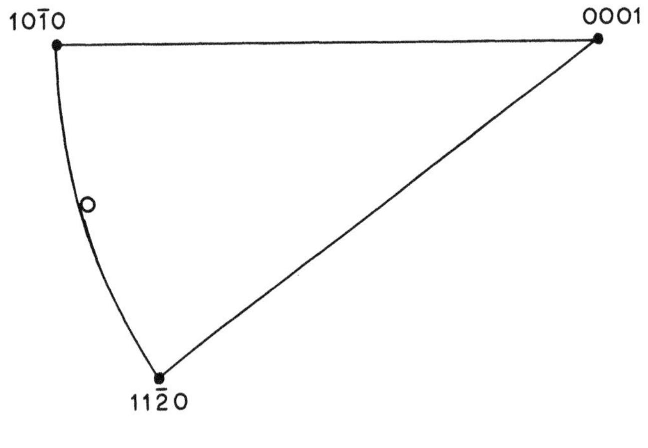

Fig. 1
The orientation of the specimen rod axis used for samples 6NB and 5N5

The ends of the specimens, 50 mm in length and 5 mm square in section, were compressed normal to the c-axis in a hand operated vice. Recrystallization was spontaneous at room temperature and the (0002) pole figure for this structure is illustrated in Fig. 2. This was obtained using a conventional Schultz goniometer in the reflection mode. The center of the pole figure corresponds to the compression axis and is therefore parallel to the c-axis of the undeformed single crystal matrix.

After the spontaneous recrystallization of the heavily deformed end of the crystal a strain gradient exists ahead of the recrystallized grains, and it is this feature that provides the driving force for grain boundary migration during subsequent growth selection anneals.

Specimens were annealed in batches of 40-50 at constant temperature in silicone oil baths. Annealing temperatures ranged from 160°C to 315°C and were maintained to within ± 1°C of the desired temperature. The annealing times required for the development of dominant grains during growth selection ranged from one hour at 315°C to two months at 160°C.

The orientations of the dominant grains and the matrix were determined by the back reflection Laué technique. The accuracy of the resultant orientation relationships was estimated at ± 2°.

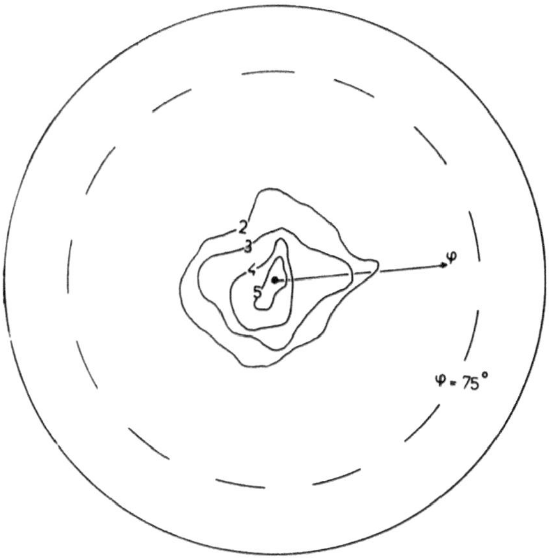

Fig. 2
(0002) pole figure taken normal to the surface at the deformed end

RESULTS AND DISCUSSION

The simplest way of presenting growth selection data is in the form of a pole figure with the matrix in standard (0001) projection. An example is given in Fig. 3 where the c-axis and three [10$\bar{1}$0] directions are used to define the selected orientation. Since the c-axis is uniquely defined it is possible to minimize confusion in the pole figures by rotation of all the basal poles into a single unit triangle.

A one dimensional description of the data is given in Fig. 4, which shows the total percentage of basal poles as a function of the angular deviation ϕ between the hexagonal axes of the matrix and selected orientations. Each figure shows an inflection in the region $\phi = 45^{\circ}$, an observation which is consistent with the work of Klar and Lücke (1) and Godwin (2). The inflection implies that there are at least two separate groupings of selected orientations; one for $\phi > 45^{\circ}$ and the other for $\phi < 45^{\circ}$. Table I shows the percentage of basal poles contained in both groupings. For both 6NB and 5N5 there is a marginal trend for the preference for $\phi < 45^{\circ}$ to decrease with increasing temperature.

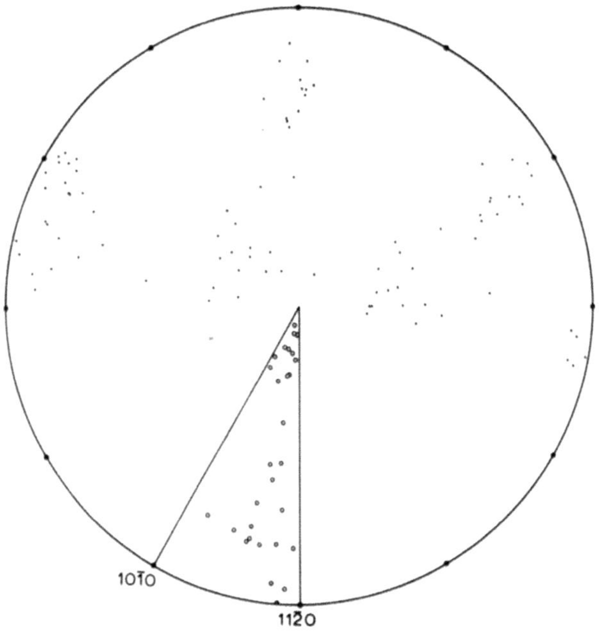

Fig. 3
Example of the distribution of selected orientations displayed in a stereographic projection

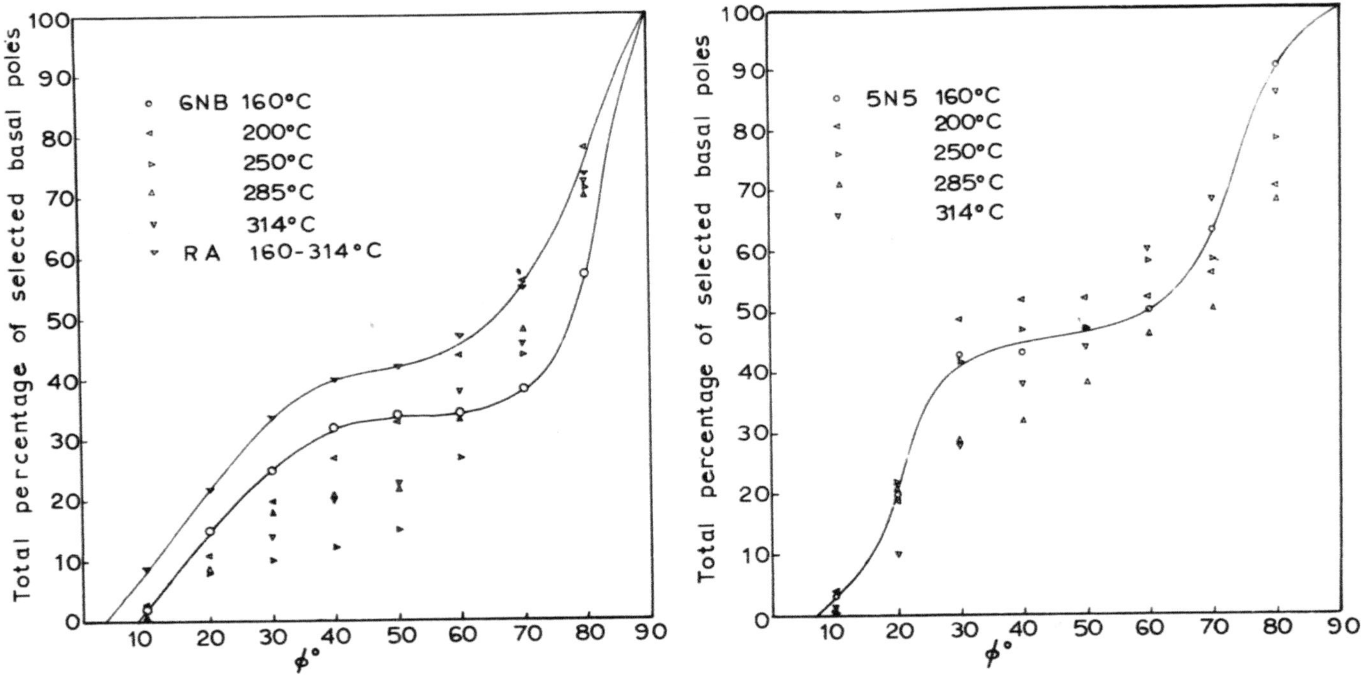

Fig. 4
The variation of the total percentage of selected orientations as a function of the angular deviation ϕ between the c-axis of matrix and selected orientations

TABLE I. Percentage of Basal Poles in Specimens

	Temp. °C	Percentage of Poles in Range ϕ = 0-45°	ϕ = 45°-90°
6NB	160	34	66
	200	33	67
	250	13	87
	285	22	78
	314	20	80
5N5	160	47	53
	200	52	48
	250	47	55
	285	42	58
	314	36	64

Percentage of selected orientations with ϕ in the ranges 0-45° and 45°-90° where ϕ is the angle between the hexagonal axes of the matrix and selected orientations.

The effect of solutes is more pronounced. The addition of 5 atom p.p.m. lead to 6NB almost doubles the preference for the range $\phi < 45°$ at each annealing temperature.

Although the preceding analysis does indicate that the annealing temperature and solute concentration are important parameters in modifying the relative preference for various orientation relationships, it is severely limited in that only one dimension (ϕ) is considered. Since the analysis of an orientation distribution is a three dimensional (3-D) problem, the ambiguities arising out of the use of a stereographic projection can be eliminated by the use of a 3-D representation of the growth selection data.

The definition of an orientation relationship requires at least three independent parameters, and these may be expressed in a variety of different ways. In the past a number of authors (3,4) have used a representation in which two parameters define a direction common to both orientations, and a third a rotation about this direction. However, in the hexagonal system 6, and in the cubic system 24, such common rotation axis/angle pairs exist for each orientation relationship. Since there is no justification for choosing any one of these pairs and because an analysis which is required to handle all of them would be unnecessarily complex, an alternative approach is needed.

In view of the recent success of the 3-D representation of textures by Bunge and Haessner (5) and Morris and Heckler (6), using the three Euler angles as the basis of a representation, a similar approach is adopted here.

The Euler angles are defined in the following manner (see Fig.5a). Firstly, two Cartesian coordinate systems XYZ and X'Y'Z' are positioned in the matrix and selected orientations respectively. The axes X, Y and Z correspond to the directions $[11\bar{2}0]$, $[10\bar{1}0]$ and $[0001]$ respectively. The Euler angles defining the orientation relationship can be obtained by first bringing the two Cartesian coordinate systems into coincidence and then rotating successively by the angles ϕ_1, ϕ and ϕ_2 about the axes Z, X" and Z' respectively. The Euler angles can be read directly from a stereographic projection in the manner described in Fig. 5b. The angles ϕ_1 and ϕ define the position of the c-axis of the selected orientation and ϕ_2 defines a rotation about this axis. Since symmetrically equivalent points occur at 60° intervals ϕ_2 can be limited to the range 0°-60°.

The three angles ϕ_1, ϕ and ϕ_2 can be used as the basis of a linear orthogonal 3-D space in which the matrix orientation is at the origin (i.e., (0,0,0)). The entire orientation space is defined by ϕ_1, ϕ and ϕ_2 in the limits:

$$\phi_1 \in (0,360)$$

$$\phi \in (0,90)$$

and

$$\phi_2 \in (0,60)$$

When the data is considered in the unit triangle of the stereographic projection the limits of ϕ_1 becomes:

$$\phi_1 \in (0,30)$$

When the data for each temperature and purity combination is represented in Euler space the scatter in the selected orientations depicted in Fig. 2 is again evident. This large scatter and the

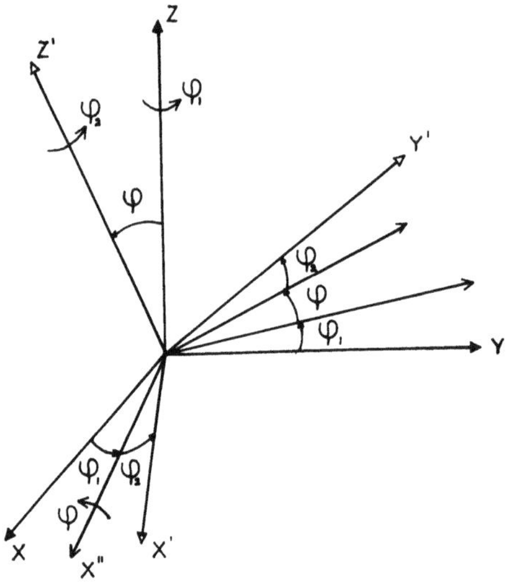

Fig. 5a
The definition of the Euler angles relating the selected and matrix orientations.

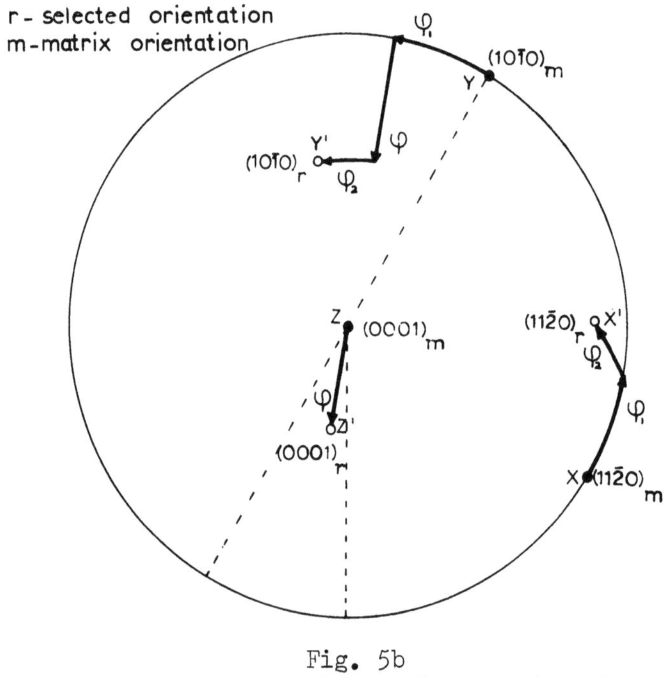

Fig. 5b
The relationship between the Euler angles and the stereographic projection.

relatively small quantity of data available in each case makes it impractical to determine preferred orientations. To determine the effects of temperature thoroughly much more data are required at each temperature. A concentration of the present data can be effected by combining all of the data available at each temperature for each of the samples 6NB, 5N5 and RA. This procedure requires the assumption that the relative positions of any preferred relationships are unaffected by temperature. It is not possible from the available data to prove the validity of this assumption, however, some evidence in support of this is offered later.

In Figs. 6, 7 and 8 density contour maps are given for the density of selected orientation in Euler space for the three samples 6NB, 5N5 and RA respectively. The contour maps are shown at three degree intervals for constant ϕ_1 sections.

These maps were determined with the aid of a computer analysis which considered the space as a cubic lattice with axes parallel to the space axes ϕ_1, ϕ and ϕ_2 and with a basic lattice unit of three degrees. Each lattice site was designated by the number of selected orientations occurring within a six degree cube centered on the lattice site. The numbers on the contours correspond to the number of times the random density, and the maps were determined from 239, 186 and 94 orientation relationships for 6NB, 5N5 and RA respectively.

The three maps show marked similarities. An analysis to determine the centers of the intensity peaks reveals that those peaks occurring in two or all three of the maps agree to within six degrees. Within the limits of resolution of the present analysis this indicates that although the relative preference for the various preferred relationships may be a function of solute concentration and rod-axis the positions of the preferred orientations are not. This observation lends some support to the earlier assumption that the positions of the preferred orientations are temperature independent.

On the basis of the foregoing observations, it is useful to combine all 519 selected orientation measurements, and as a result determine more accurate estimates for the positions of the preferred orientations.

Figs. 9, 10 and 11 show the distribution and density of selected orientations obtained by combining all of the data, for constant ϕ_1, ϕ and ϕ_2 sections respectively. Orientation densities ≥ 9 times the random density were considered to be resolved deviations from a random distribution. Six intensity groups satisfying this criterion were observed. The coordinates of the peak intensities of these six groups ($G_1, G_2 - G_6$), are listed in Table II.

TABLE II

Coordinates of Peak Intensities

$(\phi_1 \ \phi \ \phi_2)$

G_1	(9, 15, 27)	G_4	(15, 78, 24)
G_2	(9-21, 24, 17)	G_5	(27, 81, 24)
G_3	(9, 78, 9)	G_6	(21, 87, 38)

Coordinates of the centers of the high intensity peaks of the orientation space plots in figures 9-11.

In order to observe whether any of the symmetrically equivalent points of the preferred orientations are preferred, the distribution of the selected orientations was examined in the entire orientation space. In this case ϕ_1 has the range 0°-360°. There was no statistically significant preference for any of the symmetrically equivalent positions. That grain boundary orientation is important in influencing the observed distribution has been demonstrated by Klar and Lücke(1), who noted that the distribution was a function of the specimen rod axis. A similar conclusion may be drawn from Fig. 4 where it can be seen that the relative preference for the ranges $\phi < $ and $> 45°$ in the two high purity cadmium samples (6NB and RA) is modified by the orientation of the specimen rod axis.

The structural factors which give rise to the special properties of these preferred orientations in hexagonal metals are unknown. All six of the resolved relationships are irrational, none being ideal coincidence relationships. For hexagonal symmetry no ideal coincident relationships exist when the basal planes are non-coplanar. However, it has been noted by Bishop, Hartt and Bruggeman (9) that approximate coincidence relationships exhibit low energy characteristics in zinc and cadmium, indicating that they retain a structural significance. The possibility exists therefore, that the observed preferred orientations correspond to boundaries with special structural features, but at this time no evidence is available to support this conjecture.

It may be argued that the preference for certain orientations and the avoidance of others is a reflection of the original nucleus distribution and does not arise because of growth selection. Unambiguous resolution between these alternatives would require a knowledge of the original nucleus distribution. The measurement of such a distribution is generally difficult and the inaccessibility of the nuclei of interest in the present experiment makes such measurement impracticable.

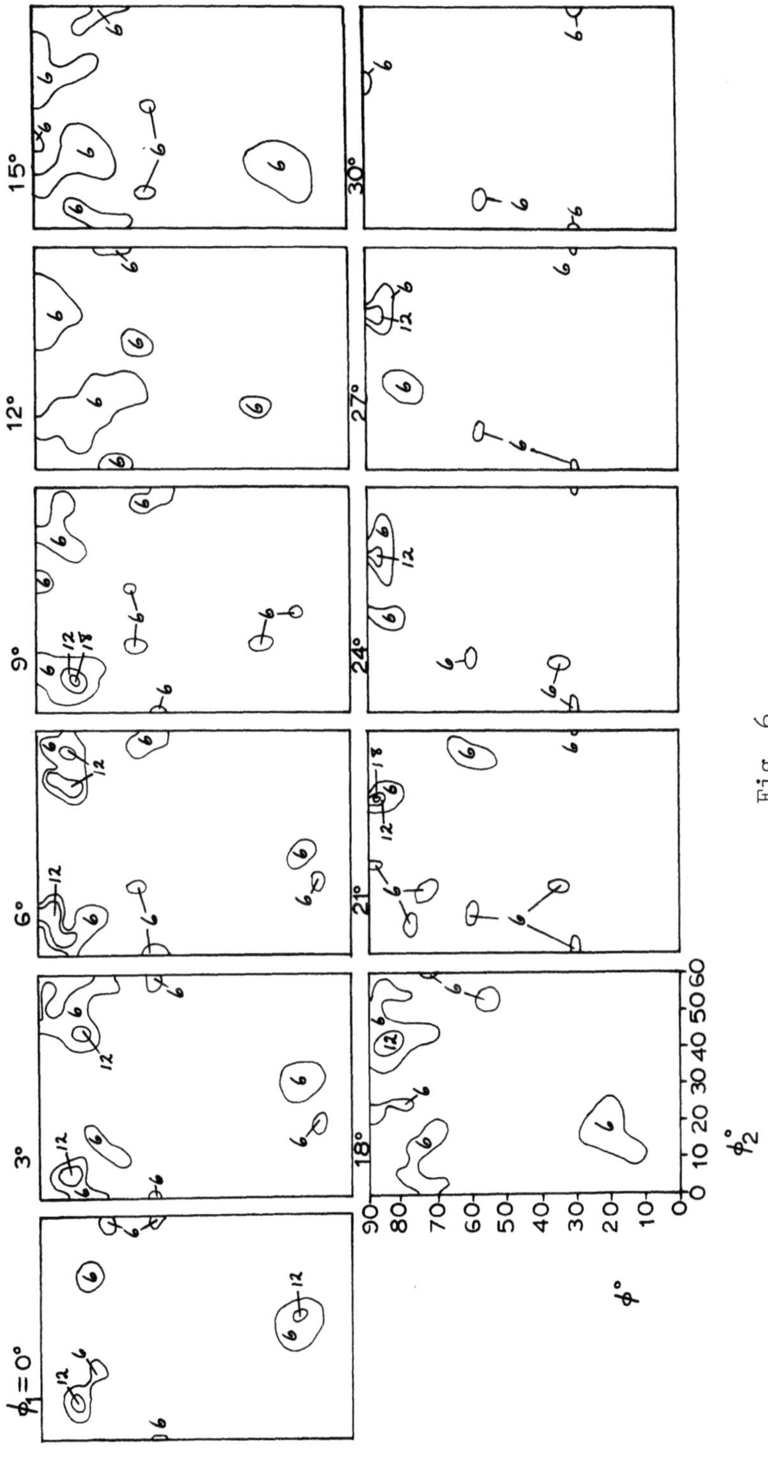

Fig. 6
The density and distribution of selected orientations in constant Φ_1 sections for 6NB.

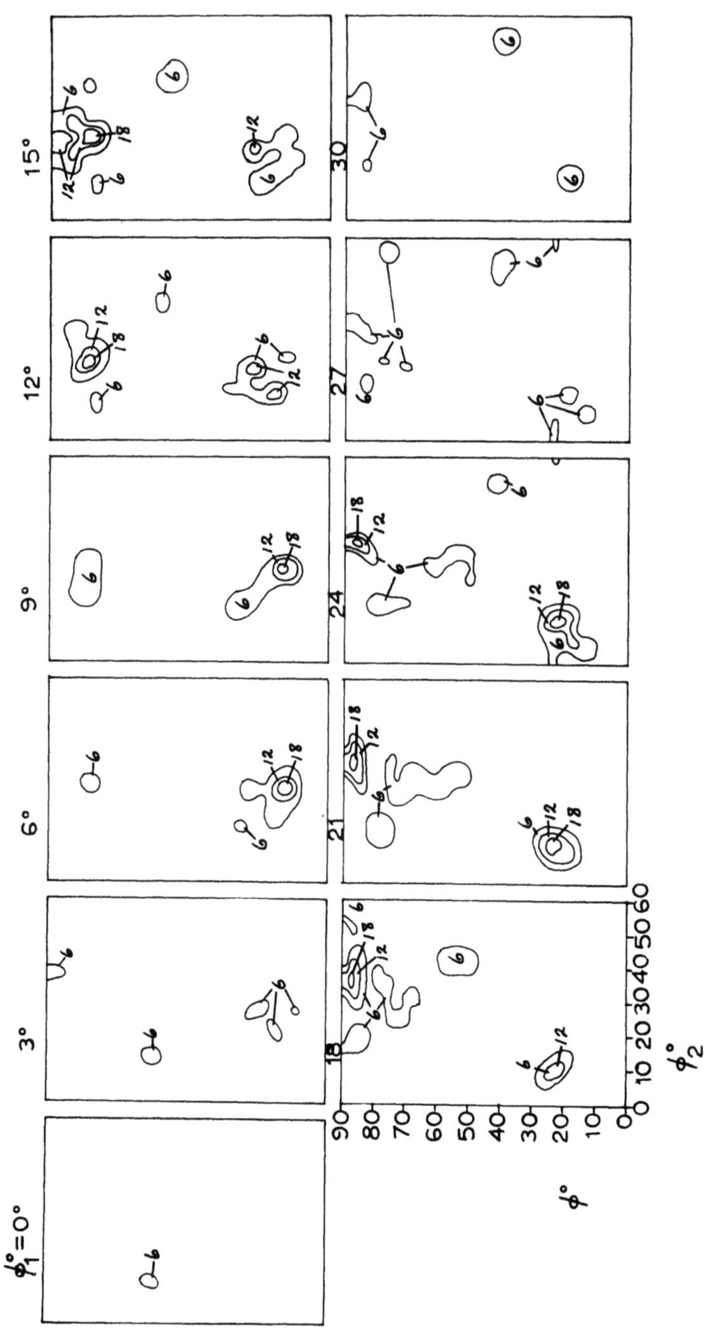

Fig. 7
The density and distribution of selected orientations in constant ϕ_1 section for 5N5.

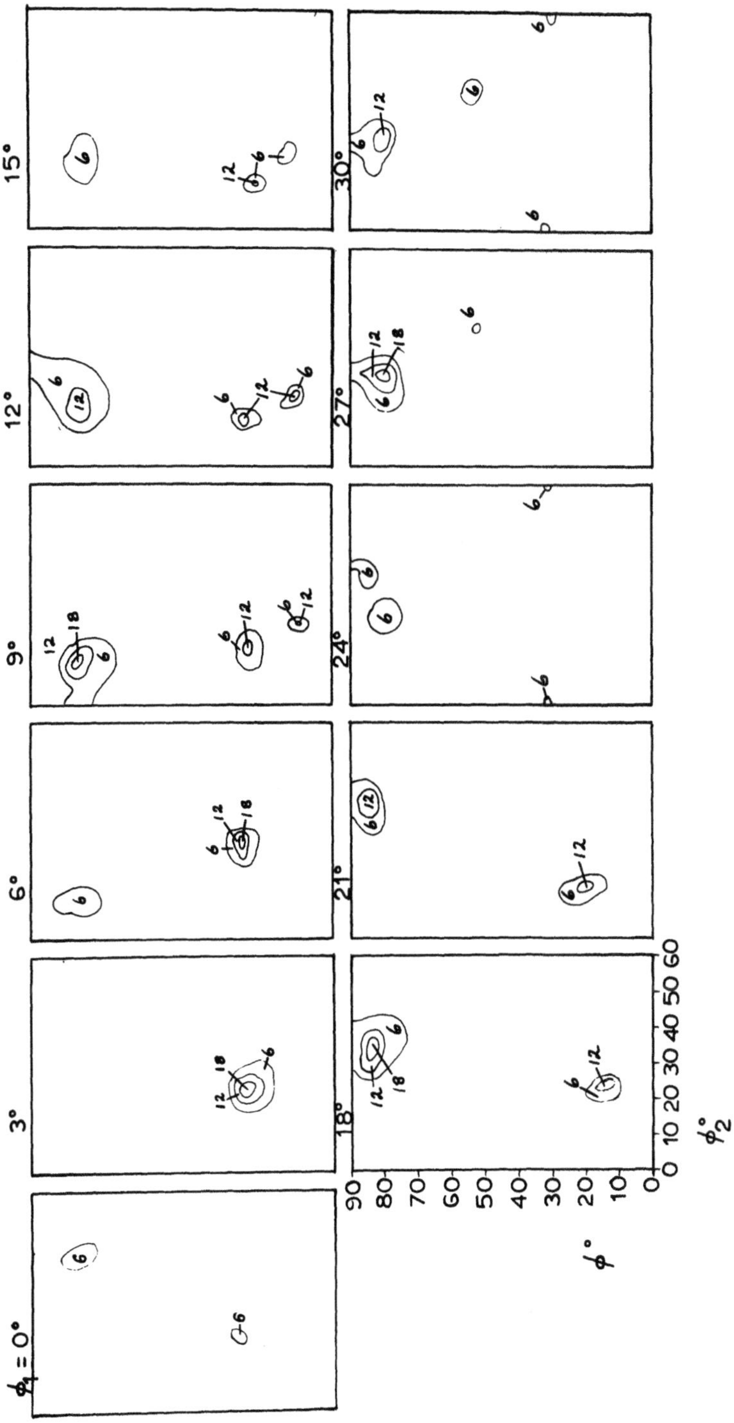

Fig. 8
The density and distribution of selected orientations in constant ϕ_1 section for RA.

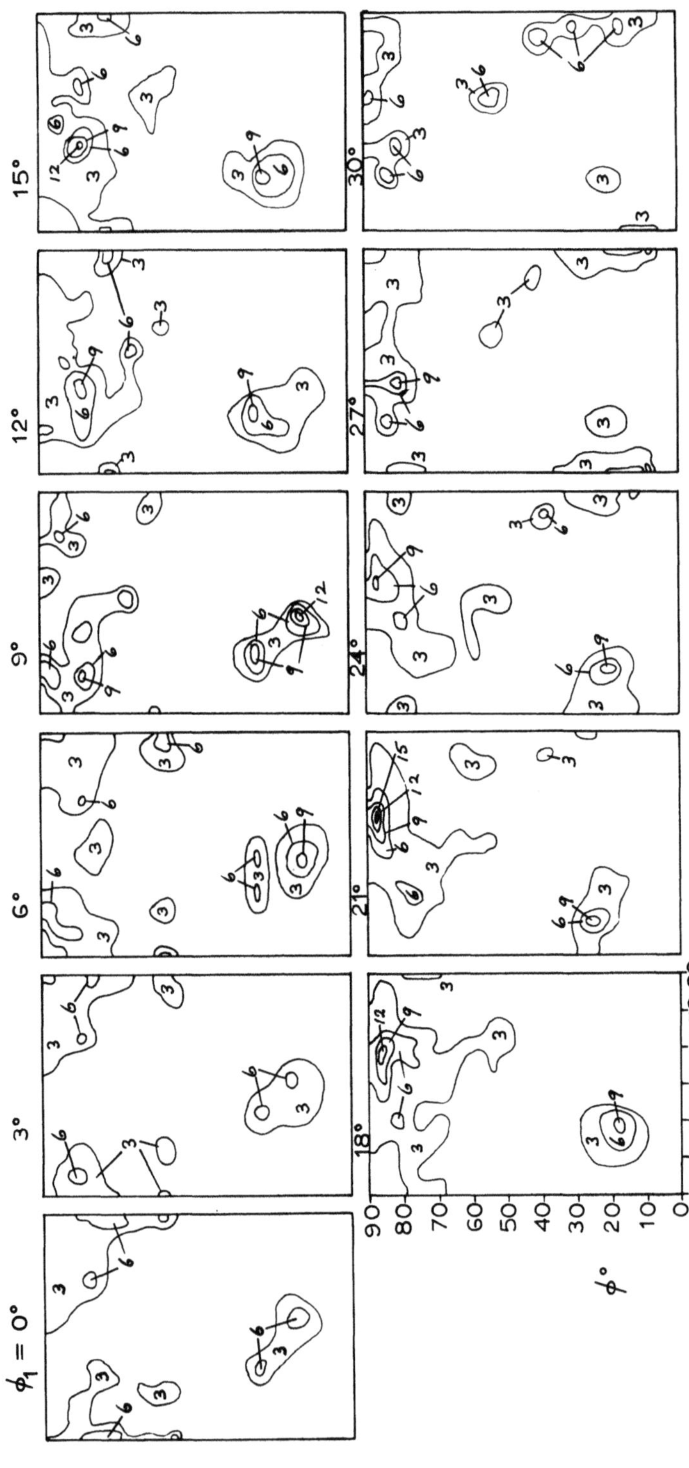

Fig. 9
The density and distribution of selected orientations in constant ϕ_1 section for all data taken together.

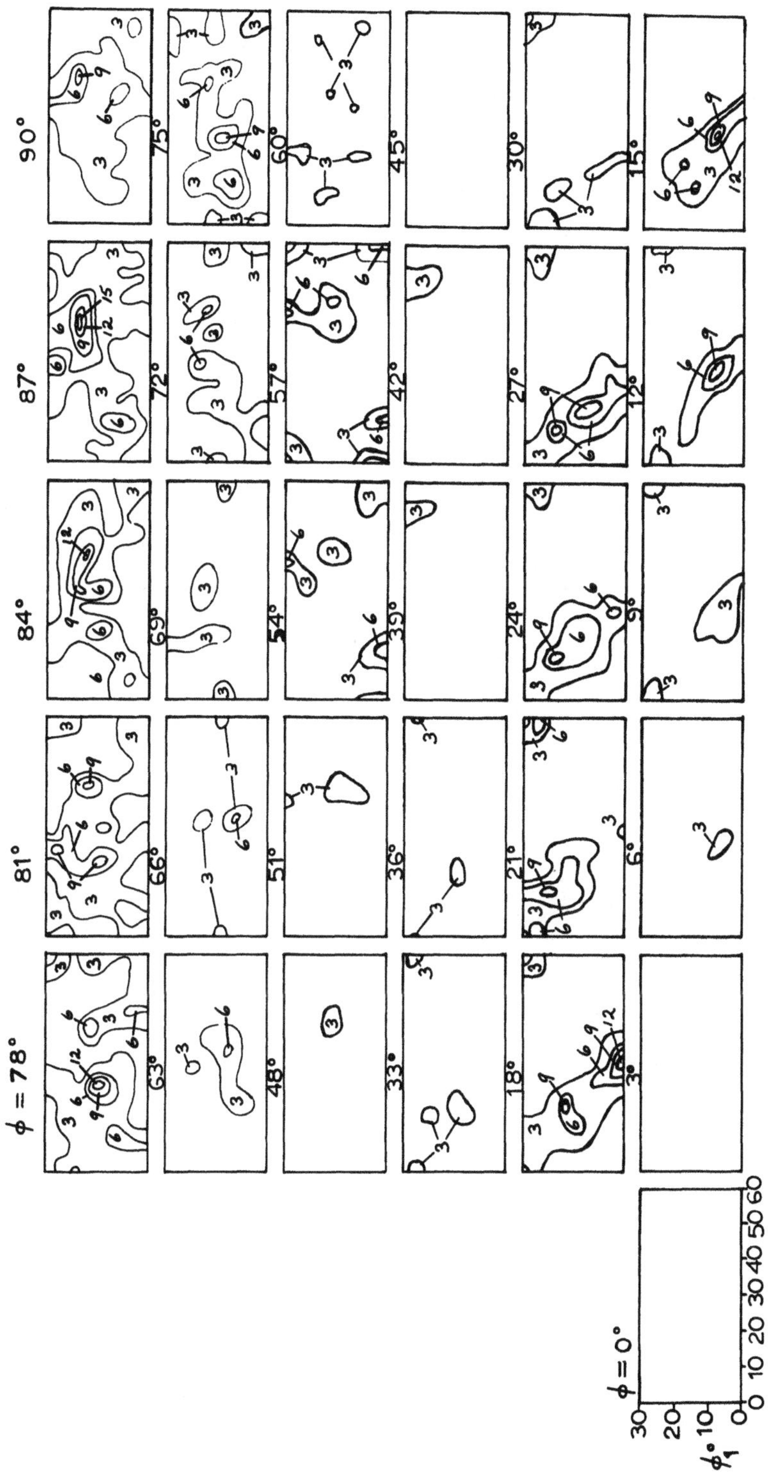

Fig. 10
The density and distribution of selected orientations in constant φ section for all the data taken together.

Fig. 11

The density and distribution of selected orientations in constant ϕ_2 section for all data taken together.

A comparison (8) between the observed distribution of selected orientations about the preferred orientations with the predictions of the Ibe-Lücke statistical theory of growth selection supports the growth selection hypothesis.

It should also be pointed out that the grain boundary energy may play a role in growth selection, in the sense that a grain at the advancing interface, with a low boundary energy with respect to the matrix, can develop a geometrical configuration that favours its development. It is of interest to note that the agreement with the statistical theory referred to above could result from either 'migration' selection or 'energy' selection.

However, in the final analysis the effectiveness of the growth selection experiments as a tool for determining high mobility can only be measured by examining the mobilities of individual grain boundaries. In this regard, we are currently employing bicrystal techniques to investigate the orientation dependence of grain boundary migration rates in cadmium.

CONCLUSIONS

(1) The distribution of grain orientations observed during growth selection can be represented satisfactorily in a 3-D orientation space.
(2) Six preferred orientation relationships have been resolved.
(3) The relative preference for these six relationships is a function of solute concentration, annealing temperature and rod axis.
(4) The positions of the preferred orientations are invariant under solute concentration, specimen rod axis. This also appears to be true for temperature but has not been resolved unambiguously.
(5) None of the preferred orientations corresponds or even approximates to the classical [0001] axis coincidence relationships.

ACKNOWLEDGEMENTS

We would like to acknowledge Dr. A. J. Heckler who carried out the pole figure determination, and the University of Liverpool for the provision of facilities during the course of this work. One of us (EAG) also acknowledges financial support received from the University of Liverpool.

REFERENCES

1. R. Klar and K. Lücke, Z. Metallk., 99, 194, (1968).
2. A. W. Godwin, Ph.D. Thesis, University of Liverpool, (1966).

3. K. T. Aust and J. W. Rutter, Trans. TMS-AIME, 218, 50, (1960).
4. Y. C. Liu, Trans TMS-AIME, 230, 1513, (1964).
5. H. J. Bunge and F. Haessner, J. Appl. Phys., 39, 5503, (1968).
6. P. R. Morris and A. J. Heckler, Trans. TMS-AIME, 245, 1877, (1969).
7. G. Ibe and K. Lücke, Recrystallization and Grain Growth and Textures, ASM, Metals Park, Ohio, p. 434, (1966).
8. E. A. Grey, Ph.D. Thesis, University of Liverpool, (1971); to be published.
9. G. H. Bishop, W. H. Hartt and G. A. Bruggeman, Acta. Met., 19, 37, (1971); and this volume.

A CRYSTALLOGRAPHIC ALTERNATIVE TO THE COINCIDENCE RELATIONSHIPS IN COPPER

Y. C. Liu

Scientific Research Staff, Ford Motor Company

Dearborn, Michigan 48121

ABSTRACT

The preferred orientation relationships between growing grains and strained (4-8 pct in tension) single crystal matrices of 99.999 pct purity copper were observed to be $\langle 001 \rangle$ 19° and $\langle 111 \rangle$ $\sim 30°$, identical to those previously found after secondary recrystallization. Nearly all of the growing grains contained an annealing twin. By considering the orientation of the parent (or twinned) grains, the preferred relationships may be expressed, approximately, in terms of parallelism between the following pairs of orthogonal directions:

P-relationship:	Growing Grain	Matrix
	$\langle 001 \rangle$	$\langle 111 \rangle$
	$\langle 010 \rangle$	$\langle \bar{1}10 \rangle$
	$\langle \bar{1}00 \rangle$	$\langle 112 \rangle$
Q-relationship:	Matrix	Growing Grain

By considering the matrix during the secondary recrystallization in copper as (001)[100] and its four twin orientations, the present alternative suggests that it is the twin matrices, not the (001)[100] orientation, which support the growth of both $\langle 001 \rangle$ 19° and $\langle 111 \rangle$ 22° secondary grains. When the occurrence of annealing twins is rare, such as in aluminum, it is therefore expected that these two relationships should be absent, as observed experimentally. The findings of only two preferred relationships in the present investigation is in agreement with the alternative offered, but not with the coincidence model, which yields many relationships with a high coincidence density.

INTRODUCTION

Growth-selection experiments have been used in the past to study the orientation dependence of grain boundary migration. The driving force in these experiments is either a network of stable grain or subgrain boundaries, or strain energy. The growing grains are either artificially nucleated or are pre-existing grains in the matrix. In many cases, the experimental results indicate the existence of preferred orientation relationships between the fastest growing grains and their matrices. This phenomenon was first reported by Kronberg and Wilson in copper[1] and by Beck and Hu in aluminum.[2] Both groups of investigators described the preferred relationships in rotational terms: ⟨001⟩ 19° and ⟨111⟩ 22° and 38° in copper, and ⟨111⟩ 40° in aluminum. These relationships agree with the coincidence model of Kronberg and Wilson.[1] Although Kronberg and Wilson viewed their model as a possible nucleation mechanism, it was subsequently considered as a likely structure for grain boundaries.[3] The latter view was supported experimentally by the work of Aust and Rutter on zone-refined lead[4-6] and of Rath and Hu on zone-refined aluminum.[7] The coincidence model also plays an important role in the structure of large angle boundaries described by the theory of Brandon, Ralph, Ranganathan and Wald;[8] this theory is supported by observations provided by field-ion microscopy.

On the other hand, arguments against the concept of the coincidence model as a grain boundary structure have been raised from time to time. Li pointed out that grain boundaries of this type should have low mobility, simply because of their "good fit" and low "porosity."[9] They should therefore be of low energy[10] and relatively immune to the segregation of impurities.[4] In order to explain the high mobility of these boundaries which is experimentally observed, a deviation from the ideal coincidence relationship is needed to provide the necessary "porosity."[9,11] This suggestion, however, is not substantiated by the experimental results of Rutter and Aust, in which the apparent activation energy for boundary migration was found to increase with even a 1° deviation from an ideal coincidence relationship.[6] In contrast with Rutter and Aust's results, a much greater deviation from ideal coincidence relationships was generally observed in the growth-selection experiments. For instance, based on the observation of 1200 growing grains in deformed single crystal matrices of aluminum and its alloys, Lücke and Ibe[12] reported a preferred orientation relationship of ⟨111⟩ 40°, but the rotational axes deviated by 12° on the average from the ideal coincidence positions. With such a large deviation from the ideal relationship, it is doubtful that coincidence model could still have physical significance assigned to it.[13] The difference in the experimental results obtained from the two types of experiment suggests that they may not be dealing with the same phenomenon.

From the observations on secondary recrystallization in copper and aluminum, one may raise three fundamentally important questions. (a) Although many coincidence relationships are available, the experimental results show that surprisingly few are actually observed. (b) The ⟨001⟩ 19° relationship appears frequently. However, this relationship corresponds to only one coincidence site in 37, whereas ⟨001⟩ 16½° and ⟨001⟩ 22½°, which are not found experimentally, correspond to the higher densities of 1 in 25 and 1 in 13, respectively. (c) Since the coincidence model deals only with the crystallographic factor, it is rather curious that both the ⟨001⟩ 19° and the ⟨111⟩ 22° relationships are observed in copper but not in aluminum.

Based upon the observations on growth selection in copper single crystals with strain energy as the driving force, an attempt will be made to answer these questions in the present paper.

EXPERIMENTAL PROCEDURE

The copper used was ASARCO grade with a quoted purity of 99.999 pct. Single crystals, 19 x 3.2 x 110 mm., were grown from the melt under a vacuum of 5×10^{-4} torr. All crystals were strained in tension at room temperature. The two grip-ends provided ample artificial nucleation sites from which strain-free grains could grow upon heating. Specimens were encapsulated under a partial pressure of purified helium; after annealing, these capsules were water quenched and broken. Orientations of the growing grains and matrices were determined using the Laue back-reflection technique.

Three heating schedules were followed. In the first, the capsules were heated at a rate of about 170° C/hr from room temperature to 850°C and held at this temperature for 10 hrs. The second schedule was the same as the first but with a final temperature of 650°C. In the third schedule, the specimens were rapidly heated to 850°C (in 145 ± 3sec.) by heating to near 850°C in a furnace maintained at 1100°C and then transferring them to another furnace held at 850°C; the annealing time at this temperature was again 10 hrs.

The range of shear strain (and corresponding shear stress) was 5-8 pct (120-180 g/mm^2), 6-8 pct (150-230 g/mm^2) and 4-7 pct (150-260 g/mm^2), respectively, for the three heating schedules. Orientation data were obtained on 54 successfully growing grains whose length was at least equal to the width of the specimens.

EXPERIMENTAL DATA IN TERMS OF THE COINCIDENCE MODEL

Summarizing the present results in terms of the coincidence model, the preferred orientation relationships between the growing grains and their single crystal matrix are of two types: ⟨001⟩ 19° and ⟨111⟩ 23-33°. Of the 54 growing grains, 34 grains were ⟨001⟩-related and 11 grains ⟨111⟩-related. The remaining 9 grains, which shared no low index poles with their matrices, were classified as "random." Fig. 1 shows the tensile axes of the matrix grains in a unit stereographic triangle. There is no obvious dependence of the preferred relationships upon matrix orientation. In five cases, two grains grew from opposite ends of the same specimen but exhibited different relationships with the matrix. Also, as listed in Table I, there is little indication that either the heating rate or annealing temperature affected the occurrence of the observed relationships.

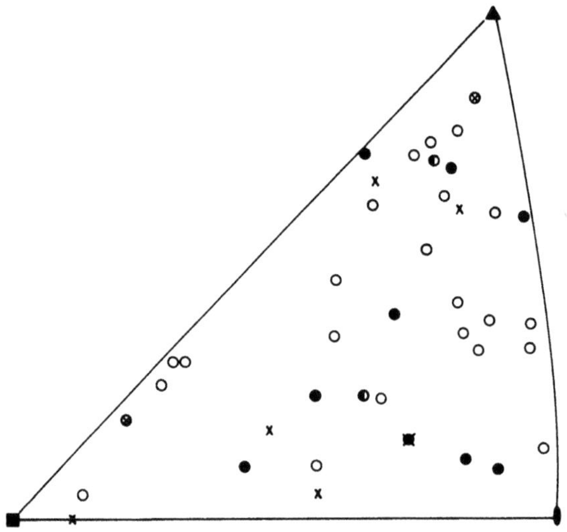

Fig. 1. The stress axis of the single crystal matrices in which o - the ⟨001⟩-related grains, • - the ⟨111⟩-related grains, and x - grains of random orientation relationships were observed.

Figure 2 shows a quadrant plot of the cube poles of ⟨001⟩-related growing grains, with the matrix orientation in the standard projection. The broken circles are of 10° radius. However, there is much greater scatter if the same data are expressed in terms of axes of minimum rotation, Fig. 3a. A return to the original amount of scattering is possible if the axes of the complementary angle of minimum rotation, i.e., 71°, are plotted, as shown in Fig. 3b.

TABLE I. Frequency of Preferred Orientation Relationships Observed after Various Heating Schedules

	⟨001⟩ 19°	⟨111⟩ ~30°	Random
Slow Heating to 850°C	18	8	4
Slow Heating to 650°C	7	1	4
Fast Heating to 850°C	9	2	1
Total	34	11	9

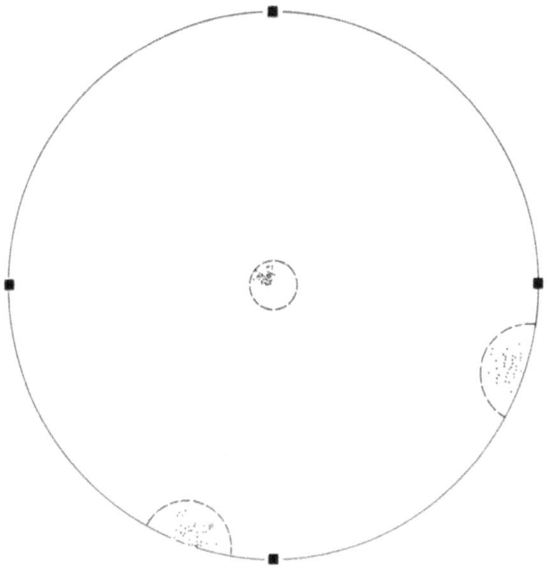

Fig. 2. Quadrant plot of ⟨001⟩ poles of the ⟨001⟩-related grains.

As previously observed,[14] such behavior is due primarily to the properties of the stereographic projection.

Figure 4 shows the frequency distribution of misorientations in the observed ⟨001⟩ relationship. Although the spread in misorientation, θ, ranges from 13 to 24°, the designation of ⟨001⟩ 19° by Kronberg and Wilson[1] is a good description of this group of orientation relationships.

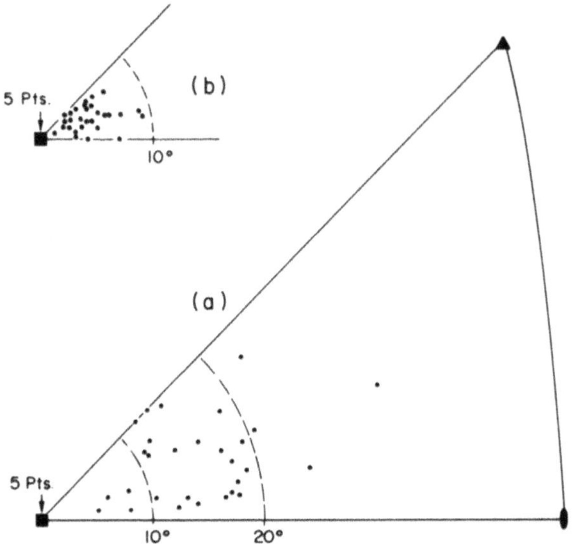

Fig. 3. The rotational axes of same data used in Fig. 2. (a) Rotational Angle - 19°, (b) Rotational Angle - 71°.

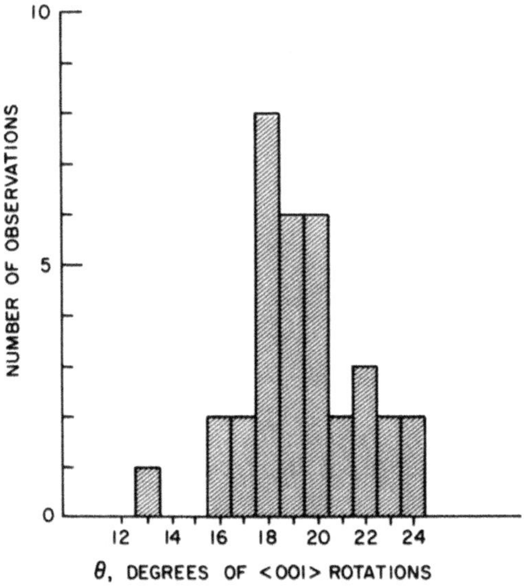

Fig. 4. Frequency plot of amount of rotation of ⟨001⟩-related grains.

Of the 11 grains which are ⟨111⟩-related to the matrix orientations, the amount of rotation ranges from 23 to 33°, with 6 determinations segregated between 28 and 30°.

Thus, the present experimental results are in concordance with the results of Kronberg and Wilson[1] on secondary recrystallization in copper and are in good agreement with Sharp and Dunn[15] who summarized their results as ⟨001⟩ 19° and ⟨111⟩ 30°. Differences in experimental results appear, however, with respect to the growth-selection studies of Aust, Ferran and Cizeron[16] and of Parkinson, Drunen and Saimoto[17] in high purity copper. In both of these investigations, the ⟨001⟩ 19° relationship was not mentioned.

MICROSTRUCTURE RESULTING FROM GROWTH-SELECTION

The microstructure of growing grains was dominated by the presence of annealing twins. It is rather unusual to find a growing grain did not contain an annealing twin, Fig. 5A. In contrast to its usual straight line appearance, sometimes a twin boundary would exhibit a jagged morphology, Fig. 5B.

The grain boundary between the growing grain and the matrix was observed in quite a few specimens. The boundary of an ⟨001⟩ related grain was usually free of intersections with twin boundaries. This could occur either during the early stages of growth, Fig. 5C, or after the twin boundary had grown out of the specimen, Fig. 5D. In eight out of twelve cases, the orientation of these boundaries could be analyzed; all were found to be ⟨001⟩ tilt boundaries. In a ⟨111⟩-related grain, boundaries of both tilt and twist type were observed, Fig. 5E.

A repeated twinning sequence, as reported by Sharp and Dunn,[18] was also observed in a ⟨111⟩-related grain, Fig. 5F. The straight boundary between A and C is not a coherent twin boundary, but it bears a relationship of the type $\{111\}_A // \{115\}_C$.

EXPERIMENTAL DATA IN TERMS OF P- AND Q- RELATIONSHIPS

Although the present experimental results are in good agreement with the general concept of the coincidence model, the same data can be analyzed in terms of an alternative set of relationships. This was first noticed on the repeated occurrence of approximate parallelism between certain low index poles of the twin of the ⟨001⟩-related grain and those of the matrix. These relationships may be expressed as:

P-relationship	Growing Grain	Matrix
	$\langle 001\rangle$	$\langle 111\rangle$
	$\langle 010\rangle$	$\langle 110\rangle$
	$\langle 100\rangle$	$\langle 112\rangle$
Q-relationship	Matrix	Growing Grain

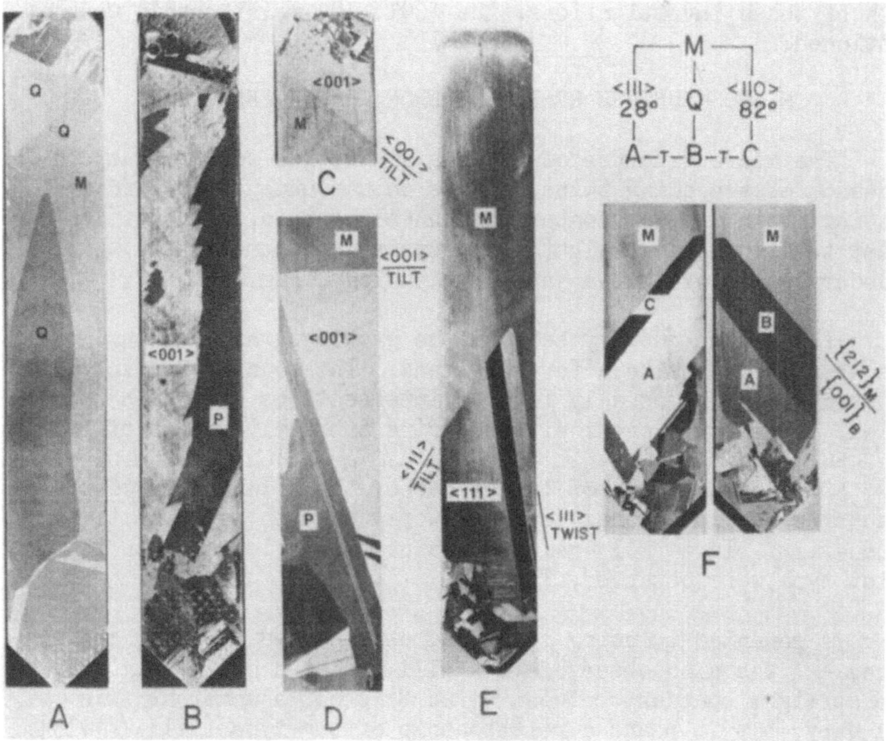

Fig. 5. Microstructures of Growing Grains and Matrices.

In the P-relationship, the $\langle 001\rangle$ poles of the growing grains are approximately parallel to the orthogonal directions, $\langle 111\rangle$, $\langle 110\rangle$ and $\langle 112\rangle$, of the matrix. Fig. 6 shows the experimental data of such a relationship with the matrix orientation in the standard projection. A 6°-radius circle suffices to enclose most of the data points. From this, the experimental P-relationship may be written as $\langle 001\rangle_G \sim\langle 111\rangle_M\text{-}6^\circ$, $\langle 010\rangle_G \sim\langle 110\rangle_M\text{-}6^\circ$ and $\langle 100\rangle_G//\langle 112\rangle_M$. Also included in Fig. 6 is the experimental data

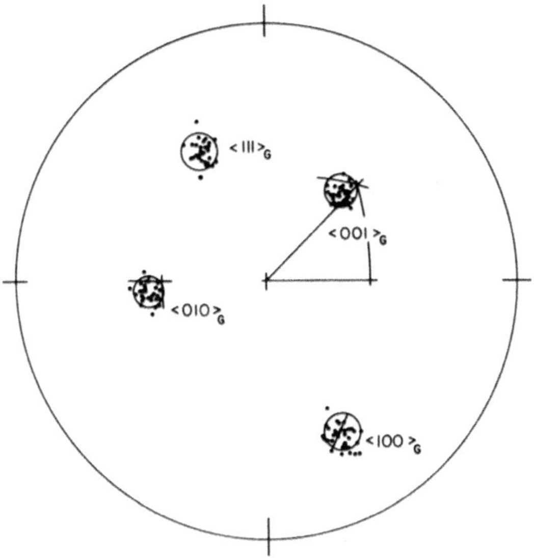

Fig. 6. Observed ⟨001⟩ poles of P-related growing grains.

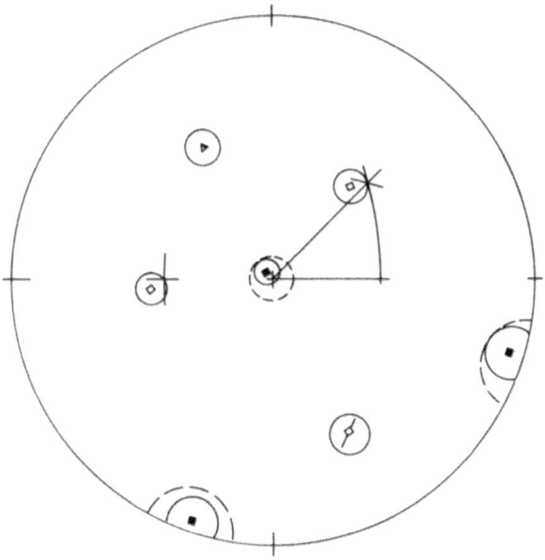

Fig. 7. Stereographic analysis of P and its twin orientations. □ - P-orientation; ■ - ⟨001⟩ 19°; solid line - 6°-radius; dotted line - 10°-radius.

on the $\langle 111\rangle_G$ pole of the twin habit plane. Fig. 7 shows the stereographic analysis of the twin orientation of P-related grains, which bear a $\langle 001\rangle$ 19° coincidence relationship with matrix. The experimental data so derived for $\langle 001\rangle$ 19° grains is comparable to that shown in Fig. 2.

By exchanging the respective directions in the growing grains and matrices, one obtains a Q-relationship. This relationship was held by the parent (or twin) of those growing grains which bear a $\langle 111\rangle$ relationship according to the coincidence model. This is shown in Fig. 8. As in the case of P-relationship, a slight deviation from parallelism between the respective directions was noticed. The experimentally observed relationship is as follows: $\langle 001\rangle_M \sim \langle 111\rangle_G$-6°, $\langle 010\rangle_M \sim \langle 110\rangle_G$-4° and $\langle 100\rangle_M \sim \langle 112\rangle_G$-4°. Fig. 9 shows the relationship of the Q-orientation and its twin with respect to the matrix orientation in the standard projection. Fig. 10 shows the superimposition of the orientations of the twin of the Q-orientation and of the ideal $\langle 111\rangle$ 22° coincidence relationship on the experimental data. Both descriptions seem to fit the data well.

Table II shows the re-tabulation of the experimental data in terms of both the coincidence model and the present alternative. Some of the growing grains which were classified as "random" according to the coincidence model now bear a Q-relationship. On the other hand, not all the $\langle 001\rangle$- and $\langle 111\rangle$-related grains show a twin orientation which is either P- or Q-related to the matrix. The percentage of growing grains which bears a preferred orientation relationship with respect to the matrix is almost identical in the two treatments.

TABLE II. Number of Growing Grains in Terms of Both Coincidence and Present P and Q Relationships.

		P	Q	Random
$\langle 001\rangle \sim$ 19°	34	32		2
$\langle 111\rangle \sim$ 30°	11		8	3
Random	9		4	5
	54	32	12	10

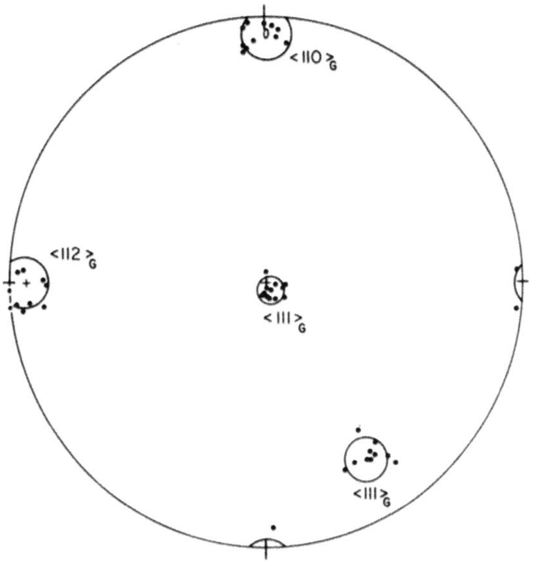

Fig. 8. Experimental data on Q-related growing grains.

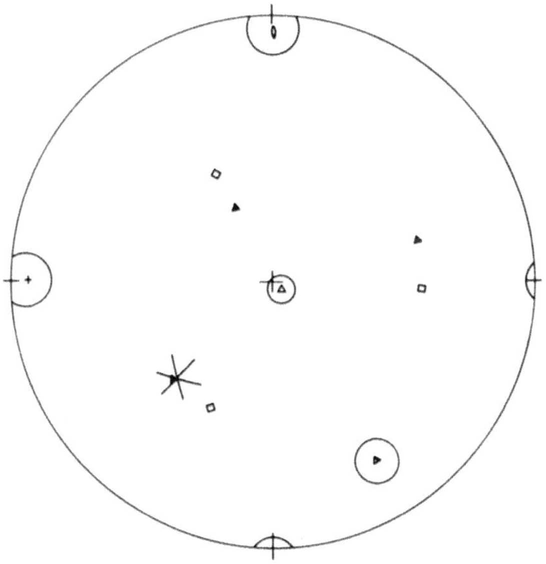

Fig. 9. Stereographic analysis of Q and its twin orientations. Open symbols - Q-orientation; filled symbols - twin of Q-orientation.

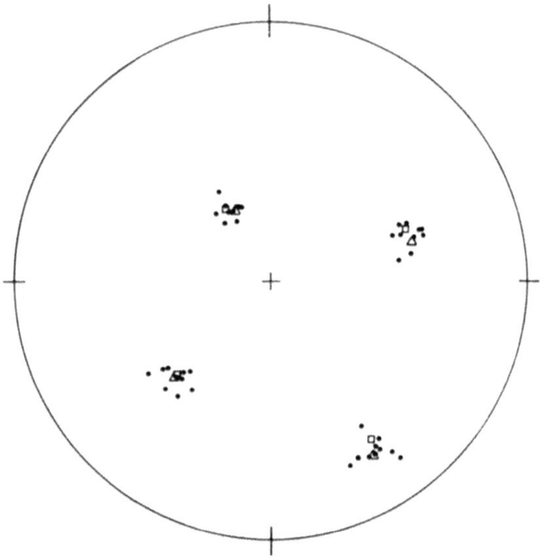

Fig. 10. Octahedral poles of ⟨111⟩-related growing grains. ● - ⟨111⟩-related grains; Δ - a twin of Q-orientation, (Fig. 9); and □ - ideal ⟨111⟩ 22° orientation.

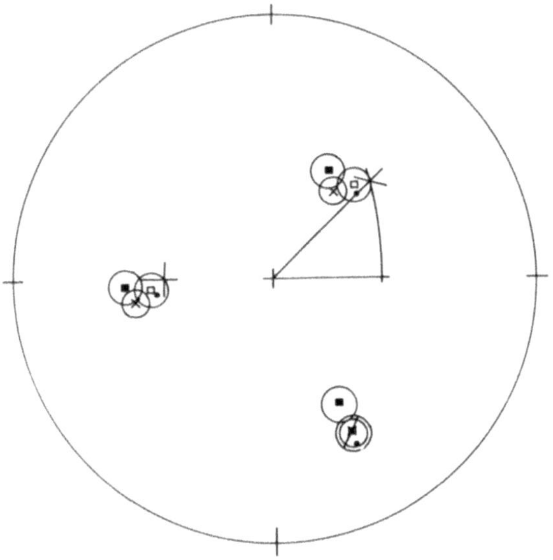

Fig. 11. Comparison of various orientation relationships, ⟨100⟩ poles. ■ - Q-orientation; □ - P-orientation; ● - {122} ⟨457⟩ orientation; X - ⟨012⟩ 114° orientation.

As shown in Fig. 11, the difference between the P and Q orientations is not too great. Parkinson, Van Drunen and Saimoto summarized their results on the growth-selection experiments in high-purity copper in terms of an ⟨012⟩ 114° rotation within a 5° scatter.[17] This description fits closely the suggested P-orientation. Also the P-orientation agrees well with Sharp and Dunn's {122}⟨457⟩ type orientations resulting from secondary recrystallization in copper.[15]

The preference of a particular twin habit such that the orientation of this annealing twin of a P- or Q-related parent grain would have a ⟨001⟩ 19° or ⟨111⟩ 22° coincidence relationship with the matrix is of interest. It is true that annealing twins on any one of the four {111} planes of a ⟨001⟩ 19° grain would bear a P-relationship with matrix, but only one of the twins of a P-related grain would have a ⟨001⟩ 19° relationship with the matrix. This observation is a first indication that the grain boundary of a P- or Q-related grain may not be a low energy boundary. As shown by Aust and Rutter,[19] the occurrence of annealing twins results in the replacement of the original boundary of a grain boundary of coincidence relationship with a lower energy. Since neither a P- nor a Q-related grain is in a coincidence relationship with the matrix, a twin event on a particular habit which will bear a coincidence relationship with matrix is therefore favored.

DISCUSSION

Based on the present experimental results, we have shown that by considering the parent of an annealing twin (or vice versa), an alternative to the coincidence relationship can be develoepd. We shall extend this alternative to the observations reported in the literature on the secondary recrystallization of copper.

The primary matrix during the secondary recrystallization of copper has previously been treated as a pseudo single crystal of (001)[100] orientation. Since it is also known that such a matrix free of annealing twins would not undergo secondary recrystallization,[1] it is therefore imperative to treat the matrix not only as a (001)[100] grain but also as its four twin orientations.

Kronberg and Wilson reported that secondary grains which are ⟨001⟩ 19°-related to the matrix are "frequently" free of twins.[1] However, Sharp and Dunn[15] noticed that about one-fifth of the secondary grains, which are twin related to the ⟨001⟩ 19° grains and which bear {122}⟨457⟩ type orientations, grew more rapidly than their parent grains. According to the coincidence model, when they meet the matrices of annealing twins, the ⟨001⟩ 19° secondary grain would have no advantage over grains of other orientations.

This is because that there is no coincidence relationship between ⟨001⟩ 19° grains and twin orientations of the (001)[100] matrix.

In view of the present alternative, it is the twin matrices of (001)[100] orientation, bearing a Q-relationship with the secondary grains, which are responsible for the growth of the ⟨001⟩ 19° grains. A similar situation can also be found for ⟨111⟩ 22° secondary grains. Fig. 12 illustrates the inter-relationships between growing grains and matrices during secondary recrystallization of copper.

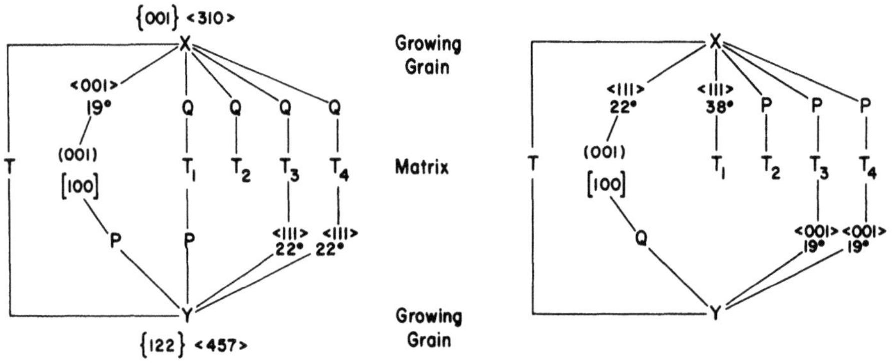

Fig. 12. The inter-relationships between orientations of secondary grains and matrices during secondary recrystallization of copper.

In metals in which the occurrence of annealing twins is rare, such as aluminum, both ⟨001⟩ 19° and ⟨111⟩ 22° are therefore considered not to be favorable relationships. This would explain the absence of these two relationships in the secondary recrystallization of aluminum.(2)

In applying the coincidence model to the results of a growth selection experiment, it leads to an inevitable question as to why orientation relationships of high coincidence density, such as 1 in 5 of ⟨001⟩ 37°, are not observed experimentally. The limited number and lack of variety in the observed relationships during both growth selection and secondary recrystallization experiments, on the other hand, are in excellent agreement with the present alternative, in which the number of permissible relationships is restricted to only two.

The crystallography of the P- and Q-relationships corresponds well with the result of the Rowland transformation.(20) Verbraak has adopted this transformation as a possible nucleation mechanism for a strain-free grain in copper.(21) The basic process of the Rowland transformation, however, involves a martensite-like shear mechanism.(21) At present, it is difficult to perceive the operation of such a mechanism during the migration of a grain boundary.

ACKNOWLEDGMENTS

Appreciation is expressed to Mr. W. S. Stewart for carrying out the experimental work. Discussions with Drs. H. I. Aaronson and N. A. Gjostein during the course of writing were appreciated.

REFERENCES

1. M. L. Kronberg and F. H. Wilson, Trans. TMS-AIME, 185, 501 (1949).

2. P. A. Beck and H. Hu, Trans. TMS-AIME, 185, 627 (1949).

3. F. C. Frank, referred by ref. 5.

4. K. T. Aust and J. W. Rutter, Trans. TMS-AIME, 215, 119 (1959).

5. K. T. Aust and J. W. Rutter, Trans. TMS-AIME, 215, 820 (1959).

6. J. W. Rutter and K. T. Aust, Acta Met., 13, 181 (1965).

7. B. B. Rath and H. Hu, Trans. TMS-AIME, 245, 1577 (1969)

8. D. C. Brandon, B. Ralphy, S. Ranganathan and M. S. Wald, Acta Met., 13, 813 (1964).

9. J. C. M. Li, in Recovery and Recrystallization of Metals, Interscience, New York, 1963, p. 160.

10. K. T. Aust, in Surface and Interfaces, Syracuse University Press, 1967, p. 435.

11. P. Gordon and R. A. Vandermeer, in Recrystallization, Grain Growth and Textures, ASM, Metals Park, Ohio, 1966, p. 205.

12. K. Lücke and G. Ibe, Final Technical Report, Contract N. DA-91-591-EUC-2875, Department of Army, 1964.

13. H. P. Stüwe, Trans. TMS-AIME, 221, 203 (1961).

14. Y. C. Liu, Trans. TMS-AIME, 230, 1513 (1964).

15. M. Sharp and C. G. Dunn, Trans. TMS-AIME, 194, 42 (1952).

16. K. T. Aust, G. Ferran and G. Cizeron, Comptes Rendus, 257, 3595 (1963).

17. E. C. Parkinson, G. Van Drunen and S. Saimoto, Scripta Met., 4, 1009 (1970).

18. M. Sharp and C. G. Dunn, Trans. TMS-AIME, 194, 1344 (1952).

19. K. T. Aust and J. W. Rutter, Trans. TMS-AIME, 218, 1023 (1960).

20. P. R. Rowland, J. Inst. Met., 83, 1620 (1954).

21. C. A. Verbraak, Acta Met., 6, 580 (1958).

INFLUENCE OF SOLUTES ON THE MOBILITY OF TILT BOUNDARIES

B.B. Rath* and Hsun Hu

U.S. Steel Res. Lab., Monroeville, Pa. 15146; *now with

McDonnell-Douglas Res.Lab., St. Louis, Missouri 63166

ABSTRACT

Grain boundaries have a profound effect on the strength of metals and alloys. Material properties can be changed significantly by boundary migration. Studies of grain boundary mobility in matrices of poly- and single-crystals under the conditions of a difference in stored energy across the moving boundary have provided only limited advances towards understanding the behavior of grain boundaries. Primarily, this has been due to uncertainties in the magnitude of the force acting on the boundary during its motion. This difficulty has been avoided by examining boundary motion in wedge-shaped bicrystals in which the boundary migrates under the driving force of its own interfacial energy. The driving force per unit area of the boundary is simply related, in this case, to the specific interfacial energy and to the distance between the boundary and the tip of the wedge. Results of an investigation of the velocities of pure tilt boundaries in zone-refined aluminum and dilute aluminum-magnesium alloy bicrystals of preselected orientations have been presented. The effects of varying driving force and solute concentration on thermally activated boundary motion will be discussed in relation to the postulates of current theories.

INTRODUCTION

In recent years several investigations and a number of excellent reviews concerning grain boundary migration as related to recrystallization and grain-growth in metals have been published (1,2,3). It is generally accepted that grain boundary migration is strongly influenced by: (i) the species and concentration of impurity atoms in the matrix (4,5), (ii) the volume free energy stored in the matrix (6), and (iii) the nature of the atomic disorder at the grain boundary (7).

Direct measurements of migration rates of grain boundaries as a function of crystal orientation and impurity concentration in zone-refined lead were performed by Aust and Rutter (1). The driving force for the motion of grain boundaries was derived from the grown-in lineage structure in the matrix crystals. Rath and Gordon (8) conducted similar experiments in deformed zone-refined aluminum. In both cases the driving force for migration arises from a difference in the dislocation density across the boundary. These methods neither provide a condition in which the grain boundary structure remains uniform throughout the boundary plane nor allow an evaluation of the force acting on the boundary at any instant during its migration. Investigation of grain boundary migration during normal grain growth also suffers from similar difficulties due to the uncertainties in the principal radii of the boundary curvature (9), which constitutes the driving force for the grain-growth process.

These experimental uncertainties have been removed by using wedge-shaped bicrystals, in which the force acting on the boundary as well as its resulting velocity have been determined during boundary migration. The driving force is determined simply from the gradient of the total interfacial free energy of the boundary and its corresponding velocity measured from its displacement. The effects of boundary misorientation angle and the concentration of substitutional impurities (concentration of Mg varying from 3-250 ppm) in boundary migration in zone-refined aluminum bicrystals have been investigated by this method. The effects of driving force, boundary orientation and substitutional impurities on boundary velocity have been examined in the light of current theories on grain-boundary migration.

EXPERIMENTAL METHOD

The materials used in this investigation were two batches of zone-refined aluminum. Spectrographic analysis of the as-received aluminum given in Table I indicated a purity of 99.999+%. The first batch consisted of seed-oriented bicrystals grown from the melt. These contained tilt boundaries such that a <111> axis common to both crystals was parallel to the boundary plane. Three such bicrystals, 2.5 cm wide, 0.5 cm thick and 12 cm long, containing 16°, 30° and 40° tilt boundaries were grown. The second batch of zone-refined aluminum was given three additional zone-refining passes followed by zone-leveling at a pressure of 10^{-5} mm of Hg to improve purity. Four Al-Mg alloys were prepared from

TABLE I. SPECTROGRAPHIC ANALYSIS OF ZONE-REFINED ALUMINUM CRYSTALS.

Element	ppm by wt.	Element	ppm by wt.
Cu	<0.2	Ga	<0.25
Fe	1.5	Zn	1.2
Mn	0.15	Ca	0.2
Cr	0.6	K	0.2
V	0.4	Si	0.8
Ti	0.2	Mg	0.5
S	<0.3	Na	0.5

this batch in purified H_2 atmosphere with magnesium concentrations varying from 3-250 ppm. Bicrystals containing 40° - <111> tilt boundaries (of each composition, including the re-purified zone-refined aluminum) were grown in the manner mentioned earlier. Wedge-shaped specimens were prepared from each bicrystal by spark cutting followed by electrolytic polishing. Care was taken to prevent contamination or deformation during handling. A section of an oriented wedge bicrystal is shown in Figure 1.

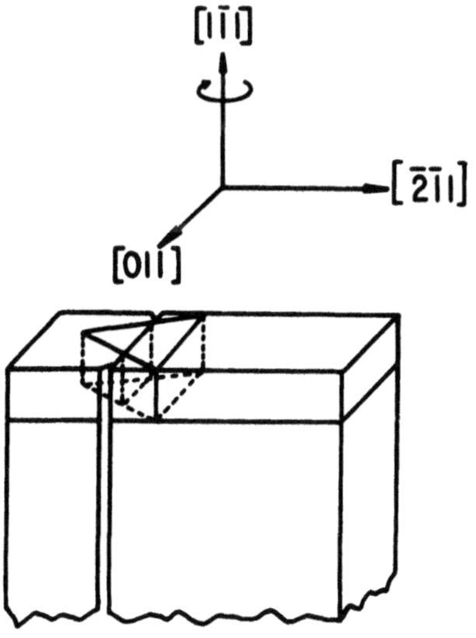

Figure 1. Geometry of an oriented bicrystal showing the outline of the wedge-shaped specimen.

The crystallographic directions shown represent the orientation of the larger section of the bicrystal. The orientations of the two sections of the bicrystal differ only by rotation around the common [1$\bar{1}$1] axis. Therefore, the top and bottom faces of the wedge always correspond to (1$\bar{1}$1) planes in both crystals, whereas the orientations of the side faces vary depending on the tilt misorientation angles between the two sections. The bicrystal orientations were determined by the Laue back-reflection method. The angle of the wedge for the bulk of this investigation was kept at 40° and the thickness of the specimens was 0.25 cm. Twenty to twenty-five wedges were cut from each bicrystal such that the distance between the intercrystalline boundary and the tip of the wedge varied over a range of 0.2 to 0.5 cm.

In order to determine whether the orientation of the inclined side faces of the wedge or its thickness had an effect on boundary migration, a few specimens were prepared with an apex angle of 80° whereas others were cut to a thickness of 0.5 cm. Yet another group of samples were prepared with the apex of the wedge in the opposite direction. This allowed measurement of the boundary velocity in opposite directions.

The wedges were annealed for various periods at temperatures ranging from 610° to 650°C in purified hydrogen. The samples were electrolytically polished between anneals. The initial and final positions of the migrating boundary as well as its distance from the tip of the wedge were determined by microscopic examination. The extent of thermal grooving during annealing was determined by interference microscopy.

RESULTS AND DISCUSSION

The grain boundary in the wedge-shaped bicrystals, when annealed at high temperatures, in the absence of any other forces, should move towards the tip of the wedge to decrease its total interfacial energy. Furthermore, the initially planar grain boundary would assume a steady-state shape during its motion, such that the force per unit area is uniform over the entire boundary. Li (10) has discussed the steady-state shape of such a boundary based both on the minimum rate of entropy production and on the minimum thermokinetic potential. Figure 2 shows the top view of a bicrystal wedge sample with the positions of the boundary before and after annealing. In all cases, the initially straight boundary quickly became curved during annealing such that the center of curvature, being independent of the boundary position, remained at the tip of the wedge. Observation of boundary migration in situ, using a hot-stage microscope, reveals that the segment of the boundary, initially straight, which intersects the inclined side faces of the bicrystal rapidly moves towards the tip, thus curving

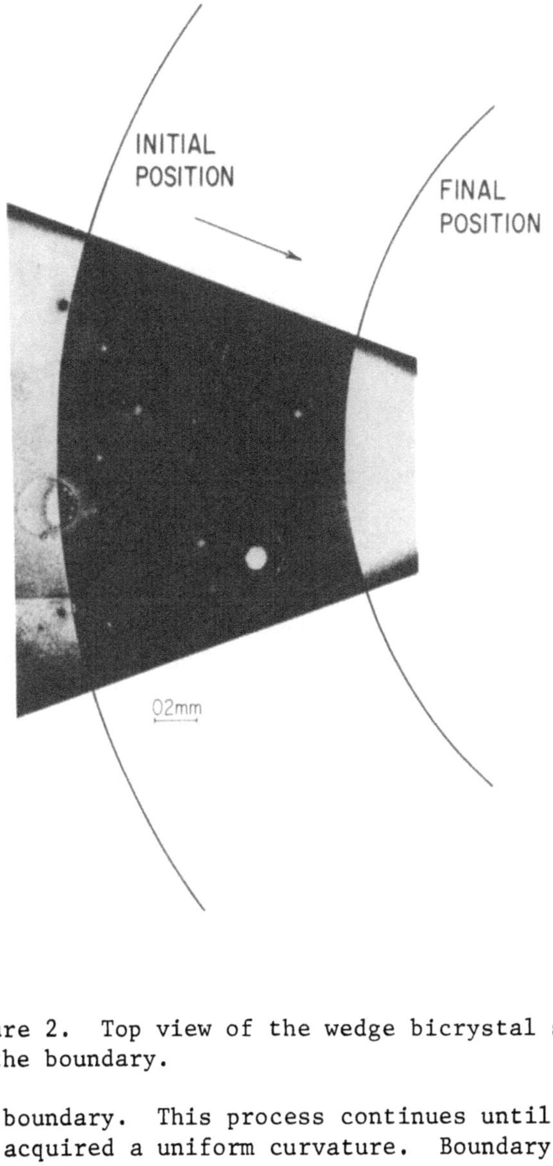

Figure 2. Top view of the wedge bicrystal showing radial migration of the boundary.

the boundary. This process continues until the planar boundary has acquired a uniform curvature. Boundary velocities were not determined until the boundary curvature was uniform as viewed on the top face of the wedge sample. The top and side views of a typical wedge sample revealing two positions of the boundary are shown in Figure 3. The cylindrical arc shape of the boundary during its migration towards the apex of the wedge is clearly evident. The fact that the boundary shape is independent of its misorientation angle and that it maintains a cylindrical shape during its migration further indicates that any extraneous forces acting on it at the inclined surfaces are negligible.

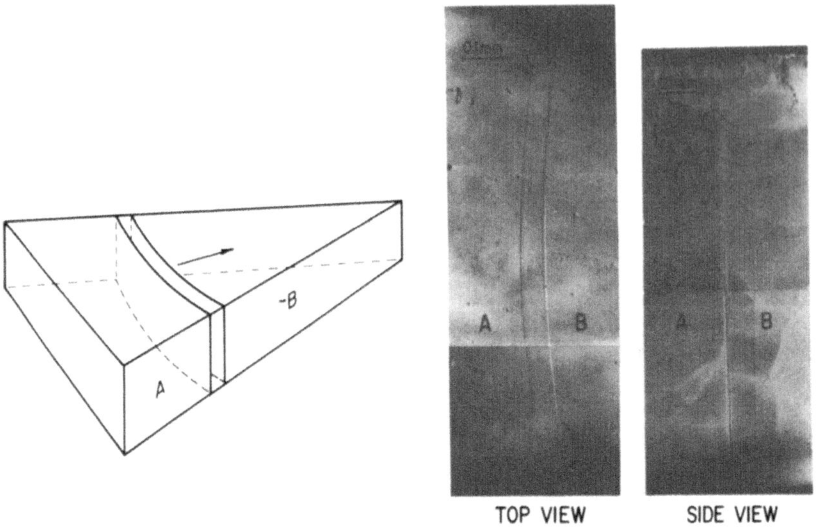

Figure 3. Boundary configuration during migration.

During cooling the bicrystal from the high-temperature anneal (0.92 to 0.98 Tm), boundary grooving occurred, presumably due to condensation of thermal vacancies on the surface during cooling. These vacancies, concentrated at the boundary, migrate to the outer surfaces, producing the grooving shown in Figure 4.

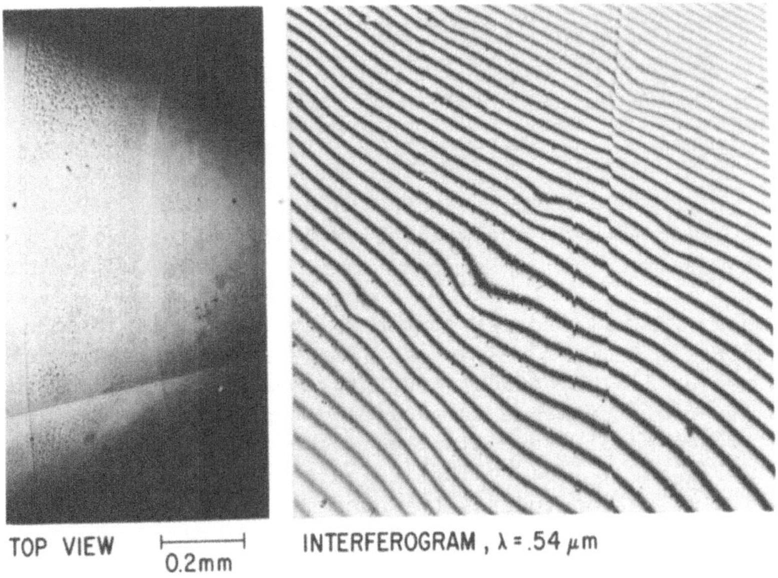

Figure 4. Thermal grooving of a boundary after cooling.

Surface pitting also occurred during cooling. The depth of the grooves, measured from interferograms, is about 350 Å and appears to be independent of the angle of misorientation of the boundary. Occasionally, very fine traces representing intermediate positions of the boundary were visible. However, the depths of these traces were below the limits of detection of an interferometer. If these surface traces exerted an appreciable drag force on the migrating boundary, it should bulge inward in the interior of the wedge. Several samples were cut and microscopically examined. As seen in Figure 5, the straight-line intersection of the boundary with the cut surface suggests that the boundary remained a segment of a cylinder during its motion, and that therefore, the drag force is negligibly small in comparison with the force derived from the gradient of boundary interfacial free energy.

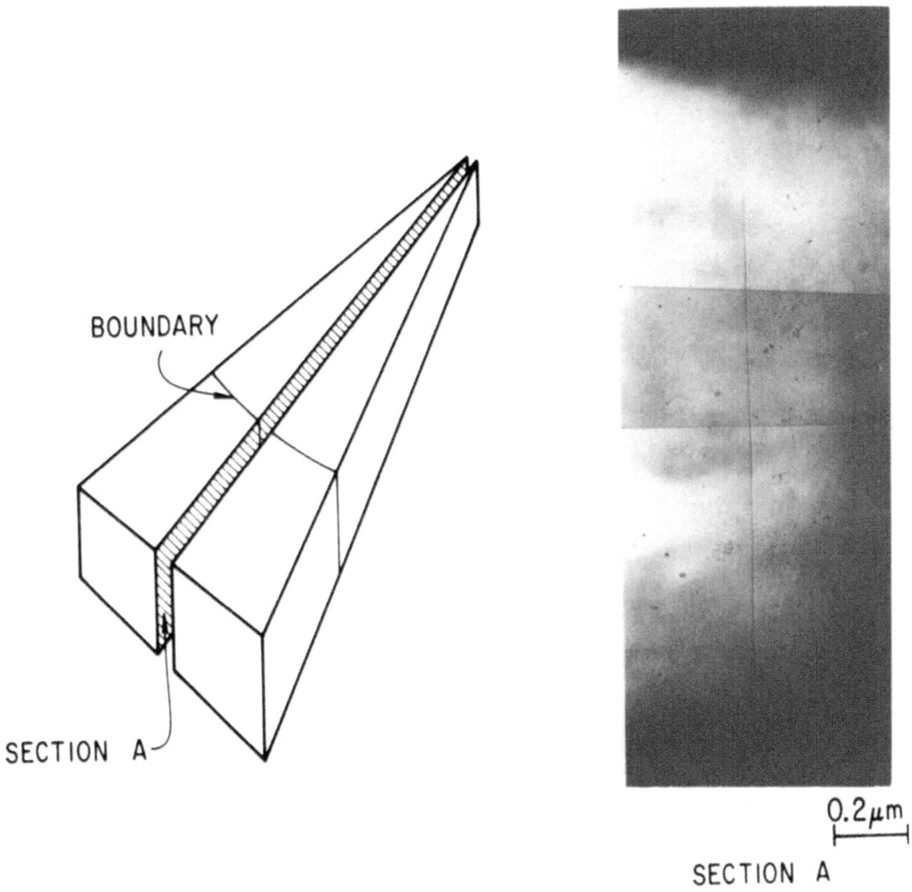

Figure 5. Section of a bicrystal specimen showing straight-line intersection of the boundary with the cut surface.

Effect of Driving Force on Boundary Velocity

In the wedge configuration the force acting on the boundary increases as the boundary approaches the tip of the wedge. For the cylindrical shape of the boundary, the center of the circular arc is at the tip; the total boundary energy is given by

$$E = \gamma X \alpha Z \qquad (1)$$

where γ is the specific boundary energy, X is the distance of the boundary from the tip of the wedge, α is the apex angle, and Z is the thickness of the specimen. This is illustrated in Figure 6.

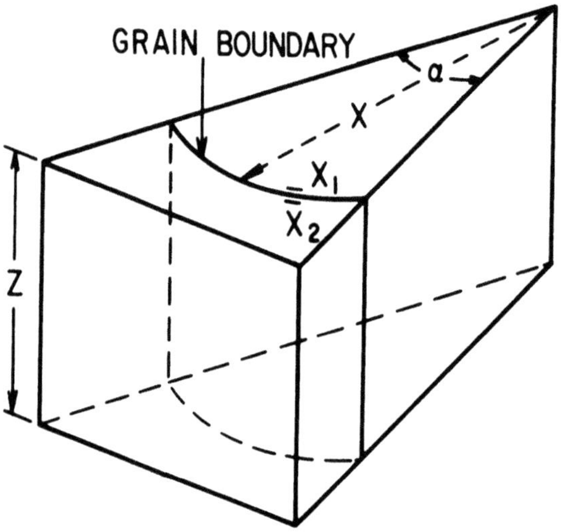

Figure 6. Schematic illustration of a wedge bicrystal specimen containing a cylindrical arc boundary.

The total driving force on the boundary, therefore, is

$$F_T = -\frac{dE}{dX} = \gamma \alpha Z \qquad (2)$$

and the force per unit area of the boundary is

$$\Delta F = \frac{\gamma}{X} \qquad (3)$$

This indicates that the driving force for boundary migration is inversely proportional to the distance of the boundary from the tip of the wedge, γ being the proportionality constant. The boundary velocity was found to be independent of (i) the thickness of the wedge, Z, when increased from 0.25 cm to 0.5 cm, or (ii) the apex angle, α, when increased from 40° to 80°.

It was found that the grain boundary velocity was highly sensitive to the distance of the boundary from the tip of the wedge. The three micrographs in Figure 7 illustrate the

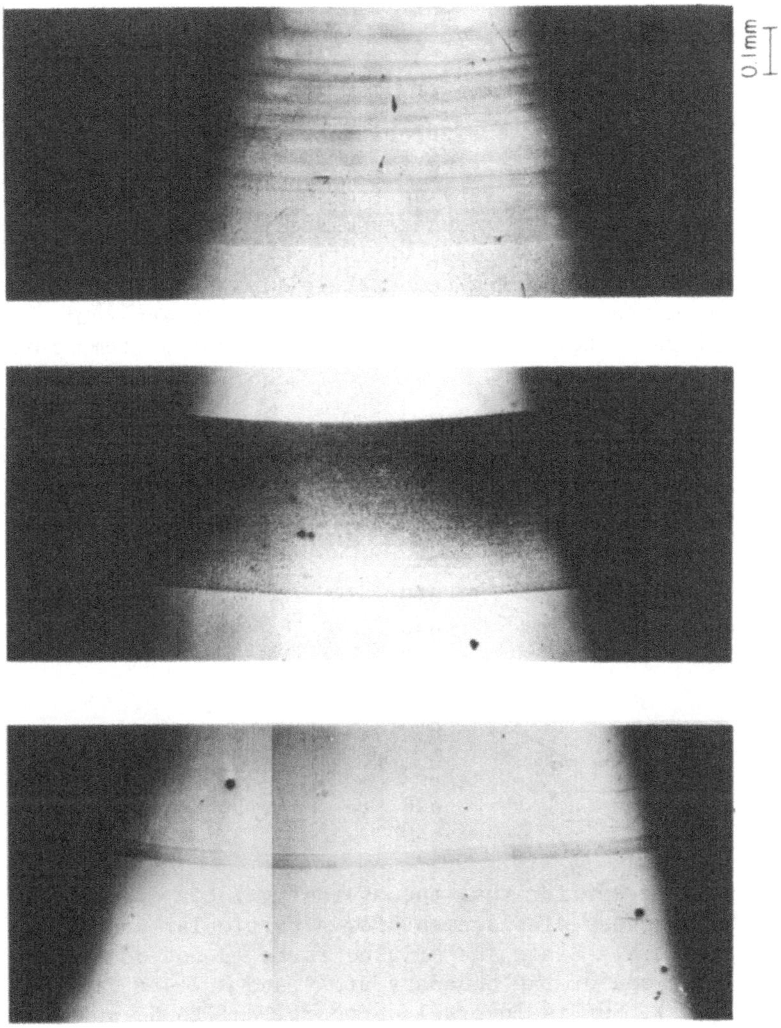

Figure 7. Effect of position of boundary with reference to the tip of the wedge bicrystal (distance X) on boundary velocity.

significant change in boundary displacement with position in the bicrystal wedge. The average boundary velocity as a function of X in the first batch of bicrystals is shown in Figure 8 for various annealing temperatures and boundary misorientation angles. The marked change in boundary velocity due to a relatively small change in X, as shown in Figure 8, indicates that the boundary velocity can be expressed by

$$V = M\ (\gamma/X)^m \qquad (4)$$

or

$$\log V = \log\ (M\gamma^m) - m \log X \qquad (4a)$$

where M is the mobility, or the velocity of the boundary at unit driving force, (γ/X). The driving force exponent, m, is dependent on the misorientation angle, varying from 4.0 to 3.2. Table II lists the values of m for the boundary misorientation angles and annealing temperatures. The small variation of m with temperature, i.e. decreasing with increasing temperature, is well within the limits of experimental scatter, and therefore, this apparent temperature effect is not conclusive within the narrow temperature range studied.

Table II. EFFECT OF BOUNDARY MISORIENTATION ON THE DRIVING FORCE EXPONENT

Misorientation Angle, θ, deg.	Temp. ^{o}C	Driving Force Exponent, m
40	610	4.0
	640	4.0
	650	3.7
30	620	3.4
	640	3.4
	650	3.3
16	620	3.4
	640	3.2

It was stated earlier that the average velocity was determined from the boundary displacement for a particular annealing time and temperature. Since the driving force is not constant between the positions of the boundary at X_1 and X_2 (see Fig. 6), where $\Delta X = (X_2 - X_1)$, but is inversely proportional to X, and furthermore, since the velocity in this region is proportional to the driving force to the power m, the effective distance, X, which corresponds to the average driving force acting on the boundary between X_1 and X_2, is given by

$$X = 1/[(\Delta X)(m+1)/(X_2^{(m+1)} - X_1^{(m+1)})]^{1/m} \qquad (5)$$

In view of these experimental conditions, ΔX was kept to less than 100μm by choosing appropriate annealing periods at all temperatures.

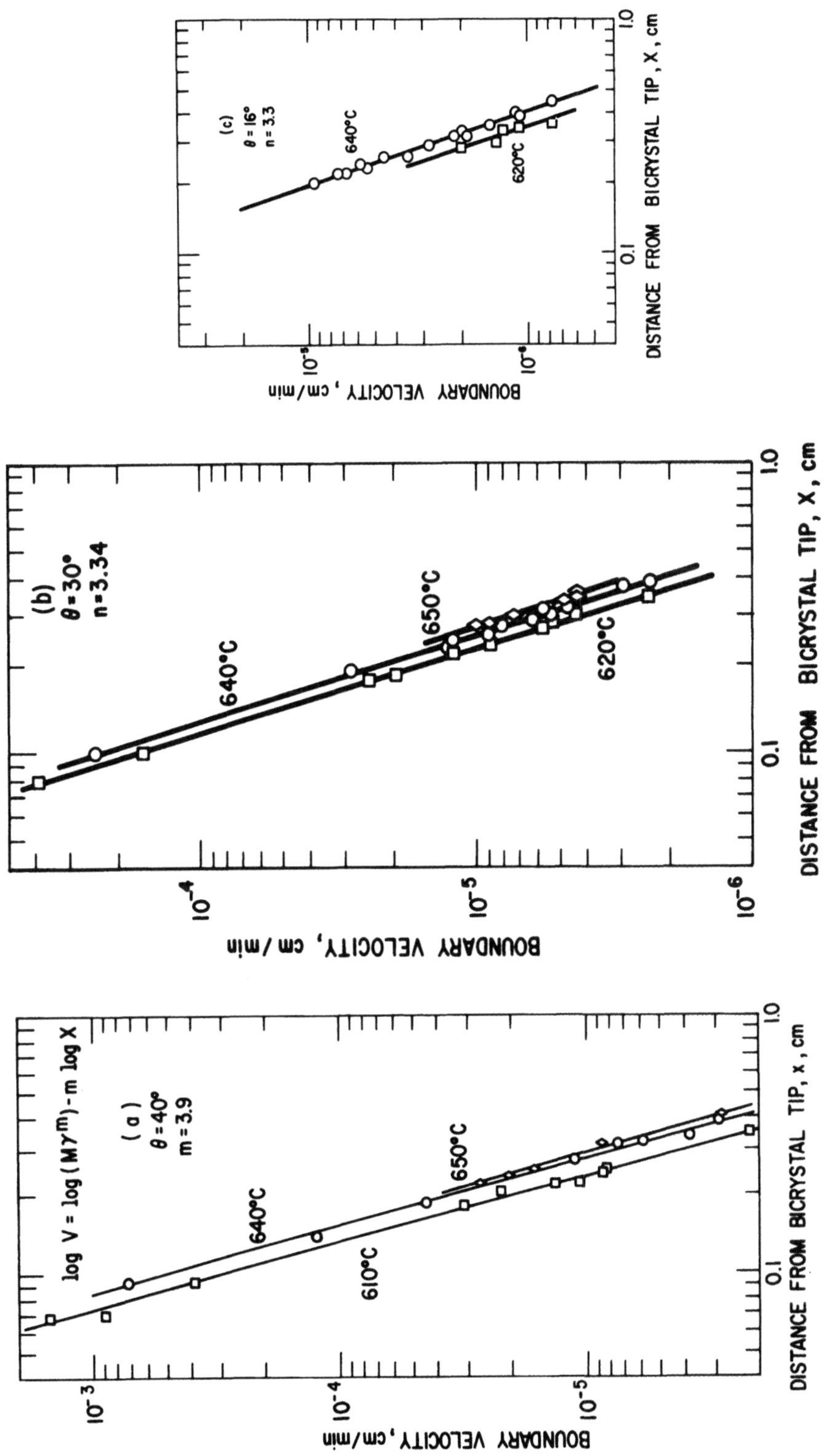

Figure 8. Boundary velocity as a function of the distance from the bicrystal tip, X, for misorientation angles (a) 40°, (b) 30°, and (c) 16°.

The distance of the boundary from the apex of the wedge, X', was taken as the average, i.e. $X' = (X_1 + X_2)/2$. From the relation between X' and V, the value of m was obtained. Subsequently, knowing m, X_1 and X_2, the effective distance X was evaluated from Eq.(5). Since electropolishing after each anneal changes the value of X, it was not possible to determine the velocity of the boundary from the slope of boundary displacement vs time plot.

For comparison, the velocities of the three tilt boundaries at 640°C as a function of X are shown in Figure 9(a). The boundary velocities decreased with decreasing angle of misorientation.

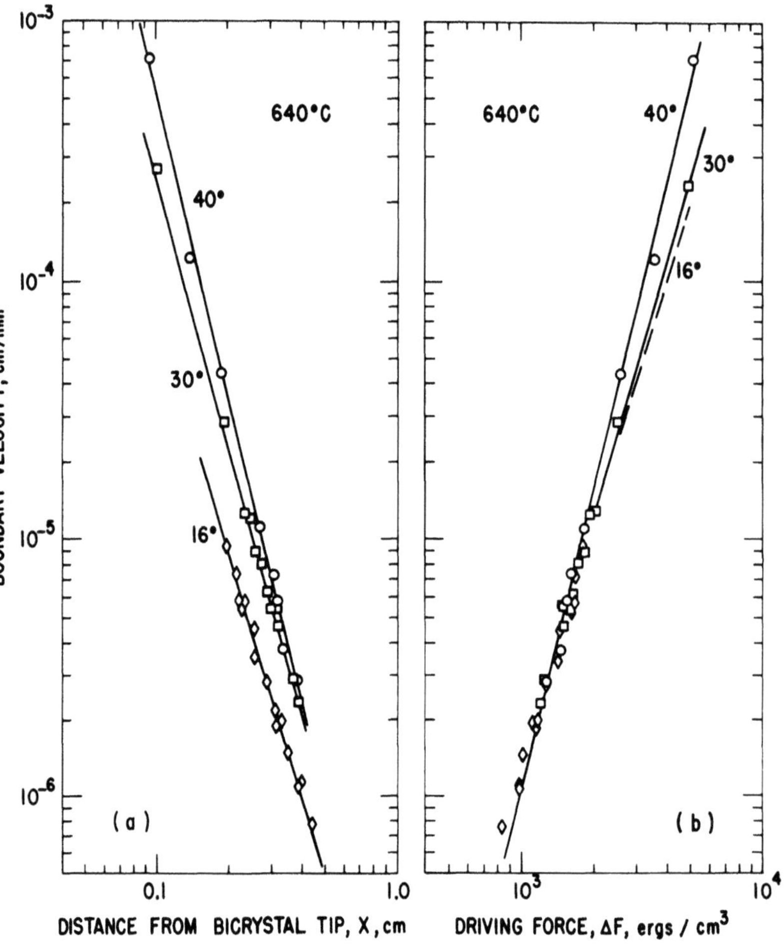

Figure 9. Effect of misorientation angle on the driving force dependence of boundary velocity.
(a) Boundary velocity vs distance X, and (b) Boundary velocity vs driving force, ΔF, calculated from distance X.

However, owing to a dependence of m on boundary misorientation, represented by the slope of each line, the velocity difference among the boundaries increased with decreasing X, or increasing driving force. In order to evaluate the driving force acting on the boundary, it is assumed that the specific energy of the 40° tilt boundary is 500 ergs/cm^2. It is further assumed that the ratios of boundary energies for different θ, the misorientation angle, are the same as those found by Gjostein and Rhines (11) in copper. Although this value is only an estimate of the energy of a 40° tilt boundary in aluminum (12), an alteration would change the observed driving force-velocity relationship. The specific energies of 30° and 16° tilt boundaries, corresponding to 480 and 370 ergs/cm^2, respectively, have been used to evaluate the driving force from Eq.(3). The effects of driving force and boundary misorientation on boundary velocity are shown in Figure 9(b). The figure unambiguously demonstrates the very large effect of driving force on grain boundary migration. Such profound effects of driving force on boundary velocity have been reported by Lücke (13) and Vandermeer (14). Due to their experimental methods the values of the driving force could not be obtained with accuracy. In an earlier investigation, Rath and Hu (15) found the boundary velocity to vary by as much as seven orders of magnitude when the recovered cell size in deformed aluminum crystals changed by a factor of only five.

The driving-force dependence of grain boundary velocity, shown in Figure 9(b) further illustrates that up to a force of 2×10^3 ergs/cm^3 the velocity is relatively independent of the angle of misorientation, whereas upon increasing the force, the boundary velocity increases with misorientation. This suggests that during recrystallization of a lightly deformed matrix with low stored energy the growth rates of "preferred" grains will be comparable to those of random types. On the other hand, when the stored energy in the matrix is high, the growth rates of "preferred" grains would predominate. This observation agrees well with the findings of Rath and Hu (6) and Green (16), that the percentage of preferred grains, which in fcc metals are described by a 40° rotation about a common <111> axis, increases with increasing prior deformation.

The temperature dependence of boundary velocity at a constant driving force decreases only slightly with the angle of misorientation. The boundary velocity at X = 0.3 cm as a function of reciprocal temperature is shown in Figure 10. The apparent activation energy, Q, decreased from 33 to 30 kcal/mol as a result of a change in the misorientation angle from 16° to 40°. This change is small in comparison with the effect of driving force on Q, shown in Figure 11, where Q decreased from 33.5 to 27.5 kcal/mol as the driving force increased from 1×10^3 to 5×10^3 ergs/cm^3. The relatively small effect of misorientation angle on Q suggests that

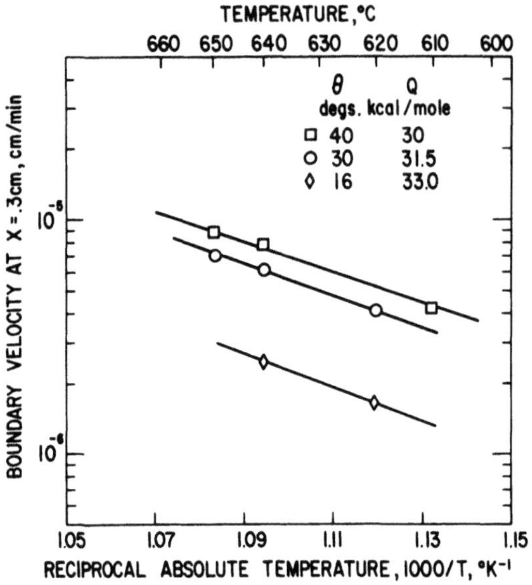

Figure 10. Temperature and misorientation dependence of boundary velocity at a distance X = 0.3 cm from the bicrystal tip.

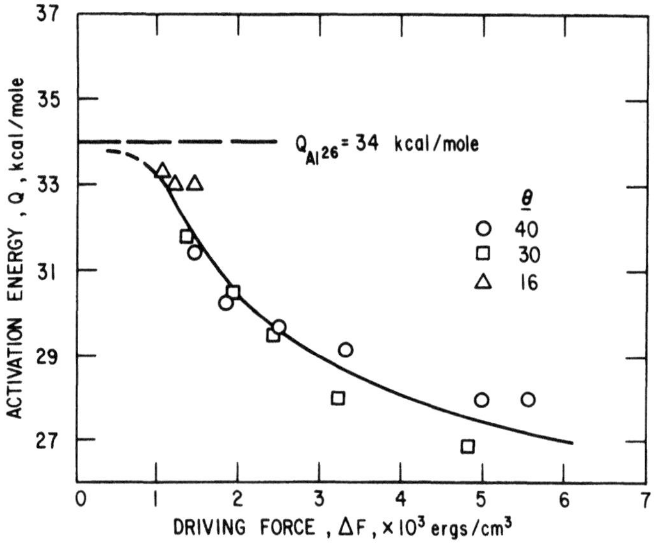

Figure 11. Effect of driving force on the activation energy for boundary migration. The dash line represents the activation energy for bulk self-diffusion in aluminum.

although for a constant driving force the number of atoms crossing the boundary per unit time per unit area may differ with misorientation; the activation barrier for a single atom jump remains almost unaffected by a change in misorientation across the boundary. On the other hand, the decrease in Q with increasing driving force could be due to an increase in porosity of the boundary because of its high velocity at higher driving forces. Based on a concept first suggested by C.S. Smith with reference to work of Mullins(17)--boundaries require vacancies in order to move--it is conceivable that at very low driving forces, hence low velocities, the boundary porosity would be minimal. Consequently, atom transfer across the boundary would be near that of bulk self-diffusion. When boundary velocity increases because of an increase in driving force, it would also result in an increase in boundary porosity. This would reduce the activation energy for atom transfer to less than that for bulk diffusion, which is about 34 kcal/mol for aluminum(18). It is therefore likely that for driving forces greater than 5.5×10^3 $ergs/cm^3$, which was the upper limit of boundary velocity measurements in this investigation, still lower values of Q would be obtained. This hypothesis is consistent with the measured Q of about 15 kcal/mol, as reported by Rath and Hu (7), Gordon and Vandermeer (2), and Frois and Dimitrov (19), for the growth of recrystallized grains in deformed aluminum with estimated stored energy of 10^7 to 10^8 $ergs/cm^3$. Aust (20) has reported an activation energy of 13 kcal/mol for boundary migration in a 12-pass zone-refined aluminum, with an estimated stored energy of only about 9×10^3 $ergs/cm^3$. This apparent discrepancy is clearly due to differences in the amount and species of trace impurities in these several zone-refined aluminums. The importance of such impurities is clearly demonstrated by our experience with zone-refined aluminum bicrystals. In the first batch, the activation energy for boundary migration was 30 kcal/mol. After three additional zone-refining passes and zone leveling, as previously described, the activation energy in our second batch was reduced to 17 kcal/mol.

The presence of these impurities at the boundary may also explain the lack of a pronounced effect of misorientation on Q. However, this does not invalidate the explanation of the driving-force dependence of the measured activation energy.

Effect of Impurities on Boundary Velocity

The effect of impurities on boundary migration was studied in the second batch of zone-refined aluminum containing 3, 10, 25, 50 or 250 ppm of magnesium. All bicrystals in this batch were grown with a 40° - $[1\bar{1}1]$ boundary. Typical boundary migration behavior in these alloys is shown in Figure 12, for Al+50 ppm Mg at $620^\circ C$. The driving force exponent, in this case m = 3.9, is nearly unaffected by the presence of Mg. It may be noted here

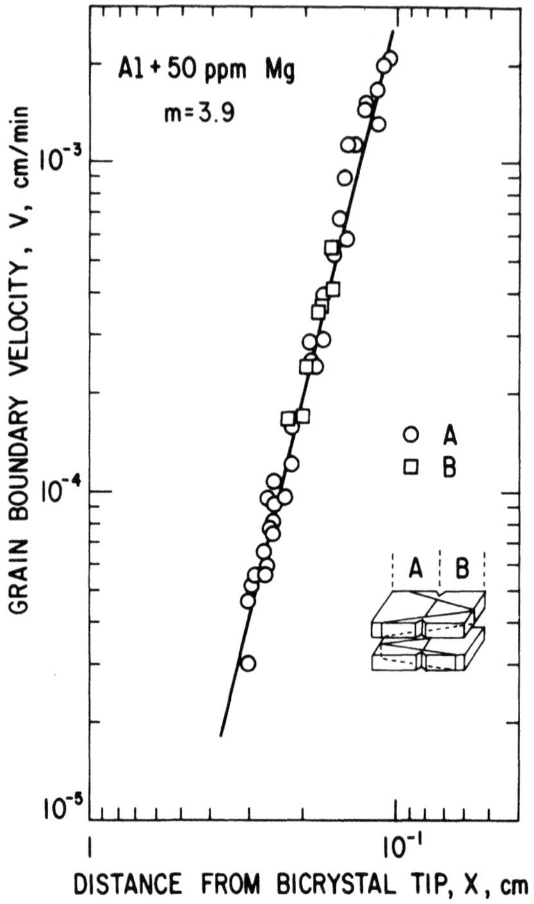

Figure 12. Boundary velocity as a function of distance X in Al + 50 ppm Mg bicrystals.

that Figure 12 is a plot similar to Figures 8 or 9, except that in Figure 12 X decreases with increasing V, thus the curve has a positive slope. To confirm that boundary velocity is not affected by the crystallographic direction of its motion, the migration of the boundary in opposite directions was determined for each alloy by cutting specimens so that the apex of the wedge was reversed. The velocities in these two directions are identical, suggesting that the rate of boundary migration is independent of the direction of its motion.

When the magnesium content was less than 250 ppm, the boundary velocity became much less sensitive to X for $X \leq 0.1$ cm or $\Delta F \geq 5 \times 10^3$ ergs/cm^3. This is shown in Figure 13. At a driving force of 6×10^3 ergs/cm^3 (X = 0.1 cm), m decreases from 4.0 to 1.2. It would seem to indicate qualitatively that at this driving force,

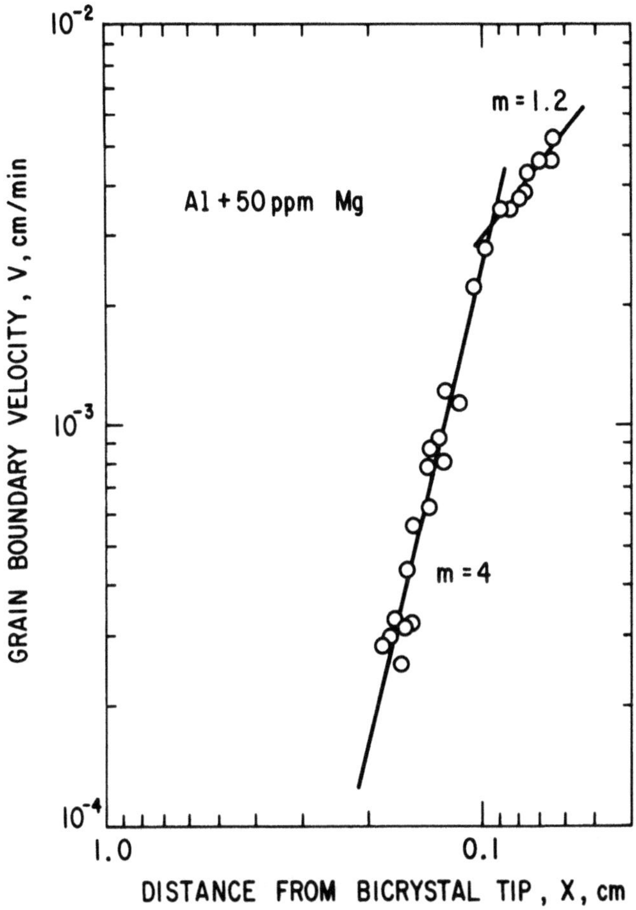

Figure 13. A change in slope of the boundary velocity vs distance, X, plot in Al + 50 ppm Mg bicrystals.

the boundary undergoes a transition from an impurity-controlled to an impurity-independent behavior, in accordance with the impurity-drag theory to be discussed later. However, velocity measurements in this region of X are inaccurate because of uncertainties in the measurements of X and of the annealing time and temperature. Because of very high boundary velocities near the apex of the wedge, the duration of annealing is only 5 to 10 min.

With the exception of the 250 ppm Mg alloy, the boundary velocity was practically unaffected by the addition of Mg to the second batch of zone-refined aluminum. For the 250 ppm Mg alloy, the boundary velocities were at all driving forces lower by one order of magnitude. This is shown in Figure 14. The 250 ppm Mg alloy did not show a transition in boundary velocity at X >0.1 cm. A comparison of the boundary velocities at 640°C in the two batches of zone-refined aluminum for the 40° tilt boundary reveals that the

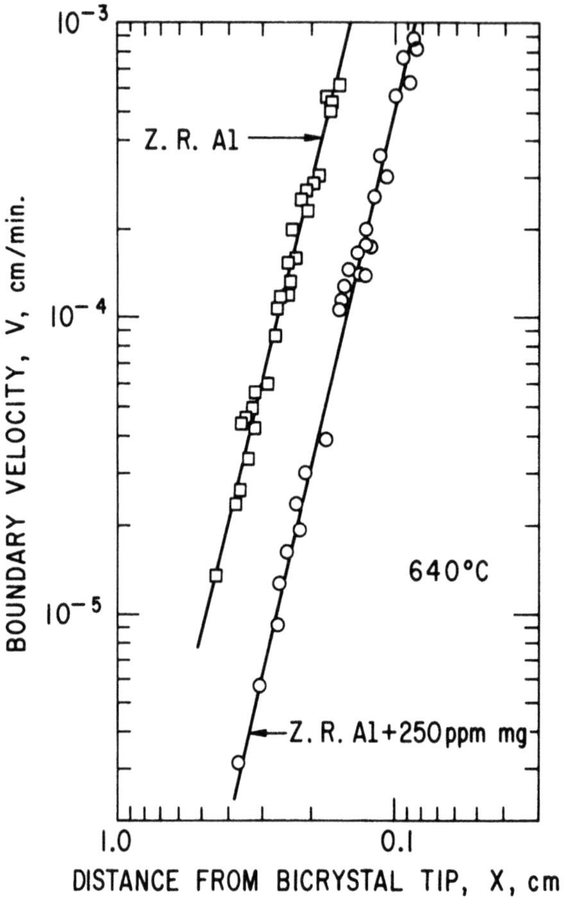

Figure 14. Boundary velocity as a function of distance X in zone-refined aluminum and Al + 250 ppm Mg bicrystals.

velocity increased appreciably by the additional zone refining treatment. For example at X = 0.3 cm, the boundary velocity in the second batch is 6 x 10^{-5} cm/min., whereas in the first, it is 5.5 x 10^{-6} cm/min. This indicates that the boundary velocity is affected by trace impurities in bicrystals of the first batch. However, it is not clear if such effect still persists in the second batch after additional zone refining.

The temperature dependence of the boundary velocity is shown in Figure 15. The activation energy for boundary migration in zone-refined aluminum and in alloys containing up to 50 ppm Mg is about 18 kcal/mol whereas the addition of 250 ppm Mg increases Q to 34 kcal/mol. Within the accuracy of the measurement of Q, it appears to decrease with increasing driving force, up to 50 ppm Mg, in a manner similar to that reported earlier for the first

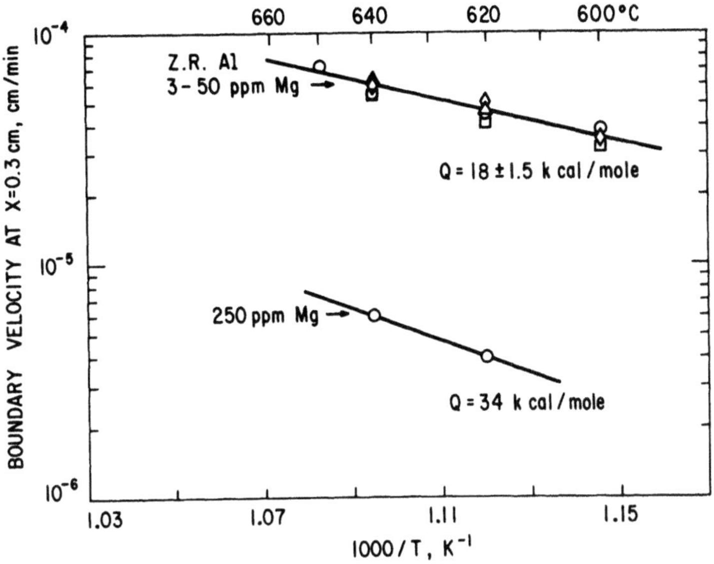

Figure 15. Temperature dependence of boundary velocity at X = 0.3 cm in Al-Mg bicrystals of various magnesium concentrations.

batch of bicrystals. This is illustrated in Figure 16 for $\Delta F = 1.7 \times 10^3$ ergs/cm^3 and 5×10^3 ergs/cm^3. However, the effect of driving force on Q appears to diminish considerably in the 250 ppm Mg alloy. It appears that within the range of driving

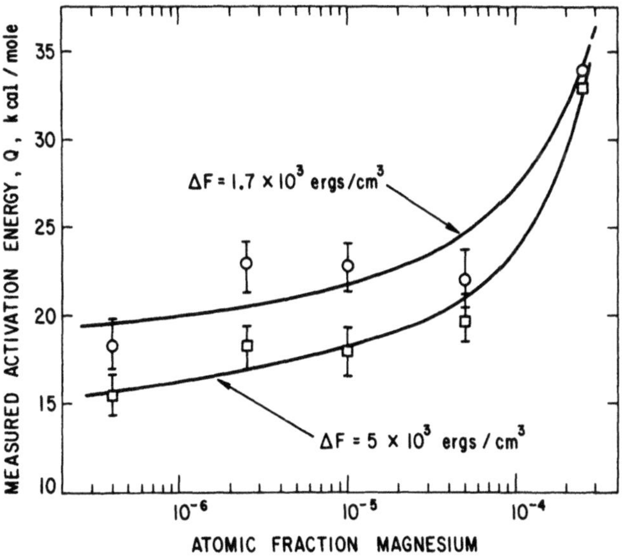

Figure 16. The dependence of apparent activation energy for boundary migration on solute concentration and on driving force.

forces and boundary velocities studied the effective porosity in the boundary does not alter during boundary migration when the bulk concentration of Mg is 250 ppm.

A comparison of boundary velocities at 640°C as a function of Mg concentration is shown in Figure 17 for $\Delta F = 5 \times 10^3$ ergs/cm^3 and $\Delta F = 2.5 \times 10^3$ ergs/cm^3. The boundary velocity remains essentially unchanged up to 50 ppm Mg, beyond which it decreases with increasing concentration of Mg. This behavior is nearly the same when the driving force is doubled.

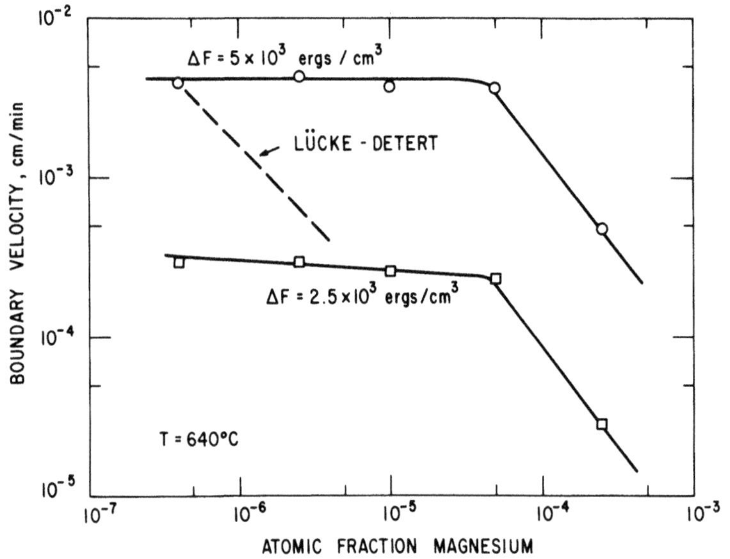

Figure 17. Effect of solute concentration and driving force on boundary velocity. Dash line, calculated from Lücke-Detert equation (27).

THEORIES OF BOUNDARY MIGRATION AND COMPARISON WITH RESULTS

Effect of Misorientation on Boundary Velocity

Beyond a minimum driving force, the velocity is affected by misorientation angle, θ. Figure 18 shows the velocities of the three boundaries as related to their misorientation angles at a constant driving force of 4×10^3 ergs/cm^3. In an attempt to predict the behavior of high-angle boundaries, Li (21) has proposed a theory based on the effects of dislocation cores on boundary properties. This theory suggests a relationship between boundary velocity and misorientation angle. At a constant driving force the velocity is given by

$$V = \frac{2}{\Pi} V_o \, \mathrm{Sin}^{-1}\left[\frac{2\Pi r_o}{b} \, \mathrm{Sin} \, \frac{1}{2}\theta\right] \qquad (6)$$

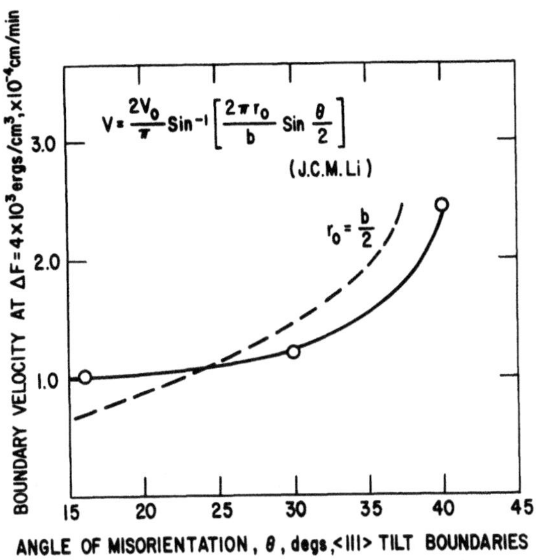

Figure 18. Effect of misorientation on boundary velocity at a constant driving force of 4 x 10^3 ergs/cm^3. Dash line, calculated from Li's equation (21).

where V_o is the velocity of the boundary in which the dislocation cores just meet, r_o, an adjustable parameter, is the radius of the core of a single dislocation, b is the Burgers vector, and θ is the boundary misorientation angle. The critical misorientation angle at which the dislocation cores meet is given by $\pi r_o/h = 1$, where h is the dislocation spacing. The boundary velocity calculated on the basis of Eq. (6) is shown by the dashed line in Figure 18, for $r_o = b/2$. There appears a reasonable agreement between theory and experiment. The assumption of $r_o = b/2$ seems appropriate, since it predicts a maximum velocity for the boundary with a misorientation angle of 37°. This prediction is in very good agreement with previous experiments (6,22), and with the coincidence site model of grain boundaries (23). The increase of velocity between 16° and 30° boundaries is somewhat less than predicted by the theory. This is due to the fact that the driving force exponent varied with the misorientation angle (see Table II). Consequently, for driving forces greater than 4 x 10^3 ergs/cm^3 the velocity increase with misorientation would be greater than that shown in Figure 18. A lack of agreement with the theory exists for driving forces below 2 x 10^3 ergs/cm^3, where the velocity appears to be independent of misorientation. Rutter and Aust (24) studied the effect of misorientation on boundary velocity in zone-refined lead. Their results at 300^oC appear to be in general agreement with the findings of this investigation.

Atomic Theories for Boundary Migration

Expressions for boundary velocities have been derived by Mott (25) and by Turnbull (26), based on absolute reaction rate theory. Whereas Turnbull treated boundary migration as a result of transfer of single atoms across the boundary, Mott suggested that boundary migration involves the transfer of groups of atoms by a process of "melting" from the shrinking crystal and solidifying onto the growing crystal. The calculated velocities were compared with our results. The velocities predicted by the group process theory were four orders of magnitude higher than those observed; whereas the single-atom process theory agreed well with experimental data. Since the comparison does not afford a rigorous test of the theory because of the use of experimental Q to obtain calculated velocities, the pre-exponential terms, V_o, of theory and experiment have been compared. Typical results together with those of Aust and Rutter (1), Gordon and Vandermeer (5), and Frois and Dimitrov (19) on zone-refined aluminum are presented in Table III. The agreement between V_o(calc) and V_o(exp) is within one order of magnitude. The discrepancy can be attributed to both the effect of orientation and to trace impurities not considered in the foregoing theories.

TABLE III. COMPARISON OF EXPERIMENTAL AND THEORETICAL V_o FOR GRAIN BOUNDARY MOTION IN ZONE-REFINED Al^o

Type of Experiment	Driving Force ergs/cm³	Log V_o, cm/sec Exp.	Log V_o, cm/sec Calc.	Orientation Relation	Ref.
Undeformed Bicrystal	3.33×10^3	1.24	1.19	40° tilt-<111>	R&H
	3.22×10^3	0.71	1.17	30° " "	
	1.23×10^3	0.53	0.76	10° " "	
Striated Single Crystal	8.6×10^3	0.18	1.57	--	A&R
Recryst. in Def. Polycrystal	1.6×10^7	6.25	5.35	--	G&V
	1.2×10^8	5.90	6.27	--	F&D

Impurity Drag Theory of Boundary Migration

Lücke and Detert (27) proposed a quantitative theory of the effects of impurities on boundary migration based on the hypothesis that impurity atoms segregate to grain boundaries because of an interaction force between the impurity atoms and the boundary. The formulation of the theory suggests that the boundary velocity should be inversely proportional to the bulk concentration of impurity atoms. Results of calculations from this theory are shown by the dash line in Figure 17. Clearly there is little agreement with the experimental findings. Aust and Rutter's results on the effect of tin on the rates of boundary migration in lead (4) were similarly incompatible with the Lücke-Detert theory. It was proposed(2) that this discrepancy is due to small orientational differences within each group of boundaries, since the theory does not consider orientational variables and their effects on grain boundary migration. This explanation is not applicable to the results of our study, since the boundary orientation in each alloy bicrystal is identical. A more probable explanation for the discrepancy may lie in our ignorance of equilibrium segregation of impurities at a moving boundary.

A more rigorous treatment of the impurity effect was subsequently presented by Cahn (28), and by Lücke and Stüwe (12), based on the same assumption that impurity atoms exert a drag force on the boundary. The equation developed by these authors relating the boundary velocity, driving force and impurity concentration is given by

$$\Delta F = \lambda V + \frac{\alpha C_o V}{1+\beta^2 V^2} \qquad (7)$$

where ΔF is the driving force, λ is the reciprocal of intrinsic boundary mobility, C_o is the bulk concentration of the impurity atoms, and α and β are impurity drag parameters related to temperature, the diffusivity profile of impurity atoms, D(x), and the profile of the boundary-impurity atom interaction energy, E(x). The theory indicates that at the lower limits of boundary velocity, the velocity is always proportional to the driving force at a constant composition, whereas at the upper limits the velocity is proportional to the driving force only when the driving force is very high or the impurity concentration is very low. Furthermore, the theory predicts regions of transitional velocities between high and low driving forces, in which as many as three velocities are possible for the same driving force, depending upon the impurity concentration. These regions may be conceived (2) as those in which the boundary velocity is highly sensitive to the driving force. In another paper in this volume, Lücke et al discuss the various ramifications of the transition region of boundary velocity.

Although the theory provides a considerable insight into the nature of boundary migration, a direct quantitative comparison of theory with experiment requires information on the atomic structure of the boundary and on the profiles D(x) and E(x). On a qualitative basis it appears that the observed relation of $V = M(\Delta F)^m$ suggests that the boundary velocity is in the transition region.

Gordon and Vandermeer (2), with several simplifying assumptions, were able to explain their results as well as the results of Holmes and Winegard (2) on the basis of the impurity-drag theory. However, this could not be done with the Aust and Rutter (4) data.

Results from several significant investigations (2,6,7,19, 20) of rates of grain boundary migration, extrapolated to 100°C, in single-crystals, bicrystals, and polycrystalline aggregates of zone-refined aluminum are shown in Figure 19. The activation energies for boundary migration (2,6,19,20) were in good agreement, varying within the narrow range of 13 to 16 kcal/mol which approximately corresponds to the activation energy for boundary self-diffusion. Consequently, one is tempted to conclude that in these

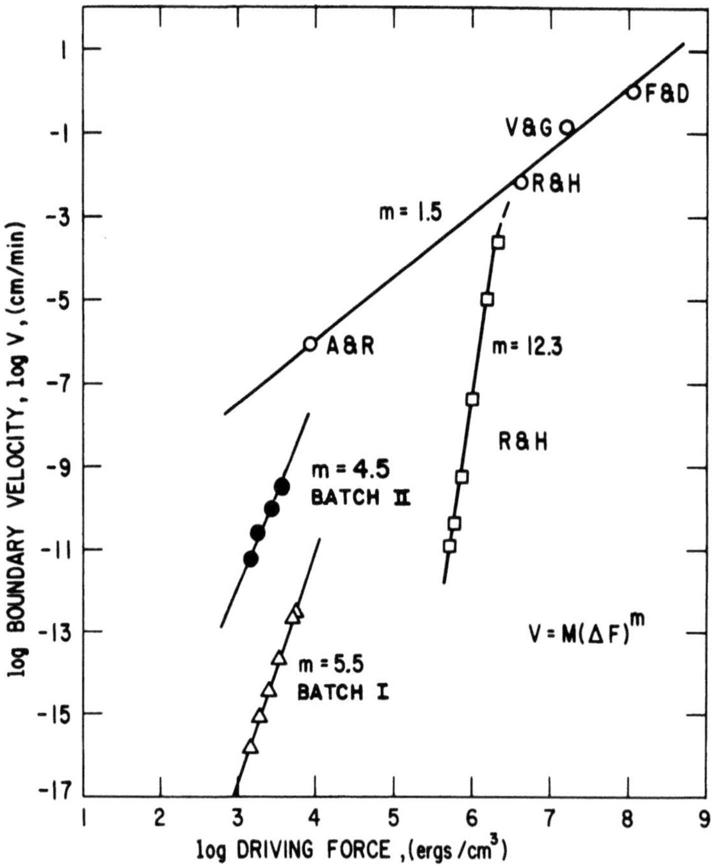

Figure 19. Dependence of boundary velocity on driving force in zone-refined aluminums extrapolated or interpolated to 100°C. (○) Frois and Dimitrov(19), Vandermeer and Gordon(2), Rath and Hu(7), and Aust and Rutter(20); (□) Rath and Hu(6); (Δ and ●) Batch I and Batch II zone-refined aluminum bicrystals, this investigation.

cases the boundaries are free of impurity effects. If so, the boundary velocity should be a linear function of the driving force, i.e. m = 1. However, as seen from Figure 19 a value of m = 1.5 is obtained. If m>1 typifies the transition region of boundary velocity, it appears that all these experimental results, which cover a wide range of velocity and driving force, fall in the transition region.

The predictions of the impurity-drag theory as given by Eq.(7), can be compared with the results of this investigation in two ways: (i) The effect of driving force on velocity, employing results from the first batch of bicrystals, and (ii) The effect of concentration of Mg in Al, using data from the second batch of bicrystals.

Differentiating Eq.(7) with respect to V^2 and rearranging terms, we get

$$\frac{\Delta F}{2V^3}\left[1 - \frac{d\ln \Delta F}{d\ln V}\right] = \left(\frac{\Delta F}{V} - \lambda\right)^2 \frac{\beta^2}{\alpha C_o} \tag{8}$$

The boundary migration behavior can be represented by

$$V = M(\Delta F)^m \tag{9}$$

Combining Eqs. (8) and (9) gives

$$\left(\frac{\Delta F}{2V^3}\right)^{\frac{1}{2}} \left(1 - \frac{1}{m}\right)^{\frac{1}{2}} = \left(\frac{\Delta F}{V} - \lambda\right) \frac{\beta}{(\alpha C_o)^{1/2}} \tag{10}$$

which suggests that on the basis of the impurity-drag theory $\left(\frac{\Delta F}{V^3}\right)^{1/2}$ should be linearly dependent on $\left(\frac{\Delta F}{V}\right)$. The plot shown in Figure 20 illustrates the incompatibility between Eq.(7) and the experimental findings. It suggests that a clearer understanding of the nature of interactions between the impurity atoms and the boundary in the transition region is needed for quantitative agreement between theory and experiments.

On the other hand, it can be assumed that the driving forces for boundary migration in this investigation fall in the low driving-force region(28) of the theory, since they are considerably smaller than those associated with recrystallization. The impurity-drag theory indicates that under this condition $V \ll 1/\beta$, hence Eq.(7) can be simplified to

$$V = \Delta F/(\lambda + \alpha\ C) \tag{11}$$

In view of the observed relation,

$$\Delta F = (V/M)^{1/m} \tag{12}$$

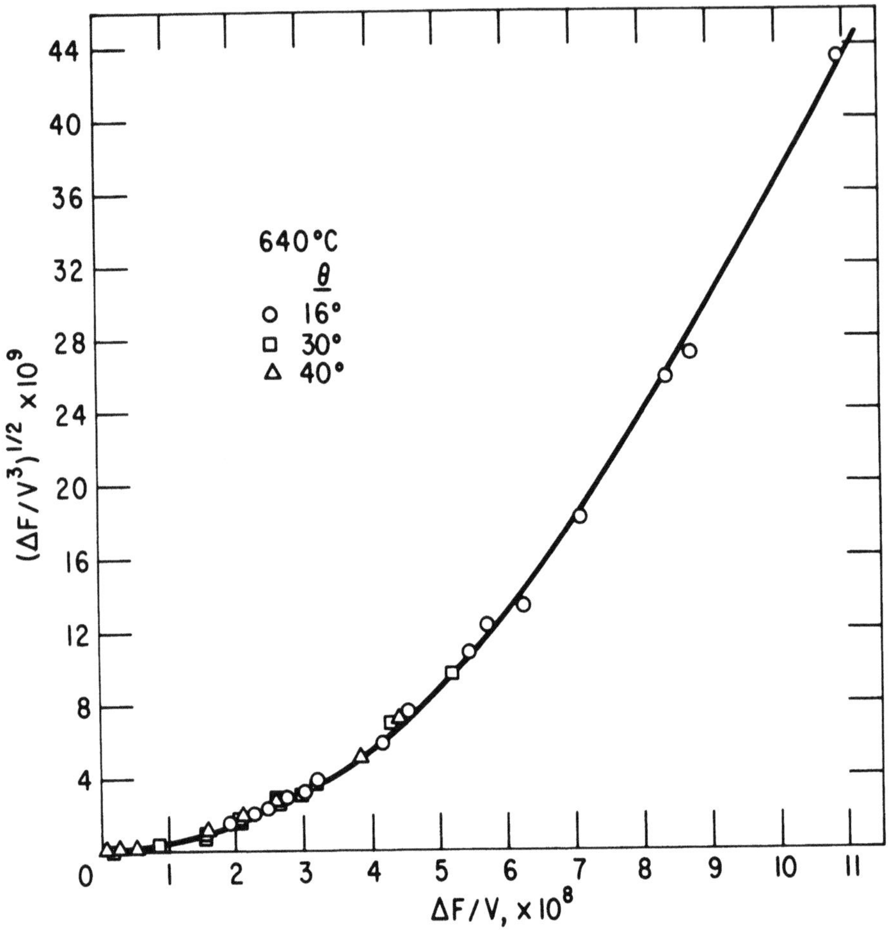

Figure 20. Comparison of experiment with impurity-drag theory(28) (See text). Data of Batch I zone-refined aluminum bicrystals.

Eq.(11) can be expressed as

$$V^{(\frac{1}{m} - 1)} = \lambda M^{\frac{1}{m}} + \alpha M^{\frac{1}{m}} C \qquad (13)$$

The consequence of this relationship is shown in Figure 21. Because of insufficient data it is difficult to ascertain the agreement between theory and experiment. However, the fact that the boundary velocity does not change significantly with increasing Mg concentration up to 50 ppm is a deviation from the theoretical prediction in this range of concentrations.

The force of interaction, E, between the impurity atom and the boundary has been considered in the development of the impurity-drag theories. This quantity was either treated as a constant(27)

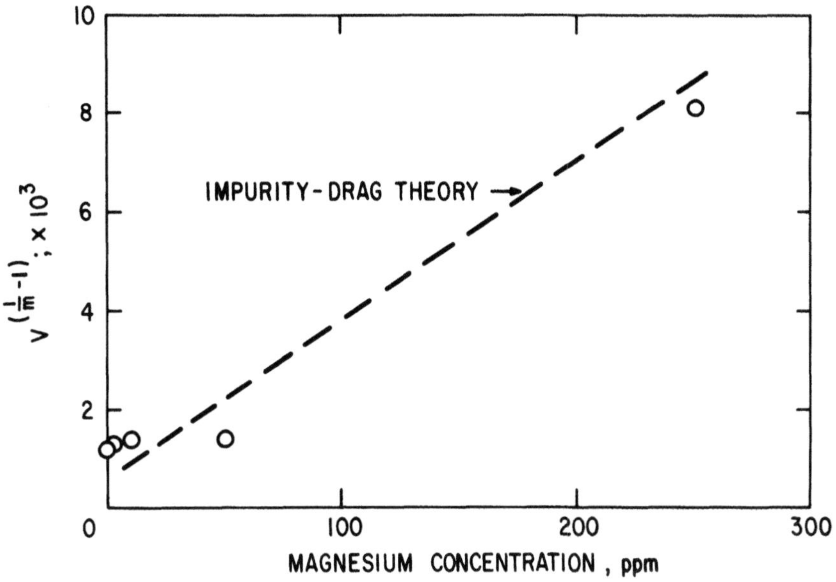

Figure 21. Comparison of experiment with impurity-drag theory(28) (See text). Data of Batch II zone-refined aluminum and Al-Mg bicrystals.

or as a function of distance, x, from the boundary, having a fixed profile with respect to x(28). It is reasonable to assume that the specific boundary energy would decrease because of impurity atom segregation at the boundary. This factor is particularly significant in the present investigation, since the driving force for boundary migration comes from the specific energy of the boundary itself. In view of this, the evaluation of driving force by taking γ as a constant is probably not valid, particularly for a moving boundary. It is likely that the boundary at its initial position would contain the equilibrium concentration of solutes. This concentration would steadily decrease, along with a possible increase of porosity, as the boundary acquires increasingly higher velocities during its migration towards the tip of the bicrystal. The consequence of this model* is presented schematically in Figure 22. If the concentration of solutes at the moving boundary at position 1 is given by C_1, then $C_1 > C_2$, and consequently the boundary energy, $\gamma_{C1} < \gamma_{C2}$. This should result in an increase in driving force as X decreases. Thus, when the modified driving force is compared with its corresponding velocity, the driving

*The increase in boundary porosity has been discussed earlier in this paper. It is assumed that the increase in porosity of a high-angle boundary during migration does not significantly affect the impurity distribution.

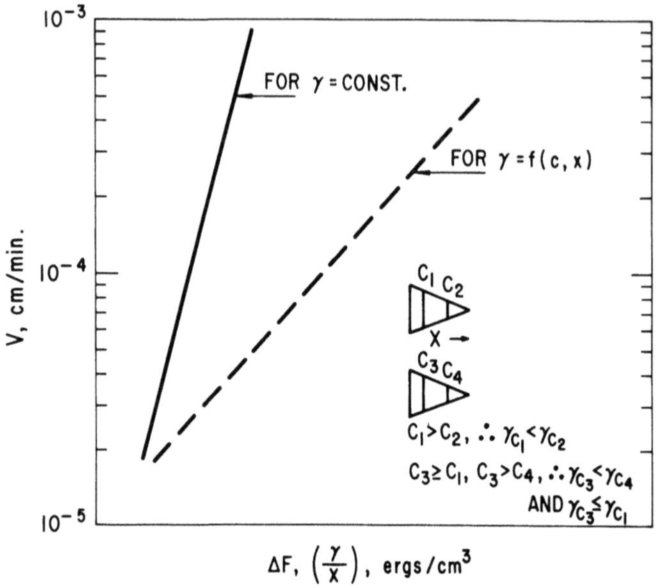

Figure 22. Schematic representation of the dependence of boundary velocity on the specific free energy of the boundary which serves as the driving force for boundary migration in pure metal and alloy bicrystals.

force exponent should decrease, going from 4 towards 1. As the impurity concentration in the alloy increases, the boundary energy may or may not continue to decrease, depending on the impurity species and the boundary saturation. However, in all cases the dynamic state of the boundary would correspond to impurity concentrations lower than equilibrium. With this modification, the impurity-drag theory may agree with the present findings.

Results of this investigation on the driving force dependence of boundary velocity, being analogous to the stress dependence of creep rate(30), can also be related to grain-growth in metals. It has been shown by Hu and Rath(31) that the driving force exponent in grain boundary migration is related to the time exponent in grain-growth, namely,

$$m = \left(\frac{1}{n}\right) - 1 \qquad (14)$$

Beck et al (32) examined grain-growth in high-purity aluminum containing varying concentration of Mg. Their results, shown in Figure 23, yield a value of $n = 0.27$ for $T/Tm = 0.9$, which remains unaffected when 0.025% Mg is added to high-purity aluminum. This value of n corresponds to a driving force exponent of $m \simeq 3$, which agrees with the results of this investigation.

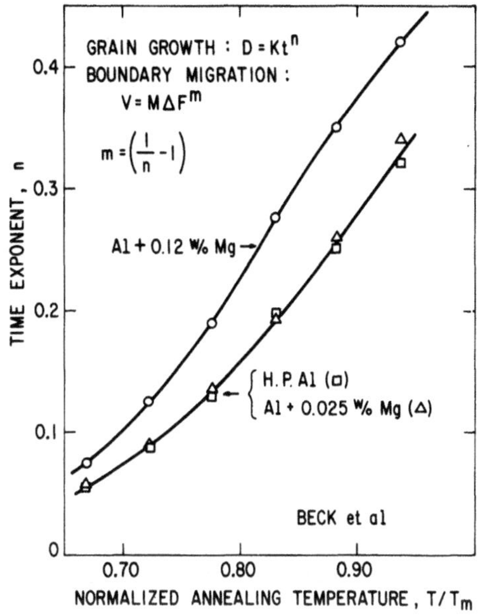

Figure 23. Correlation between the driving force exponent, m, in boundary migration, and the time exponent, n, in grain-growth as applied consistently with the grain-growth data of Beck et al(32). (See text).

CONCLUSIONS

1. High-angle tilt boundaries assume the shape of a cylindrical arc during migration in a wedge-shaped bicrystal, moving towards the center of curvature of the arc.

2. The boundary velocity increases sharply with driving force, obeying a power relation, $V = M(\Delta F)^m$; the driving force exponent m decreases with decreasing angle of misorientation, and with increasing temperature of annealing.

3. At low driving forces the boundary velocity is insensitive to misorientation angle; at high driving forces the boundary velocity increases with increasing misorientation. This is consistent with the observed effects of deformation severity on the development of recrystallization texture.

4. The activation energy for boundary migration, varying slightly with misorientation angle, approaches that for bulk self-diffusion at low driving forces. The activation energy for migration decreases with increasing driving force.

5. The addition of magnesium does not change the boundary velocity, up to a concentration of 50 ppm, beyond which the boundary velocity decreases.

6. The presence of trace impurities affects the activation energy for boundary migration. Further purification of a zone-refined aluminum decreases Q from 27 to 18 kcal/mol. With the additions of Mg up to a concentration of 50 ppm, Q remains relatively unchanged from 18 kcal/mol. The effect of driving force on Q diminishes as the Mg concentration is increased to 250 ppm.

7. The driving force dependence of the rate of boundary migration appears to be analogous to the stress dependence of creep rate, and is consistent with the correlation between boundary migration and grain growth in terms of the driving force and the time exponents, [m = (1/n) - 1].

ACKNOWLEDGMENT

The authors wish to express their gratitude to Drs. L.S. Darken and J.C.M. Li, formerly of the E.C. Bain Laboratory for Fundamental Research, for many stimulating discussions during the course of this investigation. The able assistance of D.J. Lemmon in preparation and annealing of the bicrystals is gratefully acknowledged.

REFERENCES

1. K.T. Aust and J.W. Rutter, in Recovery and Recrystallization of Metals, Interscience, New York, (1963), p.131.
2. P. Gordon and R.A. Vandermeer, in Recrystallization, Grain Growth and Textures, ASM, Metals Park, Ohio, (1965), p.205.
3. H. Gleiter, Phys. Stat. Sol., 45, 9, (1971).
4. K.T. Aust and J.W. Rutter, Trans. TMS-AIME, 215, 119, (1959), and ibid, 218, 682, (1960).
5. P. Gordon and R.A. Vandermeer, Trans. TMS-AIME, 224, 917, (1962).
6. B.B. Rath and H. Hu, Trans. TMS-AIME, 245, 1243, (1969).
7. B.B. Rath and H. Hu, Trans. TMS-AIME, 236, 1193, (1966).
8. B.B. Rath and P. Gordon, Tech. Rep. to ARO(D) and ONR, (Oct. 1962).
9. C.G. Dunn and J.L. Walter, in Recrystallization, Grain Growth and Textures, ASM, Metals Park, Ohio, (1965), p.461.
10. J.C.M. Li, Trans. TMS-AIME, 245, 1591, (1969).
11. N.A. Gjostein and F.N. Rhines, Acta Met., 7, 319, (1959).
12. K. Lücke and H. Stüwe, in Recovery and Recrystallization of Metals, Interscience, New York, (1963), p.171.
13. K. Lücke, in VII Colloque de Metallurgie, Ecroussage, Restauration, Recristallisation, Presses Universitaires de France, Paris, (1963), p.1.
14. R.A. Vandermeer, Acta Met., 15, 447, (1967).
15. B.B. Rath and H. Hu, Trans. TMS-AIME, 245, 1577, (1969).
16. R.E. Green, Jr., Trans. TMS-AIME, 233, 1954, (1965).
17. W.W. Mullins, Acta Met., 4, 421, (1956).
18. T.S. Lundy and J.F. Murdock, J. Appl. Phys., 33, 1671, (1962).

19. C. Frois and O. Dimitrov, Mem.Sci. Rev. Met., 59, 643,(1962).
20. K.T. Aust, in New Physical and Chemical Properties of High-Purity Metals, Centre National de La Recherche Scientifique, Paris, (1960), p. 99.
21. J.C.M. Li, J. Appl. Phys. 32, 525, (1961).
22. P.A. Beck, P.R. Sperry and H. Hu, J. Appl. Phys., 21, 420, (1950).
23. M.L. Kronberg and F.H. Wilson, Trans. TMS-AIME, 185, 501, (1949).
24. J.W. Rutter and K.T. Aust, Acta Met., 13, 181, (1965).
25. N.F. Mott, Proc. Phys. Soc., 60, 391, (1948).
26. D. Turnbull, Trans. TMS-AIME, 191, 661, (1951).
27. K. Lücke and K. Detert, Acta Met., 5, 628 (1957).
28. J.W. Cahn, Acta Met., 10, 789, (1962).
29. E.L. Holmes and W.C. Winegard, Trans. TMS-AIME, 224, 945, (1962).
30. J.C.M. Li, Trans. TMS-AIME, 245, 1591, (1961).
31. H. Hu and B.B. Rath, Met. Trans., 1, 3181, (1970).
32. P.A. Beck, J.C. Kremer, L.J. Demer and M.L. Holzworth, Trans. AIME, 175, 372, (1948).

SUBJECT INDEX

GPSR Compliance

The European Union's (EU) General Product Safety Regulation (GPSR) is a set of rules that requires consumer products to be safe and our obligations to ensure this.

If you have any concerns about our products, you can contact us on ProductSafety@springernature.com

In case Publisher is established outside the EU, the EU authorized representative is:

Springer Nature Customer Service Center GmbH
Europaplatz 3
69115 Heidelberg, Germany

Batch number: 10615115

Printed by Printforce, the Netherlands